Drivers of Landscape Change in the Northwest Boreal Region

DRIVERS OF LANDSCAPE CHANGE IN THE NORTHWEST BOREAL REGION

Edited by Amanda L. Sesser, Aimee P. Rockhill, Dawn R. Magness, Donald Reid, John DeLapp, Philip J. Burton, Eric Schroff, Valerie Barber, and Carl Markon

Published by
University of Alaska Press
P.O. Box 756240
Fairbanks, AK 99775-6240

Cover and interior design by UA Press.
Interior layout by Rachel Fudge.

Cover and opening image: courtesy of Steve Hillebrand, U.S. Fish and Wildlife Service. Upper Tanana River Valley in the Tetlin National Wildlife Refuge.

Contents page image from iStock, Stock photo ID:541868434.

Library of Congress Cataloging in Publication Data

Names: Sesser, Amanda L., editor.
Title: Drivers of landscape change in the northwest boreal region: a synthesis of information for policy and land management /
edited by Amanda L. Sesser, Aimee P. Rockhill, Dawn R. Magness, Donald Reid, John DeLapp, Philip J. Burton, Eric Schroff, Valerie Barber, and Carl Markon.
Description: Fairbanks, AK : University of Alaska Press, 2019. | Includes bibliographical references and index. |
Identifiers: LCCN 2019003141 (print) | LCCN 2019017003 (e-book) | ISBN 9781602233980 (e-book) | ISBN 9781602233973 (paperback)
Subjects: LCSH: Northwest Boreal Landscape Conservation Cooperative. | Landscape protection—Alaska. | Landscape protection—Alberta—Northwest Boreal Region. | Environmental protection—Alaska. | Environmental protection—Alberta—Northwest Boreal Region. | BISAC: SCIENCE / Earth Sciences / General.
Classification: LCC QH76.5.A4 (e-book) | LCC QH76.5.A4 D75 2019 (print) | DDC 304.209798—dc23
LCC record available at https://lccn.loc.gov/2019003141

This book is dedicated to the people who live and work in the Northwest Boreal region. Their perseverance, strength, and connection to the land, people, and resources are the best hope for the resiliency and sustainability of this iconic northern landscape.

And to Rick Janowicz, who dedicated his career to better understanding the dynamics of boreal systems in pursuit of sustainability.

CONTENTS

EXTENDED CONTENTS

PREFACE

The Northwest Boreal (NWB) Landscape Conservation Cooperative (LCC) was formed in 2012 to bring together conservation and resource managers in boreal Alaska and northwest Canada with the central purpose of sharing interests and resources to collectively address landscape-level issues that are common among resource managers, educators, and the public. As outlined as part of the NWB LCC's strategic plan (http://nwblcc.org/wp-content/uploads/2015/05/NWB-LCC-Strategic-Plan-V1.pdf), this document will provide a foundation by which members of the LCC and other interested parties can find baseline information regarding physical and ecological aspects of the region, as well as potential drivers of landscape change in an area that spans over 1.3 million square kilometers (321 million acres) across Alaska and Canada. This work is the culmination of more than 60 contributing authors engaged in a wide spectrum of study, including physical and biological aspects of the region; natural disturbances; social-ecological drivers; interactions among the natural and anthropogenic characteristics; and engagement among and between scientists and the public. This document may well serve as a baseline through which future meaningful engagement with the LCC can take place.

ACKNOWLEDGMENTS

The editors thank each author for their outstanding contributions to this publication. Their interdisciplinary and international collaborative efforts have made this a significant resource for both present and future members of the Northwest Boreal (NWB) Landscape Conservation Cooperative (LCC) and residents of the boreal region. We specifically thank Kristin Timm (University of Alaska Fairbanks), Keith LaBay (U.S. Geological Survey, Alaska Science Center), and Benjamin Matheson (U.S. Fish and Wildlife Service) for creating a variety of figures. Additionally, we extend our appreciation to the peer reviewers, including Jamie Kenyon (Ducks Unlimited, Canada), Mari Reeves and Dawn Magness (U.S. Fish and Wildlife Service), and Debra Grillo (U.S. Geological Survey Technical Publications Unit), for manuscript edits.

The staff at the NWB LCC were extremely helpful in coordinating all stages of the process, from inception to print. The NWB LCC, the U.S. Geological Survey, and all participating members of the NWB LCC provided financial and personnel support. Initial production of this report was provided by the U.S. Geological Survey, Alaska Regional Office. The findings and conclusions in this report are those of the authors and do not necessarily represent the views of the U.S. Fish and Wildlife Service, U.S. Geological Survey, or other organizations with which authors are affiliated.

1
LANDSCAPE CHANGE IN THE NORTHWEST BOREAL REGION

By Amanda L. Sesser and Aimee P. Rockhill[1]

The Northwest Boreal region (NWB) of North America is a land of extremes. Extending more than 1.3 million square kilometers (321 million acres), the NWB encompasses the entire spectrum between inundated wetlands below sea level to the tallest peak in North America.

Extreme low temperatures and snow dominate winters, and summers are mild and warm. Topography influences precipitation, so annual means vary over the region, but average annual precipitation in Fairbanks, Alaska, is only 280 mm (11 in) (Wendler and Shulski, 2009; Western Region Climate Center, 2018a). Permafrost gradients span from nearly continuous to absent and soils vary from ancient wind-blown loess deposits more than 100 m (328 ft) in depth (Beget and others, 2006) to shallow mineral soils that cover rocky colluvial deposits of mountain slopes.

Boreal ecosystems are inherently dynamic and continually change over decades to millennia. Kilometers-thick sheets of ice formed during repeated ice ages have left their legacies on the current landscape; Pleistocene glaciation, unglaciated Beringial refugia, and repeated colonization of boreal ecosystems shape the social and biological makeup of the NWB. The braided rivers that shape the valleys and wetlands continually change course, creating and removing vast wetlands and peatlands. Glacial melt, erosion, fires, permafrost dynamics, and wind-blown loess are among the shaping forces of the landscape. As a result, species interactions and ecosystem processes are shifting across time. Similar to this landscape change, the Alaska Native and Canada's First Nations peoples have adapted to and shaped landscape change across the boreal region for thousands of years.

Despite the stochastic nature of natural disturbance regimes, the NWB is currently undergoing directional change. Climate is projected to warm by 2–8°C by the end of

1 Northwest Boreal Landscape Conservation Cooperative, U.S. Fish and Wildlife Service, Anchorage, Alaska, USA.

the 21st century (3–14°F; Chapter 3.1). Biophysical feedbacks, or interactions among climate, disturbance, and biological composition, are moving northern regions into a new era. These changes, combined with growing populations, increased demand for globally important resources, and the need for sustainable energy and economic growth, are ushering rapid socioeconomic change away from historical ranges of variability. Not all aspects of the landscape are subject to change; at least not outside the geologic timescales. Enduring features are those landscape attributes that are relatively fixed, such as bedrock geology and topography (Chapter 3.5), yet they are important for determining which and how many species certain areas can support. As such, both drivers of change and enduring features may be important components of a diverse and comprehensive landscape management program.

With unprecedented rates of change in the NWB, there is an enormous collective opportunity to anticipate these changes and manage the system proactively. Conservation science is a discipline that reacts to ongoing or historical environmental damage. The NWB has relatively intact and functioning landscapes, which provide opportunity for proactive conservation, and some ask if economic growth and access to resources can continue while maintaining ecosystem integrity (Schmiegelow and others, 2014). A question of the Northwest Boreal Landscape Conservation Cooperative (NWB LCC) is, "Rather than repeat the typical trajectory of landscape degradation, can we execute sustainable management?"

The NWB is a data-poor region, and the intention of the NWB LCC is to determine what data *are not* available and what data *are* available. For instance, historical baseline data describing the economic and social relationships in association with the ecological condition of the NWB landscape are often lacking. Likewise, the size and remoteness of this region make it challenging to measure basic biological information, such as species population sizes or trends. The paucity of weather and climate monitoring stations also compound the ability to model future climate trends and impacts, which is part of the nature of working in the north.

Conservation communities have a choice to wait for complete information or to begin the dialogue necessary to fill the information gaps. The unparalleled opportunity available is to work collectively toward a future of supporting the needs of society while maintaining ecological integrity.

The enormous scale and remoteness of the region pose a challenge to participatory science and management; however, the collaborative and cooperative relationships of the NWB LCC provide many avenues to do meaningful work at this landscape scale. Governance structure itself can determine success or failure in these attempts; for example, current laws that were written during a time when observable climate was less variable over the long term could impede the ability to be proactive about landscape change. Likewise, success may be determined by who is included in the process and who has a voice. Empowering local decision makers requires that they have the right information at the right time.

Communication can be thought of as a driver[2] of landscape change itself, the same as wildfire, permafrost thaw, and insects and disease. Information will help inform the policy and management decisions and ultimately the ability for the boreal region to achieve sustainability.

The purpose of this volume is to create a resource for regional land and resource managers and researchers by synthesizing the latest research on the (1) historical/current status of landscape-scale drivers (including anthropogenic activities) and ecosystem processes, (2) future projected changes of each, and (3) the effects of changes on important resources. The individual chapters can be informative alone, but when combined, a holistic picture of the drivers of landscape change in the region can be seen. The chapters are short but contain a wealth of information and resources for more in-depth knowledge, and they highlight key findings and information gaps so the most important information is easy to find and digest. Generally, each chapter is coauthored by researchers and land and natural resource managers from the United States and Canada. This design serves two purposes. First, it ensures that the issues, information, and needs are expressed from both Alaska and northwest Canada. Second, it facilitates cross-border cooperation—one of the central themes of the NWB LCC.

This volume covers natural disturbance drivers of change, including wildfire, forest insect and pathogen epidemics, and invasive species. Physical drivers include climate and climate change, hydrologic response, permafrost, growing season length, and enduring features. Biological drivers of change described are vegetation composition change, novel community and trophic assemblages, wildlife parasite and pathogen life cycles, and marine-derived nutrients.

Socioeconomic drivers of change include land-use change and resource extraction, rural and Indigenous livelihoods, contaminants, international agreements, law and policy, and values and ethics. Chapter 6 explores the intricacies and uncertainty of interactions among natural disturbance and physical drivers and cumulative effects of multiple drivers of change. When multiple drivers of landscape change are integrated, a more complete picture of the complexity and cumulative change emerges. How these various drivers of change interact and enter into feedbacks is still poorly understood, but this interaction is currently the subject of intense research interest.

Somewhat different from traditional assessments, this volume explores social drivers of change and puts them at the same level as the traditional disturbance, physical, and biological drivers of change, and urges readers to consider the models under which science is conducted and how land/resource management is practiced as a driver of change. Practices of coproduction[3] suggest that meaningfully engaging communities, evolving roles of scientists and managers, assessing and projecting change, data

2 Drivers are those factors that bring change to the environment or social system.

3 Delivering public services in an equal and reciprocal relationship between professionals, people using services, their families, and their neighbors.

management, and dissemination and outreach are themselves drivers of landscape change (Chapter 7.5). If local decision makers have access to the right information at the right time, they have the ability to make science-informed decisions through local and regional policies that may ultimately help determine the sustainability of the NWB. This may then allow for ecosystem stewardship principles as an integrative idea to pull together uncertainty and trends described throughout the various chapters, and to place them in the context of actions.

INTRODUCTION TO THE NORTHWEST BOREAL LANDSCAPE CONSERVATION COOPERATIVE

The Northwest Boreal Landscape Conservation Cooperative (NWB LCC) is made up of diverse partner organizations (Table 1.1) from across more than 133 million hectares (321 million acres) of boreal forests, alpine habitat, wetlands, and rivers, spanning an altitudinal range from sea level to the highest point in North America. One of the largest LCCs in the LCC Network (https://lccnetwork.org), the NWB partnership includes the major metropolitan hubs and transportation infrastructure of the region, including the two largest cities in Alaska and the largest city in the Yukon Territory. The geographic region of the NWB LCC includes the boreal and boreal transition zones of Alaska, Yukon, northern British Columbia, and westernmost Northwest Territories (Fig. 1.1). As a true international collaboration, the NWB LCC is a growing partnership among more than 25 United States and Canadian federal and provincial/territorial agencies, nongovernmental organizations, tribes/First Nations, and institutions of higher education.

Over the last 100 years, recorded air temperature within the NWB LCC region has warmed 1.4°C (2.52°F), twice the global average, and the growing season length has increased by 50 percent (Wendler and Schulsky, 2009). At the same time, global demand for natural resources such as oil and gas, rare-earth minerals, and timber is increasing exponentially. The combination of climate and land-use change may lead to drastic alterations of these largely intact ecosystems in 50–100 years.

The need for effective, science-based, and targeted conservation at landscape scales to inform management decisions within the NWB LCC region has never been greater. Land and resource managers need decision-support tools, coordinated monitoring, and the latest results from climate and ecosystem models. By agreeing on a common vision of "a dynamic landscape that maintains functioning, resilient boreal ecosystems and associated cultural resources (NWB LCC Strategic Plan, 2015)," partners of the NWB LCC are embarking on a landscape-conservation design process to guide collaboration to achieve that vision in the context of an uncertain future.

In 2014, the NWB LCC partners completed a comprehensive information needs assessment to determine commonalities in what science and management information is needed, at what scale, and in what format to inform local and landscape-level conservation and sustainable management across the region. Biological and cultural

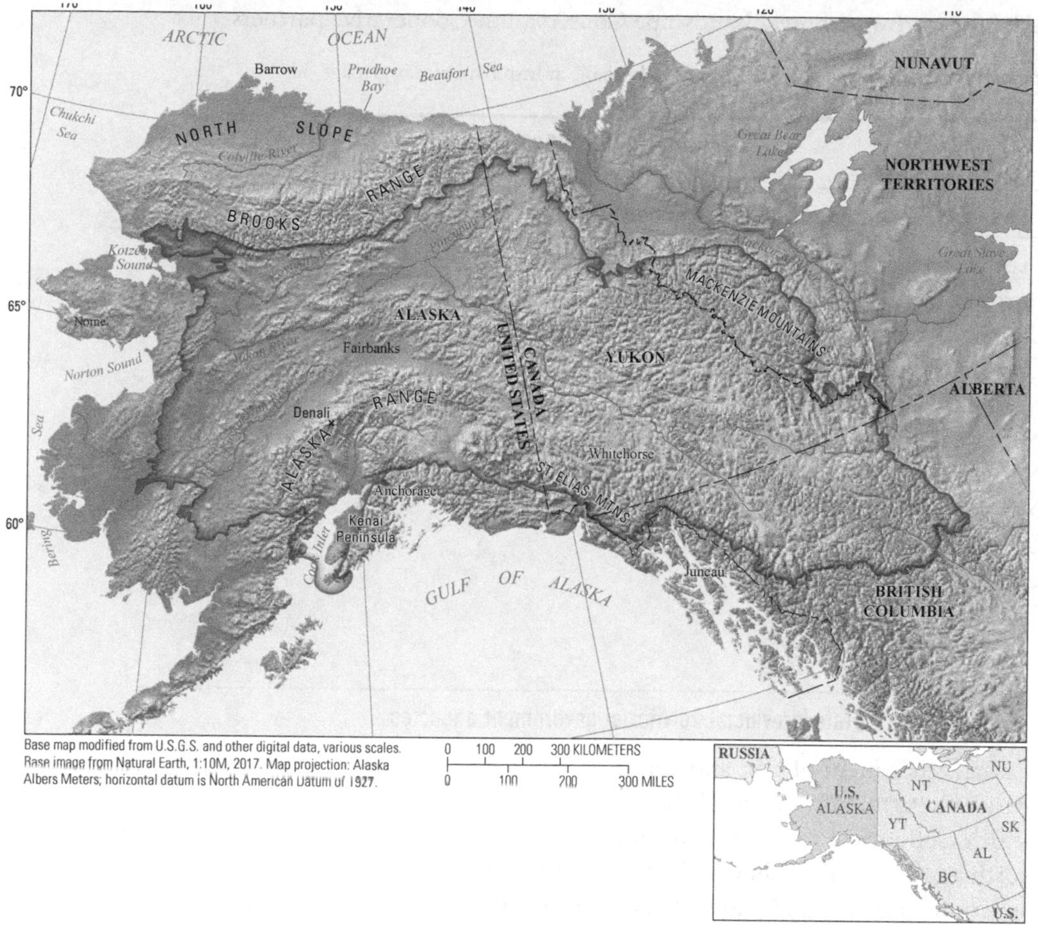

Figure 1.1. Map showing geographic area of the Northwest Boreal (NWB) Landscape Conservation Cooperative (LCC), Alaska and Canada. The LCC includes the boreal forests of Alaska, Yukon, northern British Columbia, and the western mountains of the Northwest Territories.

conservation priorities were identified as primary drivers of landscape change in NWB ecosystems.

POTENTIAL FOR LANDSCAPE-SCALE CONSERVATION IN THE NORTHWEST BOREAL REGION

Collaborative conservation that takes a holistic view of ecosystems and society across large landscapes is emerging as a 21st-century model. People and nature are not separate; society depends on ecosystem services for economic and cultural well-being. To achieve sustainable landscapes, it is important to foster resilient economies and ecosystem services. The term "cultural resources" encompasses the human components of socioecological systems. The challenge is to provoke a new way of thinking about conservation, explore the potential of LCCs, and work not only outside local jurisdictions but also outside the typical ways of doing business.

Conservation challenges facing the region today and in the future will likely

Table 1.1. Northwest Boreal Landscape Conservation Cooperative partners.

Information about partner organizations is available at http://nwblcc.org/?page_id=19

Partner organizations
United States and Canada government agencies
Army Corps of Engineers
Bureau of Indian Affairs
Bureau of Land Management
Canadian Forest Service
National Park Service
Natural Resources Conservation Service
National Oceanic and Atmospheric Administration
National Weather Service
U.S. Army
U.S. Fish and Wildlife Service
U.S. Forest Service
U.S. Geological Survey
State/provincial/territorial government agencies
Alaska Department of Fish and Game
Alaska Department of Natural Resources
British Columbia Provincial Government
Northwest Territories Climate Change Secretariat
Government of Yukon Climate Change Secretariat
Government of Yukon Community Development
Government of Yukon Energy Mines and Resources
Government of Yukon Fish and Wildlife Branch
Yukon Parks
Universities, tribal, indigenous, nongovernmental organizations, and partnerships
Council of Athabascan Tribal Governments
Ducks Unlimited Canada
Kenai and Mat-Su Basin Fish Habitat Partnerships
Kenai Watershed Forum
Pacific Birds Habitat Joint Venture
Tanana Chiefs Conference
University of Alaska Fairbanks Cooperative Extension Service
University of Northern British Columbia
Wildlife Conservation Society Canada
Yukon Research Center of Yukon College
Yukon River Intertribal Watershed Council

require collaborative efforts. Each organization in the NWB partnership has unique strengths and capacities; by working together assets can be leveraged to address landscape conservation challenges at the scale at which they occur. When working collaboratively, the NWB LCC diversity becomes an advantage. Diverse partnerships are more resilient in the face of change and can meet multifaceted and large-scale conservation challenges by bridging diverse perspectives, missions, and institutions. Local and aboriginal perspectives are particularly important to include in collective learning and adaptive management of our landscapes. By working together, we can achieve more than we can individually.

An integral part of every activity within the NWB LCC is communication, which provides an opportunity to be creative and think from a new perspective. The need to incorporate science into policy and decision-making has never been greater given high uncertainties and rates of change. Translation of new and existing information from technical sources into easily understandable and accessible forms targeted toward diverse audiences is a key role that the NWB LCC can play.

To successfully enhance the capacity of organizations and communities to respond to and adapt to change, the NWB LCC needs to be flexible and adaptive. This includes taking advantage of new opportunities as they arise and updating priorities and goals as more is learned about landscapes and trajectories of change.

LCCs are more than science partnerships. Although informed by science and traditional knowledge, LCCs are the social platforms where partners convene to discuss values, articulate shared visions of future landscapes, and plan ways to work hand-in-hand to achieve those visions. LCC functions that are not strictly based on science include coordinating existing efforts; identifying priority conservation features or areas; setting shared goals and objectives; supporting climate adaptation; identifying social and institutional constructs and barriers to change; and co-creation of knowledge with end users. For a plan to encompass all the potential functions of an LCC partnership, it is necessary to incorporate science activities within a larger agenda. This volume highlights important information gaps and science needs as they pertain specifically to the broader mission of the NWB LCC, targeting science that is necessary to move landscape conservation forward in new directions and with the largest on-the-ground conservation effects.

2
NATURAL DRIVERS

By Eric A. Schroff[1]

INTRODUCTION

Natural and human-caused disturbance affects forest and other plant community structure and ecosystem function, and disturbances often trigger ecological change. Wildfire, forest insect and pathogen outbreaks, and the effects of introduced invasive species are important drivers that act independently, or in concert, across the landscape and over time to shape the forests and the cultural landscape of the boreal region.

Whether the climate turns out to be warmer/cooler and wetter/drier in any given area within the region, there will be changes in the nature and effect of the natural disturbance drivers on successional pathways and the resultant structure of forest within the boreal region. Given the complex nature of forest systems and the uncertainty associated with how they may respond to changing climatic conditions, it is important that resource managers have access to information to guide management decisions.

This chapter consists of three subchapters:

- Wildfire in the Northwest Boreal Region
- Forest Insect and Pathogen Epidemics in the Northwest Boreal Region
- Invasive Species in the Northwest Boreal Region

Individually, the subchapters provide a quick peek into the window of each topic, inviting the reader to learn more about the topic and providing access to references cited to enhance the opportunity for resource managers to expand their knowledge base.

1 Yukon Government, Whitehorse, Canada, Retired.

2
NATURAL DRIVERS

2.1
WILDFIRE IN THE NORTHWEST BOREAL REGION

By Jill Johnstone[1], Xanthe Walker[2], and Teresa Hollingsworth[3]

KEY FINDINGS

- Fire is a major factor that has shaped the boreal forest for millennia.
- Changes in the fire regime (fire size, frequency, seasonality, and severity) that are associated with recent and projected future climate warming will almost certainly affect patterns of forest succession and the mosaic of forest cover types in boreal landscapes.
- The stability of forest cover types is maintained by feedbacks between plants and their environment that affect the way fires burn as well as other ecosystem processes. Changing fire conditions can initiate a new set of feedbacks, causing a change in forest vegetation that may persist across multiple fire cycles.
- At broad scales, conversion of spruce or pine forests to forests dominated by aspen and birch may increase the resilience of boreal forest landscapes to climate change by reducing some of the amplifying feedbacks between fire and a warming climate.

WILDFIRES IN THE NORTHWEST BOREAL FOREST

Wildfire is the primary agent of natural disturbance to boreal forests in Alaska and Northwest Canada, and fires play a central role in organizing the physical and biological attributes of the biome. Similar to currently occurring fires, wildfires have been a regular occurrence in the Northwest Boreal (NWB) region for at least 5,500 years (Hu and others, 2006). Important attributes of fires include size, frequency, seasonality, and severity; together these characteristics define the "fire regime" (Weber and Flannigan, 1997; see Box 2.1, "Characterizing the Fire Regime of Alaska and

1 University of Saskatchewan, Saskatoon, Saskatchewan, Canada.
2 Northern Arizona University, Flagstaff, Arizona, USA.
3 U.S. Forest Service, Pacific Northwest Research Station, Fairbanks, Alaska, USA.

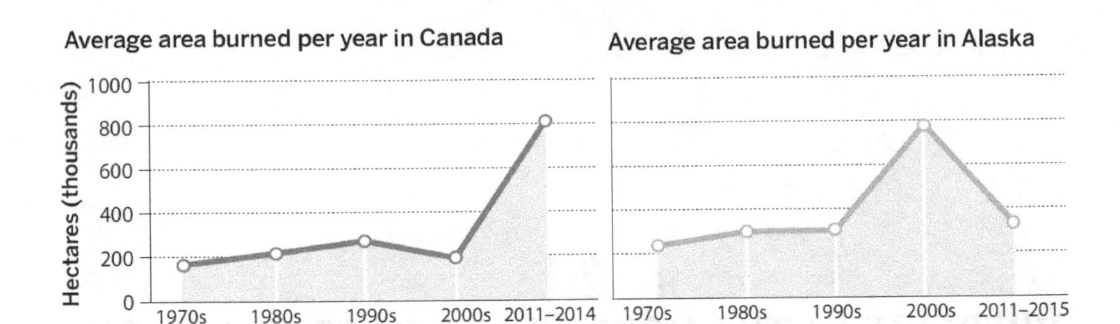

Figure 2.1.1. Average area burned annually in Canada and Alaska since the 1970s. Fire activity in the first half of the 2010 decade is already on the same level of peak fire activity seen during previous decades. Reprinted from Appenzeller (2015), used with permission.

Yukon Boreal Forests"). Fire characteristics are strongly influenced by the effects of weather on fuel moisture, lightning ignitions, and wind speed and direction. With such strong links between fire and weather, it makes sense that changes in climate will affect the fire regime in an area. Recent climate change has been linked to an increase in fire activity in the NWB forest over the past several decades (Podur and others, 2002; Kasischke and Turetsky, 2006; Soja and others, 2007; Kelly and others, 2013; Fig. 2.1.1). Continued increases in fire frequency, extent, and severity and fire season length are a likely outcome of further temperature warming and more frequent droughts during the fire season (Flannigan and others, 2005; Xiao and Zhuang, 2007; Balshi and others, 2009; Wotton and others, 2010). Balshi and others (2009) suggest that the total area burned in North America's boreal forest will double by 2050 and increase 3.5–5.5 times by the end of the century.

FIRE AND FOREST SUCCESSION

Fire-adapted conifers such as black spruce (*Picea mariana*) and lodgepole pine (*Pinus contorta*) that use aerial seedbanks stored in sealed, serotinous cones to ensure ample regeneration after fire, dominate much of the NWB region (Buma and others, 2013; Fig. 2.1.2). Studies of forest succession suggest that increases in fire severity and frequency brought on by climate warming are likely to shift boreal forests of Alaska and Yukon from a spruce- to a broadleaf-dominated landscape (Rupp and others, 2006). This is because the regeneration advantage maintained by serotinous conifers under historical fire conditions is disrupted when fires are unusually severe or if they recur after a short time. Reproductive legacies and fire effects on soil organic layer depths are key factors underlying the stability and resilience ofblack spruce forests across multiple fire cycles. Feedbacks between vascular plants, mosses, and microbial decomposition maintain deep organic soils in black spruce forests and wetlands of Interior Alaska

BOX 2.1: CHARACTERIZING THE FIRE REGIME OF ALASKA AND YUKON BOREAL FORESTS

Fire size is the area burned by an individual fire. Fire size and spatial configuration determine the extent and pattern of fire effects. As in most boreal forests, the majority of fires that occur in the NWB are small, but the few large fires (>10,000 ha) that occur account for most fire effects (Weber and Flannigan, 1997).

Fire frequency is the rate at which fires return to a given area and is often expressed as the fire cycle (the expected amount of time for an area equal to the size of the study area to burn) or fire return interval (the number of years between successive fire events at a specific location). In the NWB forest, stand-replacing fires have a typical fire cycle of 75–150 years (Viereck, 1982; Larsen, 1997).

Fire seasonality refers to the period of the year during which fires occur. Most fires in the NWB occur in spring and summer. In areas where permafrost is present, fires that burn in late summer often consume more of the forest biomass than those in early summer, when the organic soil is wet or frozen (Kasischke and Turetsky, 2006).

Fire severity is the amount of vegetation mortality and biomass consumption caused by a fire, which may be operationally defined as the proportion of tree canopy mortality or depth of burn in the forest floor (Weber and Flannigan, 1997; Keeley, 2009). Because crown fires burn much of the area burned in the NWB region, fires tend to kill most of the forest canopy and initiate a new phase of stand replacement. However, consumption of the soil organic layer is more variable and can have large effects on post-fire regeneration (Johnstone, Hollingsworth, and others, 2010).

and central and northern Yukon (Johnstone, Chapin, McGuire, and others, 2010). Under the historical fire regime, moist, cold soils typically burn at low severity and leave the surface organic layer largely intact. Thick organic layers burned at the surface create a poor-quality seedbed that requires high seed inputs from serotinous conifers, essentially barring other species like broadleaf trees from becoming established at a site (Johnstone and Chapin, 2006a). The low leaf litter and evaporative demands from conifer trees favor the reestablishment of a moss understory, and the system quickly returns to a spruce-dominated forest similar to that of the pre-fire forest.

These stabilizing feedbacks between plants and soil in boreal conifer forests can be disrupted by unusual fire events that then lead to altered patterns of succession. Severe fires that consume much of the soil organic layer create seedbed conditions that allow deciduous broadleaf species such as aspen and birch to successfully recruit from seed and dominate the post-fire regeneration phase (Johnstone, Hollingsworth, and others, 2010). Similarly, repeat fire events can disrupt the conifer regeneration cycle when the

Figure 2.1.2. Unburned clusters of cones in the canopy of a burned black spruce stand provide an aerial seedbank for post-fire regeneration, hastening the recovery of spruce forest after a fire. In this photograph, scientists take measurements in permanent plots to record vegetation recovery after a fire that burned near Chicken, Alaska, in 2004. Photograph by Jill Johnstone.

fire-free interval is too short for serotinous conifers to develop their aerial seedbank and re-seed after fire (Buma and others, 2013). In these situations, a shift from stable cycles of conifer dominance to successional cycles dominated by deciduous broadleaf trees (Johnstone and Chapin, 2006b), or disruption of forest cover to non-forest ecosystems dominated by shrubs or graminoids (Brown and Johnstone, 2012; Chapin and others, 2013), can occur.

Once broadleaf species such as aspen and birch become established, they can regenerate after fire by resprouting from belowground roots and stems and maintain their dominance across multiple fire cycles. As a result, changes in fire activity that alter successional patterns between broadleaf and conifer species can affect forest landscape composition over long periods (Kelly and others, 2013). A shift from conifer to birch or aspen canopies increases evaporation of moisture and reflectance of radiation from the forest canopy, enough that a shift to broadleaf dominance after fire can counteract the warming effect of carbon emitted from burning the conifer forest (Randerson and others, 2006). Deciduous broadleaf forests tend to have higher fuel moisture and lower rates of fire spread, and thus conversion of conifer forests to broadleaf-dominated stands may help make boreal landscapes more resilient to the effects of climate warming on fire activity (Johnstone and others, 2011). The many and complex interactions between fire, weather, and forest structure mean that understanding and anticipating landscape responses to changing wildfire characteristics requires an integrative systems approach that builds on our understanding of ecosystem succession and fire behavior.

IMPLICATIONS FOR MANAGEMENT

Changes to the fire regime caused by a change in climate and human activity are likely to alter the distribution and relative dominance of forest types within the NWB forest. Releases of carbon from wildfires and subsequent permafrost degradation are likely to cause positive feedbacks to climate warming, but shifts from conifer to deciduous forest cover may compensate for these changes by altering landscape flammability and energy transfer. Climate-driven changes in the fire regime are likely to overwhelm human capacity to manage fire in sparsely populated areas (Wotton and others, 2010), and landscape-scale conservation strategies will likely need to accommodate changing fire and vegetation mosaics.

INFORMATION GAPS

- Little is known about how fire and vegetation succession are likely to interact with permafrost in lowland forests and peatlands or in tundra environments. These interactions may be important to forest carbon dynamics over the coming century.
- Fire-initiated changes in forest vegetation may affect large mammals, like moose, caribou, elk, and bison. However, there is not yet a good understanding of how a shifting forest mosaic may lead to increases in some mammal populations and decreases in others. Additionally, little is understood about how changes in the distribution and abundance of large mammals will affect patterns of forest succession.
- Although it is known that fire may initiate the conversion of conifer forests to deciduous broadleaf forests, the degree to which these changes are likely to occur at landscape and regional scales is not known. What are the implications of such a conversion for landscape and regional processes of water, energy, and carbon cycling?
- Further research is needed on how shifts in forest succession patterns and the mosaic of forest types will influence future fire regimes and climate effects on boreal forests.

2.2
FOREST INSECT AND PATHOGEN EPIDEMICS IN THE NORTHWEST BOREAL REGION

By Matt Bowser[1] and Alex Woods[2]

KEY FINDINGS

- Character, composition, and distribution of the boreal forest have been and will continue to be shaped by irruptive outbreaks of forest insects and pathogens.
- Local climate variation, by contributing to susceptibility of trees and the success of their parasites, is a primary driver of outbreaks.
- Introductions of exotic insects and pathogens threaten to alter the boreal forest irreversibly through novel outbreak relationships.

INSECTS

The Northwest Boreal (NWB) region supports a variety of insects and pathogens that can affect the overall health, age, and structure of the forests. Outbreaks of forest insects affect vast areas of the boreal forest, altering forest composition and structure. Insects cause less total damage than fire in the drier, continental regions of the NWB; the reverse is true nearer to the coastal limits of the boreal forest (Malmström and Raffa, 2000). Broadleaf forests of birch (*Betula* spp.) and poplars (*Populus* spp.) are susceptible to outbreaks of leaf-feeding insects. Conifer forests of spruces (*Picea* spp.), larch (*Larix*), and pines (*Pinus* spp.) support outbreaks of both leaf-feeding and phloem-feeding (the tissue under the bark that conducts food produced by photosynthesis to all parts of the plant) insects. In conifers, defoliation by leaf-feeding insects often predisposes trees to attack by phloem-feeding insects (Werner, Holsten, and others, 2006).

As with fire, insect outbreaks in the boreal forest tend to be cyclical and patchy, recurring in 10–30-year cycles for leaf-feeders (Werner, Raffa, and Illman, 2006) and

1 Kenai National Wildlife Refuge, Soldotna, Alaska, USA.
2 Ministry of Forests, Lands and Natural Resource Operations, Smithers, British Columbia, Canada.

BOX 2.2.1: SPRUCE BEETLE

In south-central Alaska, regional scale outbreaks of spruce bark beetles (*Dendroctonus rufipennis*) in white and Lutz spruce (*Picea glauca* and *Picea* x *lutzii*) have been occurring from the oldest available dendrochronology records in the mid-1700s to the present. Spruce beetle outbreaks coincided with warmer phases of El Niño Southern Oscillation and the Pacific Decadal Oscillation (Sherriff and others, 2011). Warmer, drier summers lead to water stress in spruce trees while quickening the development of spruce beetles (Berg, 2000). From 1990 to 2000, 1.19 million hectares (2.94 million acres) were infested in south-central Alaska, in many stands killing more than 90 percent of spruce trees greater than 11 cm (4.3 in) diameter at breast height (Werner, Holsten, and others, 2006), an outbreak apparently unprecedented in terms of regional synchrony within the last 250 years (Sherriff and others, 2011). Potential regeneration of the affected areas as spruce forest or as a more open, grassy forest is a current topic of debate and research (Berg, 2000; Werner, Holsten, and others, 2006; Boggs and others, 2008).

as much as more than 100-year cycles for phloem feeders (Sherriff and others, 2011). This heterogenic of insect outbreaks in time and space contributes to variation in tree species composition and successional states of the boreal forest.

Forest insects interact with other natural and human-caused disturbance agents including fire, pathogens, and timber harvest. Insect-caused damage facilitates the ignition and spread of fires. Spruce beetles (*Dendroctonus rufipennis*) carry plant pathogenic fungi that aid in overwhelming host trees (Werner, Holsten, and others, 2006). In south-central Alaska, infection of alders (*Alnus tenuifolia*) by canker fungi appears to make these shrubs more vulnerable to severe defoliation by leaf-feeding insects (Ruess and others, 2009).

The NWB region forests are vulnerable to extensive change through the potential introduction of exotic insect species. Large-scale, cyclical outbreaks of at least some forest insects in the boreal forest are synchronized by cyclical climatic events. Cyclicity of spruce bark beetles in south-central Alaska has been linked to the Pacific Decadal Oscillation and the El Niño Southern Oscillation (see Box 2.2.1, "Spruce Beetle").

Even though outbreak cycles of irruptive forest caterpillars appear to be driven mainly by interspecies interactions among trees, herbivores, and the herbivores' natural enemies (Berryman, 1996), coincidence of outbreaks over large areas suggests that their population dynamics are synchronized by either dispersal or climate oscillations (Liebhold and Kamata, 2000).

Due to the complexity of host-disease interactions, consequences of climate change on pathogen outbreaks are difficult to project (Coakley and others, 1999), but it is expected that warming will release some pests from their temperature constraints (Woods and others, 2005). In the case of the widespread and ecologically important nitrogen-fixing shrub green alder (*Alnus viridis*), a warming climate is expected to increase susceptibility to insect herbivory (Rohrs-Richey, 2010).

Characteristic low diversity of these forests renders them less resilient than more diverse forests and contributes to disruptiveness of even native forest insects (Juday and others, 2005). For example, recent introduction of the green alder sawfly (*Monsoma pulveratum*) has contributed to the decline of thinleaf alders (*Alnus tenuifolia*) over much of south-central Alaska (Kruse and others, 2010).

PATHOGENS

Pathogens are less obvious as agents of change in the western boreal forests than insects and far less significant than wildfire. Pathogen outbreaks are arguably less predictable under various climate change scenarios than either insect or wildfire given their correlation to precipitation. The phenomenon of alder decline in southeast and south-central Alaska illustrates the extent to which a fungal pathogen in conjunction with environmental stress can rapidly lead to possible long-term ecological changes in some ecosystems (Ruess and others, 2009). For example, drought conditions in the first decade of the 21st century seem to have predisposed thinleaf alder, an important nitrogen-fixing riparian shrub species, to attack by a pathogenic fungus *Cytospora umbrina* (asexual state), the causal agent of cytospora canker. Alder decline illustrates how the combination of environmental change in association with opportunistic fungal pathogens has the potential to alter processes as fundamental as nutrient cycling in some ecosystems of the NWB region.

Although drought has been implicated in alder decline, the opposite trend in precipitation has been posited as the driver in another pathogen outbreak. An increase in summer precipitation in the neighboring region to the south has demonstrated how typically favorable climatic changes can have unexpected implications if those changes remove environmental thresholds allowing for rapid spread and intensification of a fungal pathogen. An epidemic of Dothistroma needle blight caused by the fungus *Dothistroma septosporum* has been linked to directional climate change in northwest British Columbia (Woods and others, 2005). In this area, the foliar disease has resulted in mortality in both immature and mature pine stands where pine is naturally a minor component of the forested landscape. A similar disease epidemic in the NWB region could have larger ecological consequences given the preponderance of lodgepole pine (*Pinus contorta*) and lack of species diversity in the area. The northernmost documented report of Dothistroma needle blight detected in British Columbia was within the southern boundary of the NWB Landscape Conservation Cooperative (LCC) (McCulloch and Woods, 2009) and well within the northern range of lodgepole pine (Wheeler and Critchfield, 1985), its host in this region. Depending on how climate change unfolds in the coming decades, the potential exists for foliar diseases such as Dothistroma needle blight to increase in both incidence and impact.

Rapid environmental change can upset long-established relationships between hosts and pathogens (Harvell and others, 2002). Pathogens can be particularly damaging to host populations when environmental changes expose species to novel threats for which little genetic selection pressure has previously been applied (Telford and others, 2014). In some areas, the potential exists for diseases to exploit favorable environmental change where hosts have not been exposed to such challenges due to the cold, dry climate of the NWB region. If the climate becomes too dry, still other pathogens may be able to take advantage of drought-induced stress. The magnitude and direction of changes to seasonal precipitation trends will largely dictate which threat posed by pathogens is realized.

FUTURE PROJECTIONS

Cyclical outbreaks of insects and pathogens in the boreal forest will continue, but for most species, confident projections of directionality cannot be made (Dukes and others, 2009). An exception is the spruce beetle. With a warming climate, spruce trees are expected to be severely affected by the spruce beetle as warmer temperatures stress the trees and favor rapid development of the beetles (Berg, 2000; Werner, Holsten, and others, 2006). The relationships of climate and most other irruptive agents in the NWB region are not yet understood.

The number of exotic insects and pathogens introduced to the boreal forest are likely to increase as humans continue to move materials to and within the region, and a changing climate is expected to make the region increasingly favorable for a larger number of exotic species (see Chapter 2.3). Generally, which new exotic species will be brought to the region, or what ramifications these introductions may have on the ecosystem, cannot be predicted. However, if only new phytophagous insects and pathogens are added to the regional biota and no new tree species or genotypes are added, then the projection is assumed directional, with the trees being the loser. The reverse may be true with introductions of predators, parasites, and pathogens of the irruptive agents themselves.

IMPLICATIONS FOR MANAGEMENT

Within the NWB region, managing for diverse stands should help to reduce outbreaks because mixed forests are more resistant to insects and diseases than less diverse forests (Comeau and Thomas, 1996). Compared to mixed forests, monoculture forests are at a greater risk of outbreaks of both insect (Jactel and others, 2005) and fungal disease (Pautasso and others, 2005). Every effort should be made to limit introductions of new exotic insects and pathogens into the region because the consequences of these introductions are unpredictable, potentially severe, and generally irreversible. Managers need to understand that insects and diseases are natural, inextricable parts of boreal forest systems. Even where humans contribute to changes in host-pest dynamics (due to climate change, transport of exotic species, forest management practices), managers may need to cope with these trends rather than attempt to reverse them. For example, it may be unproductive to manage toward historical climax Lutz spruce forests on the southern Kenai Peninsula where a warming climate and spruce beetles are expected to render this region less hospitable for the spruce.

INFORMATION GAPS

- Drivers of most insect and pathogen cycles are not known. Reconstruction of past climate combined with dendrochronology and life history studies are lacking for most irruptive species.
- Long-term insect and pathogen outbreak data are needed to describe irruptive cycles, including quantitative data on insect densities and outbreak incidence.

2.3
INVASIVE SPECIES IN THE NORTHWEST BOREAL REGION

By Tricia L. Wurtz[1], Bruce A. Bennett[2], and Maria C. Leung[3]

KEY FINDINGS

- To date, the ecosystems of the NWB region are less affected by invasive species than most ecosystems on Earth.
- Exotic and potentially invasive species are being introduced to the region at an increasing rate and by an increasing variety of vectors.
- To date, only a handful of the exotic species in the region have moved beyond the anthropogenic footprint and into native ecosystems.
- Little information on the ecological effects of invasive species in the NWB region is available.
- The attitudes and responses of land management agencies to invasive species vary widely.
- Once invasive species become established, they are expensive to remove and the chances of eradication are low.

The ecosystems of the Northwest Boreal (NWB) region have fewer invasive species and are less affected by invasive species than most other ecosystems on Earth. The region has been protected by its cold climate and comparative lack of roads and other development; however, these barriers are eroding. Lengthening growing seasons, warm winter temperatures (Chapter 3.1; Sanderson and others, 2012), and extensive wildland fire (Chapter 2.1.1), combined with increasing activity in agriculture, mining, oil and gas extraction, forestry, and wilderness tourism, are extending the network of travel corridors and altered landscapes for the spread and establishment of potentially invasive species.

---

1 U.S. Forest Service, Fairbanks, Alaska, USA.
2 Environment Yukon, Whitehorse, Yukon, Canada, Retired.
3 Wild Tracks Ecological Consulting, Whitehorse, Yukon, Canada.

VECTORS AND SPREAD

Vectors by which invasive species have been introduced to the NWB LCC include intentional introductions for agricultural purposes, inadvertent introductions from agricultural or forestry products, the movement of contaminated road vehicles, boats, aircraft, and animals into previously unaffected areas, and the disposal of live specimens. Introductions usually have been associated with urban centers or other sites of high human activity, but introductions into remote sites are becoming more common.

Several of the most aggressive invasive plant species in the NWB initially were introduced through agriculture, either for forage (for example, smooth brome [*Bromus inermis*], bird vetch [*Vicia cracca*]), to improve soils (for example, white sweetclover [*Melilotus albus*]), or accidentally through seed mix impurities (for example, creeping thistle [*Cirsium arvense*], field pennycress [*Thlaspi arvense*], and leafy spurge [*Euphorbia esula*]). Later, species that had proven utility in agricultural settings were used for purposes such as erosion control during road construction, mineral exploration, and mine reclamation. This enlarged the area occupied by the introduced species (Bennett and Mulder, 2009) and provided a greater source for further spread.

Horticultural activities after World War II greatly increased the various species being introduced intentionally, such as Siberian peashrub (*Caragana arborescens*) and European bird cherry (*Prunus padus*), and inadvertently, such as perennial sowthistle (*Sonchus arvensis*), in the soil of containerized imported ornamental plants (Conn, Beatty, and others, 2008; Conn, Stockdale, and Morgan, 2008).

Today, introductions are increasing in frequency and opportunity through trade (Carlson and Shephard, 2007). Imported hay and straw carry a variety of viable weed seeds (Conn and others, 2010). Shrink-wrapped bundles of firewood originating in Washington State and sold in Fairbanks harbored five species of living insects, including two exotic bark beetles (U.S. Forest Service, 2012). Gypsy moths from Europe (*Lymantria dispar dispar*) and Asia (*Lymantria dispar asiatica*) have been detected in Alaska (either captured in insect traps or intercepted in ports) seven times since 1985 (U.S. Forest Service, 2012). The European gypsy moth egg masses likely were carried north from the lower 48 states on recreational vehicles, whereas the Asian egg masses were found on cargo ships from Asia.

Once established, introduced species may maintain local populations, often without noticeably spreading, for long periods until conditions become favorable for their spread. The fruits of European bird cherry, a popular landscape tree, are readily consumed and spread by birds. The trees have gradually become widespread in natural areas within and around Whitehorse, Yukon, and in Fairbanks and Anchorage, Alaska. European bird cherry monocultures also have formed in some forested urban parklands of Anchorage (Roon, 2011; U.S. Forest Service, 2015).

Vehicles of all types spread invasive plants and animals. Highway maintenance activities, including mowing, brushing, and grading, are increasing the spread of invasive plants by transporting propagules and creating surface disturbance (Line and others, 2008). Research in both Alberta, Canada, and Alaska has examined the spread of non-native earthworms that can be released in gardens, dumped by anglers, or transported in tire treads (Cameron and Bayne, 2009; Saltmarsh and others, 2016). Boats and floatplanes have spread the invasive aquatic plant elodea (*Elodea* spp.), which was originally introduced to Alaska's waterways by aquarium release. As of this writing, elodea has been found in three lakes far off the road system, likely introduced by floatplanes traveling from infested lakes along the road system (U.S. Forest Service, 2014).

Non-native plants are common along most of the hiking trails in the Kenai Mountains, but are rare within natural vegetation communities of the area (DeVelice, 2003; Bella, 2011), suggesting that the plants were introduced by trail users or trail maintenance activities. Wind-dispersed plants (for example, common dandelion [*Taraxacum officinale*] and narrow-leafed hawksbeard [*Crepis tectorum*]) are finding their way into remote backcountry areas without trails or roads (B.A. Bennett, Environment Yukon, Whitehorse, Yukon, Canada, written commun., June 1, 2016).

ECOSYSTEM EFFECTS

Invasive species can have profound effects on native ecosystems, causing displacement or mortality of endemics, outcompeting native species for limited resources, and altering food webs. However, information on the ecological effects of invasive species in the NWB region is scant, partly because only a handful of species have moved into natural ecosystems (Rose and Hermanutz, 2004; Oswalt and others, 2015). Between 1999 and 2004, an outbreak of larch sawfly (*Pristiphora erichsonii*), an insect native to Europe, affected an estimated 240,000 hectares (600,000 acres) of interior Alaska, killing roughly 80 percent of the Alaska larch trees (*Larix laricina*) in the affected area (Burnside and others, 2010). In 2010, the green alder sawfly (*Monsoma pulveratum*), an insect native to Europe and north Africa, was determined to be widespread in south-central Alaska (Fig. 2.3.1); in some areas it completely defoliated large patches of alder (Kruse and others, 2010; U.S. Forest Service, 2011). How these sawflies originally were introduced to Alaska is not known. Although ornamental European bird cherry has densely colonized areas of the banks of two urban Anchorage streams, it does not have a strong effect on leaf litter processing by stream invertebrates (Roon and others, 2014). Models of elodea habitat suitability in Alaska showed high suitability across a large expanse of interior Alaska, an area that includes numerous Chinook salmon (*Oncorhynchus tshawytscha*) and Alaska whitefish (*Coregonus nelsonii*) spawning and rearing sites (Luizza and others, 2016); the authors concluded that elodea may pose direct negative effects on local subsistence practices related to these species.

Figure 2.3.1. Green alder sawfly (*Monsoma pulveratum*) larvae feeding on thinleaf alder (*Alnus tenuifolia*) leaves in Alaska. The species is native to Europe and North Africa. Photograph by U.S. Forest Service.

IMPLICATIONS OF MANAGEMENT

Overall, there is a need to build awareness of invasive species and their potential effects in the NWB region. The current attitudes and approaches taken by land management agencies to invasive species in the region vary widely. Some agencies seem unaware of the issue, while others conduct extensive mapping and monitoring efforts, but do not engage in control efforts, as well as those with active and effective invasive species control programs (Schwörer and others, 2014). In Alaska, several agencies have invested significant resources and effort in developing invasive species management plans, but in most cases implementation lags far behind planning. In Yukon, outreach programs to raise awareness of invasive species have begun, and public awareness of particular invasive species has been measured.

Voluntary compliance attained through public education tends to be the starting point for reducing spread of invasive species. Regulatory approaches sometimes are introduced after the effect of public education is determined to be insufficient (for example, Ontario Ministry of Natural Resources, 2012). Understanding how more experienced jurisdictions have mandated control of invasive species provides tools for managing invasive species in the NWB region. For example, Alaska adopted a ban on felt-soled waders in 2012, and in 2014 enacted a quarantine that prohibits the importation, purchase, sale, or distribution of five aquatic plant species, including elodea.

Spatial boundaries of most strategic plans are defined politically (for example, statewide or provincewide) rather than by biome. Because controlling invasive species is about working with people and their activities, networking and aligning policies among existing political boundaries would move toward landscape-scale efforts spanning several political regions. An example is the Canadian Cooperative Wildlife Health Centre, an organization that uses an "early detection, rapid response" approach to wildlife diseases with a network of federal, provincial, and territorial resources. The Alaska Committee for Noxious and Invasive Pest Management (CNIPM) is an effective network with more than 40 member organizations. It recently released a strategic plan that includes objectives in coordination, education and outreach, prevention, inventory and monitoring, control and management, and research (Alaska Committee for Noxious and Invasive Pest Management, 2016).

The pathway by which invasive species spread is one of the most influential factors in designing prevention and control measures. Pathways relate to species' life histories in a broad sense, for example, whether the species is terrestrial or aquatic, or whether it requires a host species. Likewise, the people most likely to introduce and spread invasive species are those associated with specific pathways and vectors, such as elementary school teachers (Rozell, 2009), float plane pilots, scientists (Ware and others, 2012), greenhouse and nursery owners (Reichard and White, 2001), or persons engaged in agricultural activity or transportation.

Goals of strategic planning can be different depending on whether the area has high anthropogenic development. The goal in areas of high anthropogenic development generally would be to control and reduce invasive species, whereas the goal in relatively pristine environments would be to prevent introductions and establishment.

The likelihood of invasive species reaching new areas in the NWB region is increasing with resource development and climate change. Gas lines are being developed, new and far-ranging roads are proposed in Alaska, and in recent years, mining activity has increased dramatically in Yukon (Yukon Geological Survey, 2012). The increasing extent and severity of wildfire means that more land area will be disturbed, much of it vulnerable to introduction and spread of invasive plants (Cortés-Burns and others, 2008; Spellman and others, 2014). Warming climate trends and longer shoulder seasons will reduce the climate filter that has so far prevented some invasive species from establishing in the region (Bennett and Mulder, 2009; Industrial Forestry Service, Ltd., 2011; Fig. 2.3.2).

Once invasive species become established, removal is expensive and the chances of eradication are low (Environment Yukon, 2010; Kenai Cooperative Weed Management Area, 2014; South-central Alaska Northern Pike Control Committee, 2014). For example, for the attempted eradication of elodea from three small lakes on

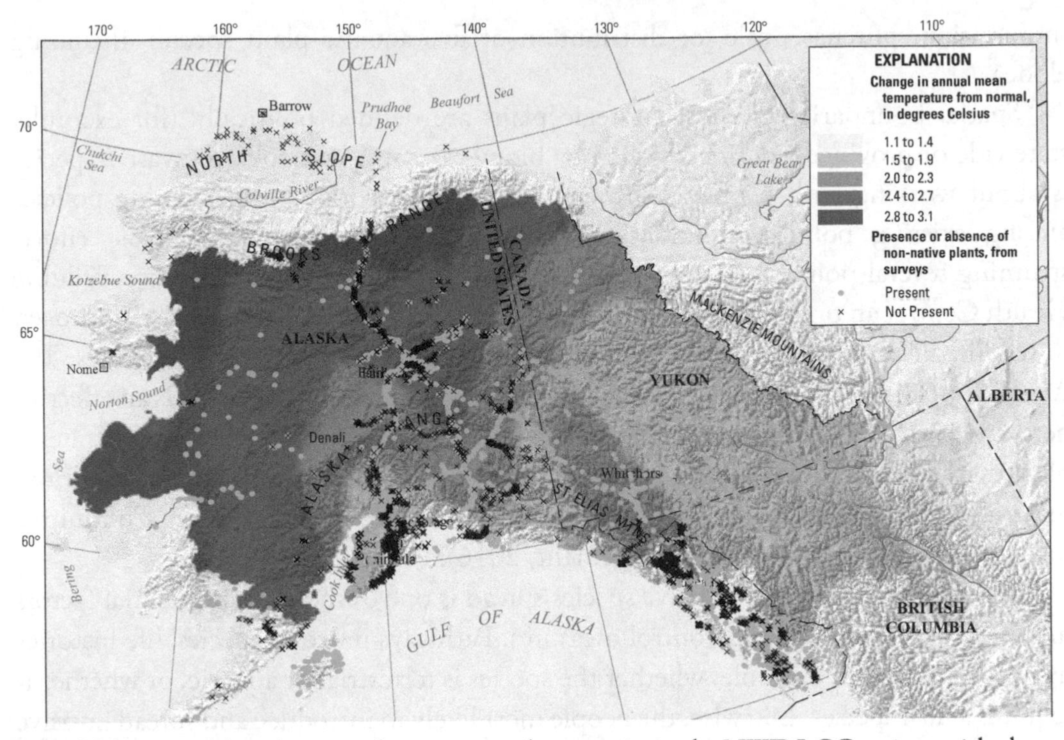

Figure 2.3.2. Known distribution of non-native plant species in the NWB LCC region, with change in annual temperature from the reference period 1970–1999 to the future period 2030–2059 for the average of five climate models. (See Chapter 3.1, Table 3.1.2, and Figure 3.1.4.) Plant data are from the Alaska Exotic Plant Information Clearinghouse Database (Alaska Center for Conservation Service, 2016); Department of Interior, Alaska Climate Science Center, data courtesy of Scenarios Network for Alaska and Arctic Planning (2014).

the Kenai Peninsula the projected cost in herbicides alone of was US$597,000 (Kenai Cooperative Weed Management Area, 2014).

All sectors of society can play a role in preventing the introduction of exotic and potentially invasive species to the NWB region. Proactive planning and risk assessment by land managers will enable them to respond rapidly and effectively when new infestations occur.

INFORMATION GAPS

- Baseline information is still needed on native species' distributions within the NWB region, especially for insects and pathogens. In some cases, determining whether a newly detected species is an overlooked native or a new invader has been difficult.
- A risk assessment of potentially invasive species that considers likelihood of introduction, potential vectors, likelihood of persistence, and effects of ecological/economic/social outcomes (Luizza and others, 2016).

- More information on the ecological effects of the most aggressive invaders already in the region (sweetclover, bird vetch, elodea, and green alder sawfly) would help land managers determine appropriate responses.
- Information about pesticide fate in the NWB region is needed to help managers evaluate pesticide utility in managing infestations because pesticide degradation rates are significantly slower in cold climates than in the climates where pesticides are typically developed and tested.

BOX 2.3.1: WHITE SWEETCLOVER IN THE NORTHWEST BOREAL REGION

White sweetclover (*Melilotus albus*) is among the most widespread invasive plants in the NWB region. It was originally introduced to increase nitrogen and organic content in agricultural soils. Later, sweetclover was planted for roadside stabilization and for reclamation projects associated with oil and gas exploration, pipeline construction, and mining. Beekeepers also have introduced it to enhance foraging opportunity for honeybees. From roadsides and reclamation project sites, sweetclover has spread onto the floodplains of at least four major rivers, where its seeds are easily carried downstream. Sweetclover spread from the town

Figure 2.3.3. White sweetclover (*Melilotus albus*) on the floodplain of the Stikine River in the Stikine-LeConte Wilderness Area of southeast Alaska. The seeds were carried downstream from the small historical settlement of Telegraph Creek, British Columbia. Photograph by U.S. Forest Service.

of Telegraph Creek in British Columbia down the Stikine River to the Stikine-LeConte Wilderness Area in southeast Alaska (Conn, Beattie, and others, 2008; Conn and others, 2008; Conn and others, 2011; Fig. 2.3.3). In Yukon, tall stands of sweetclover on roadsides compromise highway safety and require extensive and costly mowing (Keevel, 2007).

Sweetclover has moved north along the Dalton and Dempster Highways, far from areas populated by people. Eradication of this species from the NWB region is not considered a realistic goal; instead, efforts are focused on slowing its spread.

The ecosystem effects of this species have been the focus of some research. Dense stands on early successional floodplain sites create novel shade environments. Sweetclover areas had approximately 50 percent greater native seedling mortality than areas without sweetclover, suggesting that sweetclover infestations on early successional sites have the potential to change the composition of the developing plant community (Spellman and Wurtz, 2011). A study of competition for pollinators between sweetclover and native cranberry (*Vaccinium vitis-idaea*) and blueberry (*Vaccinium uliginosum*) reported a complex relationship, with negative, neutral, and positive effects on pollination and fruit set of native berries (Spellman and others, 2013).

3
PHYSICAL DRIVERS

Carl Markon[1]

KEY FINDINGS

- Regional climate trends during the latter 20th and early 21st century indicate an increase in temperature and precipitation, but the trends are small compared to observed interannual variability.
- The length of the growing season as measured by green-up and senescence generally varies across the NWB, but generally has increased, and is projected to increase.
- Higher than normal winter and spring air temperatures are producing an early onset of spring runoff and more rapid snowmelt events, resulting in a compressed runoff period with higher peak flows in some regions.
- Ground temperatures in boreal regions have remained stable or have decreased slightly and active-layer depths at undisturbed locations have remained constant overall in the same period, with some interannual variation.
- In the context of conservation planning, mapping a full suite of enduring features can provide a coarse-filter approach to habitat representation and a fine-filter approach to inclusion of special elements or habitats.

The role of key physical driving forces is increasingly gaining attention in landscape-change research, especially in light of global climate change. In this chapter, three of the most prominent and natural physical drivers of landscape change in the Northwest Boreal (NWB) region are presented—climate, hydrologic processes, and permafrost. Growing season length—a land surface quality that is a reflection of the near-term effects of those drivers—and enduring features—another land surface quality that is not as affected by those drivers but considered stable in the course of a human lifespan—also

1 U.S. Geological Survey, Homer, Alaska, USA, Retired.

are described. Each of these landscape qualities provide a better understanding of the factors that cause changes in ecosystems and ecosystem services and are fundamental to the development of landscape conservation practices.

3.1
CLIMATE AND CLIMATE CHANGE IN THE NORTHWEST BOREAL REGION

By David L. Spittlehouse[1] and Jeremy S. Littell[2]

KEY FINDINGS

- Regional climate trends in the NWB region during the latter 20th and early 21st century indicate an increase in temperature of approximately 2°C (3.6°F). Precipitation has increased, but the trends are small compared to observed interannual variability.
- Regionwide, climate model projections indicate an increase in mean annual temperature of 2.1°C (3.8°F) to 5.4°C (9.7°F) depending on time frame and emissions scenario. Annual precipitation is projected to increase by 14–34 percent depending on the time frame and emissions scenario.
- The length of the growing season varies across the NWB LCC, but generally would increase by 12–39 days.
- Interactions of precipitation and temperature determine the consequences of climate change for variables with important thresholds, such as precipitation expected to fall as snow (which increases in some locations but not in others) and climatic moisture deficit (which increases across the region, although by varying amounts) despite the increase in precipitation.

CLIMATE DRIVERS

The Northwest Boreal (NWB) region is located between 58 and 70°N, resulting in a large annual variation in solar radiation. This variation coupled with its location between the cold Arctic and the stormy north Pacific and rugged topography produces a climate of extremes (Phillips, 1990; Western Region Climate Center, 2018b). The climate is primarily continental, with an average temperature range between the

1 Climate Change and Integrated Planning Branch, British Columbia Ministry of Forests, Lands, Natural Resource Operations and Rural Development, Victoria, British Columbia, Canada.
2 U.S. Geological Survey, Anchorage, Alaska, USA.

warmest and coldest months of as much as 40°C (104°F) and snow cover lasting as many as 7 months. Parts of the NWB region in northwest British Columbia, south-central Alaska, and southwestern Alaska have stronger maritime influences (Shulski and Wendler, 2007). A large high-pressure cell over the mid-Pacific and a weak low over the northern Bering Sea control summer weather, and June–August precipitation often exceeds that of winter months. Despite this, the long days produce a substantial amount of sunshine, warm temperatures, and considerable demand for soil moisture. In autumn, the contrast between Arctic and Pacific air temperature increases storm intensity and storm tracks slowly move south. The region is cold and dry in winter because Arctic air masses dominate and can stall over the area for weeks. Radiative cooling under clear skies and long winter nights compound this coldness, and in the interior of Alaska, Yukon, and northern British Columbia, frequent to near-permanent inversions persist between October and March (Phillips, 1990; Shulski and Wendler, 2007). An almost permanent low-pressure system in the Gulf of Alaska occasionally pushes mild Pacific air over the Saint Elias and Coast ranges to offer a respite from the cold and increase precipitation in the southern part of the region.

Topography modifies the large-scale climatic influences. The southern half of the region tends to have low precipitation, being in the shadow of the Saint Elias and Coast Mountains. Farther inland, precipitation tends to increase and temperatures decrease as elevation increases in the Interior Basin of the Yukon.

RECENT CLIMATE

The climate of the NWB region is changing (Lemon and others, 2008; Markon and others, 2012; Environment Canada, 2014a; Northern Climate Exchange, 2014; Pacific Climate Impacts Consortium, 2014; Streicker, 2016). Mean annual temperatures have increased by more than 2°C (3.8°F) over the last 60 years (Stafford and others, 2000; Shulski and Wendler, 2007; Alaska Climate Research Center, 2013; Environment Canada, 2014a). Winter temperatures in much of the region have increased by twice this amount. Changes in temperature affect the amount of precipitation that falls as snow (McAfee, Guentchev, and Eischeid, 2013). The trend in precipitation is positive, but is not statistically significant and is small compared to interannual and interdecadal variability. Detection of precipitation trends in the region is complicated by the interaction of interannual climate variability and the sparse network of climate stations (McAfee, Walsh, and Rupp, 2013; Environment Canada, 2014a). Warmer temperatures mean that snowmelt has started earlier and there are fewer winter heating with more summer growing degree days. Examples of current climatic conditions (1981–2010 average) for six locations in the region are presented in Table 3.1.1. Regional estimates for the entire NWB LCC region for 1970–1999 are in Table 3.1.2.

Table 3.1.1. Current climate (1981–2010) and mean tendency in projected climates in 2080s for six locations in the Northwest Boreal region.

[Data for each emission scenarios (RCP4.5 and RCP8.5) are the average of 16 general circulation models. Data from 1981–2010 Normals: http://www.ncdc. noaa.gov/cdo-web/ and http://climate.weather.gc.ca/climate_normals/index_e.html#1981; projections: ClimateWNA v4.83. Precipitation: As snow-water equivalent. Growing Ddays: Growing degree days > 5°C. Heating Ddays: Heating degree days < 18°C. Climatic moisture deficit: Sum of reference evaporation (Erf) – precipitation for months with Eref > precipitation).

Location	Period	Mean Annual				Precipitation*		Frost-free period	Growing Ddays		Heating Ddays		Reference evaporation		Climate moisture deficit	
		Temperature		Precipitation												
		(°C)	(°F)	(mm)	(in.)	(mm)	(in.)	(days)	(°C)	(°F)	(°C)	(°F)	(mm)	(in.)	(mm)	(in.)
Dease Lake	1981–2010	-0.5	31.1	446	17.6	213	8.4	58	799	1,438	6,752	12,154	445	17.5	190	7.5
	RCP4.5	2.3	36.1	495	19.5	195	7.7	95	1,295	2,330	5,710	10,280	505	19.9	220	8.7
	RCP8.5	4.7	40.5	520	20.5	170	6.7	115	1,705	3,070	4,925	8,865	545	21.5	245	9.6
Fairbanks	1981–2010	-2.4	27.7	275	10.8	80	3.1	107	1,299	2,338	7,594	13,669	420	16.5	225	8.9
	RCP4.6	1.1	34	340	13.4	105	4.1	120	1,770	3,185	6,345	11,420	460	18.1	235	9.3
	RCP8.5	3.8	38.8	370	14.6	95	3.7	140	2,175	3,915	5,480	9,865	495	19.5	250	9.8
Fort Yukon	1981–2010	-5.5	22.1	180	7.1	75	3	98	1,171	2,108	8,603	15,485	380	15	270	10.6
	RCP4.5	-1.6	29.1	240	9.4	95	3.7	100	1,650	2,970	7,315	13,165	415	16.3	275	10.8
	RCP8.5	1.2	34.2	255	10	100	3.9	115	2,015	3,625	6,484	11,670	465	18.3	305	12
Galena	1981–2010	-3.3	26.1	314	12.4	165	6.5	122	1,126	2,027	7,903	14,225	325	12.8	135	5.3
	RCP4.5	0.4	32.7	320	12.6	190	7.5	125	1,515	2,725	6,605	11,890	360	14.2	150	5.9
	RCP8.5	3.3	37.9	350	13.8	170	6.7	140	1,895	3,410	5,620	10,115	405	15.9	165	6.5
Old Crow	1981–2010	-8.3	17.1	279	11	141	5.6	84	826	1,487	9,516	17,129	325	12.8	205	8.1
	RCP4.5	-4.5	23.9	380	15	175	6.9	100	1,165	2,095	8,160	14,690	360	14.2	195	7.7
	RCP8.5	-1.4	29.5	405	15.9	170	6.7	115	1,500	2,700	7,090	12,765	385	15.2	210	8.3
Whitehorse	1981–2010	-0.1	31.8	262	10.3	142	5.6	80	936	1,685	6,584	11,851	420	16.5	250	9.8
	RCP4.5	2.7	36.9	295	11.6	130	5.1	115	1,410	2,540	5,635	10,145	475	18.7	275	10.8
	RCP8.5	5	41	310	12.2	115	4.5	140	1,815	3,265	4,895	8,810	510	20.1	300	11.8

*As snow-water equivalent

Table 3.1.2. Historical and projected future climate variables for the NWB LCC region.

[Historical: Not all seasonal values will average/total to annual due to rounding and conversion from International System of Units to U.S. customary units; similarly with days in frost/thaw calculations. Historical climate is CRU TS 3.1 downscaled to PRISM. After Walsh and others (2008); future climate is derived from an ensemble mean of five CMIP5 general circulation models (GCMs): CCSM4, GFDL-CM3, GISS-E2_r, IPSL-CM5A-LR, and MRI-CGCM3. All original data from Scenarios Network for Alaska and Arctic Planning (http://ckan.snap. uaf.edu/).]

	Historical (1970–1999) temperature (°C [°F])	Temperature change (°C [°F])			
		2040s (2030–2059)		2080s (2070–2099)	
		RCP6.0	RCP8.5	RCP6.0	RCP8.5
Annual (January-December)	−3.0 (26.5)	2.1 (3.8)	2.8 (5.0)	3.8 (6.8)	5.4 (9.7)
December, January, February	−17.2 (1.1)	2.9 (5.2)	3.7 (6.7)	5.0 (9.0)	6.8 (12.2)
March, April, May	−2.2 (28.0)	1.3 (2.3)	2.0 (3.6)	3.1 (5.6)	4.3 (7.7)
June, July, August	11.8 (53.2)	1.6 (2.9)	2.1 (3.8)	2.9 (5.2)	4.3 (7.7)
September, October, November	−4.5 (23.9)	2.5 (4.5)	3.4 (6.1)	4.2 (7.6)	6.2 (11.2)
	Historical (1979–1999) precipitation (mm [in.])	Precipitation change (percent)			
		2040s (2030–59)		2080s (2070–99)	
		RCP6.0	RCP8.5	RCP6.0	RCP8.5
Annual (January-December)	730 (29)	14.2	18.4	23.6	34.1
December, January, February	150 (6)	17.3	16.8	23.9	32.3
March, April, May	110 (4)	19.9	19.9	25.6	39.1
June, July, August	240 (9)	13.8	20.0	24.5	33.3
September, October, November	230 (9)	11.0	17.2	21.9	35.1
	Date and Growing Season	Change in Days			
		2040s (2030–59))		2080s (2070–99)	
		RCP6.0	RCP8.5	RCP6.0	RCP8.5
Date of freeze	October 3	8	11	14	21
Date of thaw	April 22	−5	−7	−12	−18
Length of growing season (days)	165	12	18	25	39

CLIMATE PROJECTIONS

Two separate datasets were used to develop projections of climate changes in the NWB region. Climate data for Western North America (ClimateWNA; Wang and others, 2012) and the Scenarios Network for Alaska and Arctic Planning (SNAP, 2014) provide downscaled climate estimates derived from the Coupled Model Intercomparison Project Phase 5[3] (CMIP5) global climate models. This information was summarized for specific locations using a multi-model ensemble from ClimateWNA and the average over the entire region using a subset of five climate models from SNAP. Increases in temperature and precipitation are projected to continue due to global warming (Figs. 3.1.1 and 3.1.2). Mean annual temperatures are projected to increase by 1–3.5°C (1.8–6.3°F) by the 2050s and by 2–8°C (3–14°F) by the end of the 21st century.

The temperature range depends on the global climate model and emissions scenario (see Box 3.1.1, "GCMs, Emission Scenarios, and Reasons for Differences in Projections"). Precipitation is projected to increase by 2 to 18 percent and 10 to 45 percent, respectively. Projected changes in winter temperature and precipitation are greater than those for summer (Figs. 3.1.2 and 3.1.3).

Within the NWB region, spatial variability in projected temperature (Fig. 3.1.3) and precipitation (Fig. 3.1.4) is a key aspect of planning for climate change. Projected temperature increases are larger in the western and northern parts of the NWB LCC. Precipitation is projected to increase across the region, but the five-climate model average (Fig. 3.1.4) indicates increases will be more pronounced in the north-central part of the NWB LCC. Examples of projected climates in the 2080s for six locations in the region are presented in Table 3.1.1 and regionwide for the 2040s and 2080s in Table 3.1.2.

The intensity of extreme events also will increase. For example, daytime extreme maximum temperatures that occur only 2–5 percent of the time at present are projected to occur about 10 percent of the time by the 2050s (median of the projection ensemble). Variables such as frost-free days and degree-days above any defined threshold also will increase (Table 3.1.1). By the 2050s, a 10-year-return period daily precipitation event will be 10–30 percent greater than at present. This means that a 1-in-25-year event may become a 1-in-10-year event by the 2050s and a 1-in-5-year event by the 2080s (median of the projection ensemble). This follows global and regional patterns described by Sillmann and others (2013).

The greater the projection for winter warming, the greater the projected increase in winter precipitation (Fig. 3.1.2). The result is that the amount of precipitation projected to fall as snow increases in some places and decreases in others, particularly by the 2080s (Table 3.1.1). However, warming likely also will result in a later start to snowpack accumulation and an earlier start to the melt season. Increased temperature and precipitation could result in a lower snowpack and earlier snowmelt at low elevations

3 http://cmip-pcmdi.llnl.gov/cmip5/.

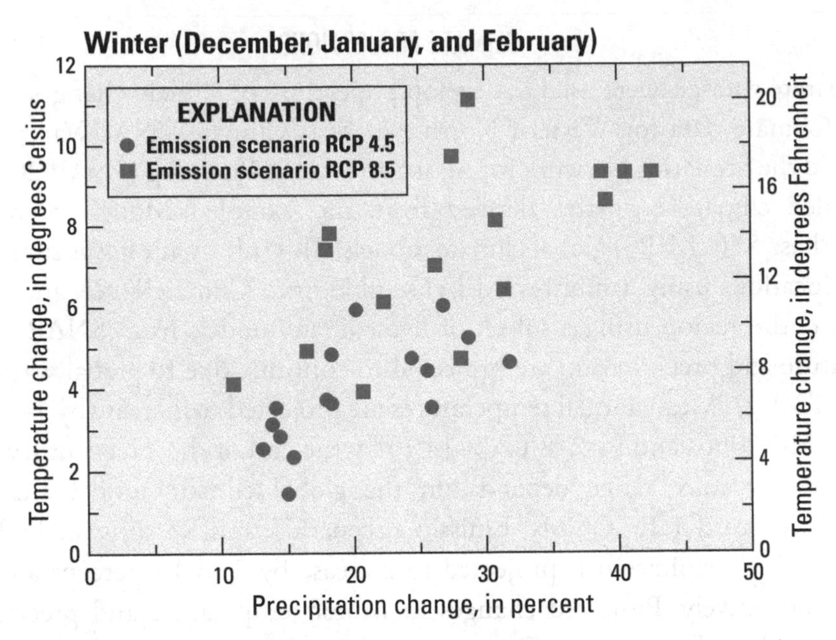

Figure 3.1.1. Change in winter (December, January, and February) temperature and precipitation for the NWB region projected by 16 general circulation models (GCMs) and 2 emission scenarios (RCP 4.5 and RCP 8.5) by the 2080s relative to 1981–2010. Data from Climate data for Western North America v4.85, accessed on April 10, 2018, at https://sites.ualberta.ca/~ahamann/data/climatewna.html.

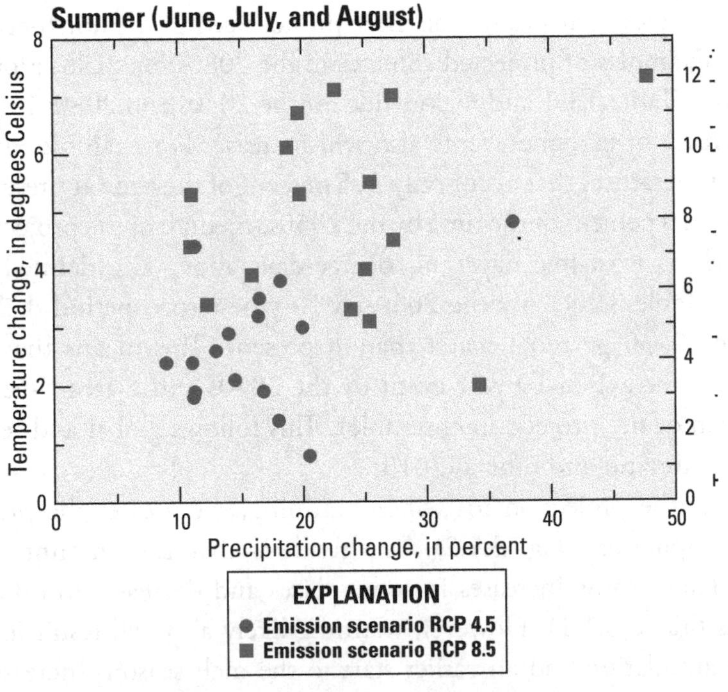

Figure 3.1.2. Change in summer (June, July, and August) temperature and precipitation for the NWBl region projected by 16 general circulation models (GCMs) and 2 emission scenarios (RCP 4.5 and RCP 8.5) by the 2080s relative to 1981–2010. Data from Climate data for Western North America, accessed October 24, 2017, at http://www.climatewna.com/.

BOX 3.1.1: GENERAL CIRCULATION MODELS, EMISSION SCENARIOS, AND REASONS FOR DIFFERENCES IN PROJECTIONS

Projections of future changes in climate are based on global climate models and plausible estimates of future greenhouse gas emissions. The models integrate physical processes of atmospheric and ocean circulation and other factors necessary to adequately simulate climate. Scenarios of greenhouse gas emissions over the 21st century used in the recent IPCC Fifth Assessment (Stocker and others, 2013) are called representative concentration pathways (RCPs) and have similar trajectories to ones used in the IPCC Fourth Assessment (Pachauri and Reisinger, 2007), which used scenario families. Both assessments project broadly similar changes in climate based on climate scenarios that explore alternative development pathways, covering a wide range of demographic, economic, and technological driving forces and resulting greenhouse gas emissions but differ in their details and are labeled differently between the two reports. The RCP4.5 (IPCC fifth assessment) is an optimistic view of the future with reduction in the rate of emissions, whereas RCP8.5 (IPCC fifth assessment) represents a business-as-usual approach with minimal reductions in emissions. RCP 6.0 (IPCC fifth assessment) is intermediate in nature and assumes a balanced use of resources not relying too heavily on one particular energy source. It provides a good midline scenario for carbon dioxide output and economic growth.

Climate models produce a range of variables for grid sizes between about 1 and 4 degrees (at grid cell areas of 10,000–150,000 km^2 in the NWB region). The projections represent area averages and are not necessarily representative of specific locations within the grid box. The projections can be downscaled to smaller areas or points of interest. A common approach uses anomalies (that is, differences between the projection for a future period and the historical baseline climatology, typically 1961–1990, 1970–1999, or 1981–2010). Analyses are usually based on 30-year windows, for example, 2020s (2011–2040), 2050s (2041–2070), and 2080s (2071–2100). In the case of temperature, the anomaly is a difference between current and future conditions; for precipitation, percentage change is used. The anomalies are applied to the baseline historical climate of the location of interest to give projections of the future climate for that location. Derived variables such as degree days and evaporative demand are then calculated for the projected climate.

To translate between the coarse scale of climate models and the local climatic conditions, a process called downscaling is used. Downscaling refers to a class of methods that range in complexity from simple statistical relationships to complicated dynamic weather modeling. Both SNAP and ClimateWNA rely on historical weather station or atmospheric data interpolated across the landscape using topography, and as such are relatively simple. Both these processes allow downscaling to scales less than

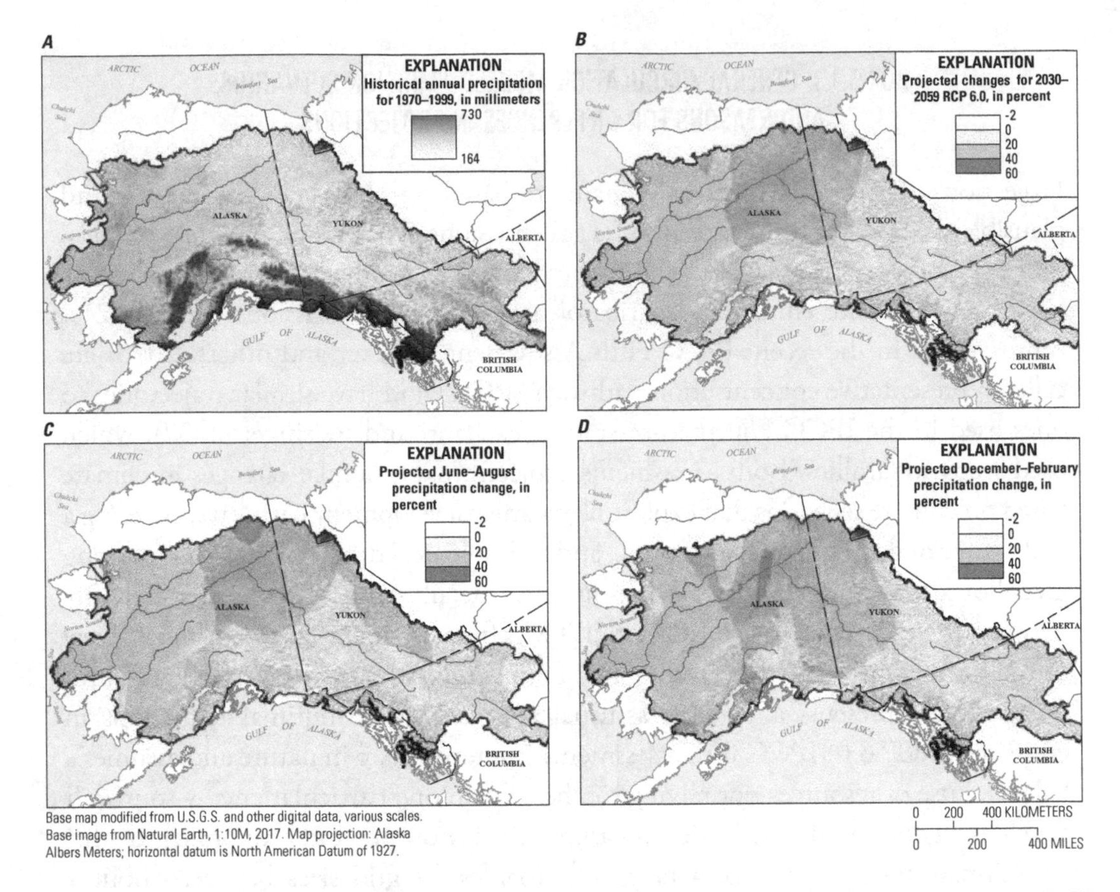

Figure 3.1.3. Historical temperature (1981–2010) and projected change data (2030–2059 RCP 6.0). (*A*) Historical mean annual temperature (CRU TS3.1 downscaled to PRISM, 1970–1999); (*B*) projected changes; (*C*) historical growing season length; (*D*) projected change in growing season length. Projections are AR5/CMIP 5 general circulation models (GCMs) downscaled to PRISM at approximately 800 m resolution and averaged for five GCMs (Table 3.1.2). Data from Scenarios Network for Alaska and Arctic Planning (http://ckan.snap.uaf.edu/dataset/projected-monthly-and-derived-temperature-products-771m-cmip5-ar5, accessed October 24, 2017).

2 km^2 (0.77 sq. miles), although in both cases it is important to remember that the finer the resolution, the more likely it is that other factors (for example, vegetation, land cover change, soils) must be considered for the downscaling to be appropriate. The main differences between the two products used here are in the historical reference stations and reference climatology used and the interpolation methods. See Scenarios Network for Alaska and Arctic Planning (2014) and Wang and others (2012) for details.

The projections have uncertainty that results from natural climate variability (the major source of uncertainty from now to about 30 years from now), how the various GCMs are parameterized (main source of uncertainty in the mid-21st century),

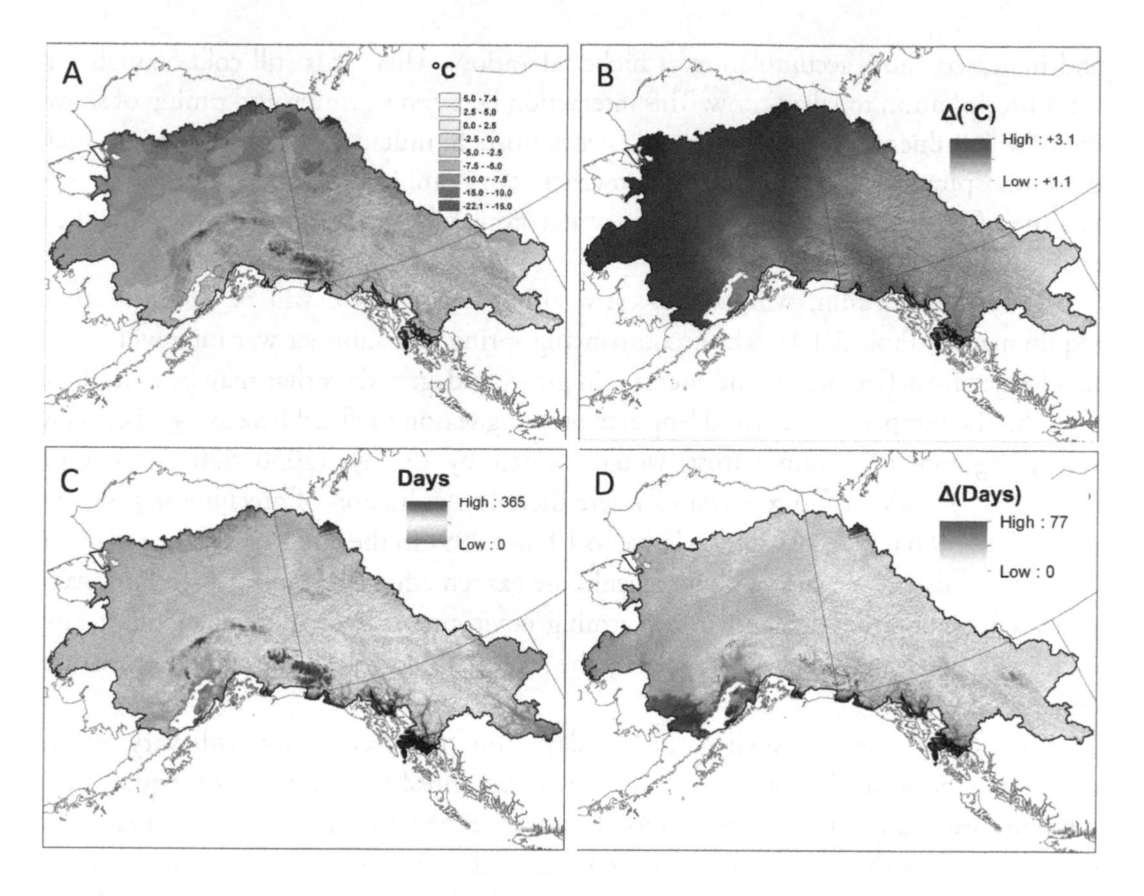

Figure 3.1.4. Historical precipitation (1981–2010) and projected change data (2030–2059 RCP 6.0. (*A*) Historical annual precipitation (1970–1999); (*B*) projected changes; (*C*) projected June–August precipitation change; (*D*) projected December–February precipitation change. Projections are AR5/CMIP 5 general circulation models (GCMs) downscaled to PRISM at approximately 800 m resolution and averaged for five GCMs (Table 3.1.2). Data from Scenarios Network for Alaska and Arctic Planning (https://www.snap.uaf.edu/).

and the emission scenarios (main uncertainty in the late 21st century). Uncertainty also can arise from the spatial and temporal scale of projections (gridded normals, annual, monthly, daily). To constrain these uncertainties, it is important to present changes in climate from multiple GCMs and scenarios and to realize that there is greater certainty in larger-scale, longer-time-frame averages than in daily projections for a point. However, it is also important to realize that the future climate will vary interannually and interdecadally, whereas averages of multiple climate models smooth out this variation. It is also possible that feedbacks between climate change and impacts (such as carbon flux from tundra, or changes in sea ice coverage) could attenuate or exacerbate impacts.

and increased snow accumulation at higher elevations where it is still cold enough for most precipitation to fall as snow. This interaction between warming and timing of snow cover will influence the occurrence of permafrost. Simultaneous changes in summer evapotranspiration and cool season snowpack can combine to affect the timing and volume of streamflow and the effective water availability for plant growth and other ecosystem processes.

A warmer autumn, winter, and early spring means there will be fewer heating requirements (Table 3.1.1). The accompanying spring and summer warming will result in a longer frost-free period and increase in growing degree days that may benefit plant growth. The temperature-derived length of growing season (defined here as days between last spring and first autumn frost) would increase by 12 days (2030–5209 relative to 1970–1999) under a future scenario where there is a reduction of greenhouse gases by as much as 39 days (2070–2099 relative to 1970–1999) in the late 21st century under a "business as usual" scenario with no greenhouse gas remediation (Table 3.1.2). However, increased evaporative demand in a warming environment will more than offset any increase in precipitation as can be seen by the projected increase in the climatic moisture deficit (Table 3.1.1).

Change in variables such as degree days and frost-free period will vary across the region because they are calculated based on threshold values. Consequently, a temperature change in a month at a location where the base temperature is equal to or greater than the threshold will affect the value of the variable, unlike another location where the increased temperature for that month does not yet reach the threshold. Thus, there will be a greater increase in lower elevations than higher elevations and southern compared with northern parts of the region. Consequently, the responses from place to place and watershed to watershed must be considered carefully. These place-specific considerations are beyond the scope of this chapter but can be addressed using downscaled products available from ClimateWNA and SNAP.

INFORMATION GAPS

- Subregional variation in historical climatology is less well understood than in comparably sized regions of southern Canada and the lower 48 United States due to the sparseness of stations with long-term data.
- High-elevation stations are poorly represented, and therefore understanding of higher elevation effects and rates of change are virtually absent.
- More active research is needed on the interactions of temperature and precipitation changes with other factors, such as permafrost, glaciers, and vegetation, which will also change with climate.
- Projecting derived climatically related variables, such as drought or runoff requires more complex considerations and should be an area of active research because they could slow or hasten the impacts of climate change.

3.2
HYDROLOGIC RESPONSE IN THE NORTHWEST BOREAL REGION
Impacts on Natural Resources, Ecosystems, and Communities

By J. Richard Janowicz[1] and Larry D. Hinzman[2]

KEY FINDINGS

- Hydrologic response in a watershed is due to the integrated effects of local meteorological events, vegetation, topography, and subsurface structure (geology, geomorphology, and permafrost). As those drivers change in some interdependent manner, the hydrological response also will change in a complex, but potentially predictable pattern. Hydrologic response (annual maximum discharge) in the region is primarily driven by snowmelt and secondarily by summer rainfall.
- Permafrost is degrading and is placing a greater importance on the interaction between surface and subsurface processes.
- Higher winter and spring air temperatures are producing an earlier onset of spring runoff and more rapid snowmelt periods, resulting in a compressed runoff period with higher than normal peak flows in some regions.
- Recorded data of the last few decades indicate that annual mean discharge has increased within continuous and discontinuous permafrost zones, with some variability within sporadic permafrost regions, whereas winter minimum discharge has increased significantly.

1 Water Resources Branch, Department of Environment, Whitehorse, Yukon, Canada.
2 Department of Civil and Environmental Engineering, University of Alaska Fairbanks, Fairbanks, Alaska, USA.

- Streamflow response trends of the glacierized drainage basins within the sporadic and discontinuous permafrost zones of southwestern Yukon indicate an increasing trend of both mean and minimum annual discharge throughout the study region.

SNOW AND RAIN

Hydrologic response in the Northwest Boreal (NWB) region is largely nival (snowmelt driven) with secondary summer rainfall influences (pluvial driven; Woo, 1986). Similar to all cold regions, spring streamflow response is characterized by a rapid rise in discharge as a result of snowmelt (Marsh, 1990). Secondary maximum discharge occurs during the summer months as a result of rainfall, whereas smaller streams experience their annual maximum discharge as a result of intense summer rainfall (Bolton and others, 2000).

Early spring snowmelt-driven discharges initiate river ice break-up that may produce ice jams and associated flooding in favorable locations (Timoney and others, 1997; Arctic Monitoring and Assessment Programme, 2011). These floods often produce the annual peak water level. Minimum annual discharge occurs in March or April, coinciding with minimum annual groundwater inputs (Janowicz, 2008). This flooding notably precedes the annual maximum discharge, which also corresponds with the maximum groundwater infiltration and recharge (Kane and Stein, 1983). Southern regions of the NWB region have substantial glacier and ice cap coverage. Peak flows in glacierized basins are delayed until later in the summer due to supplemental inputs from glacier melt, unlike nival regimes, which experience their annual maximum discharge in May or June due to snowmelt inputs (Janowicz, 2004, 2008; Fleming, 2005).

PERMAFROST

Permafrost has a dominant control over hydrologic response in northern regions by producing short pathways to the stream channel, with little interaction with subsurface processes (Hinzman and others, 2013). A thicker active layer increases residence time and promotes a longer (slower) pathway to the stream channel (as compared to a more rapid response of near surface flow). Hydrologic response in the NWB region follows this principle, and is closely tied to the underlying permafrost (Bolton and others, 2004; Janowicz, 2008). Although precipitation decreases in higher latitudes, the ratio of runoff to precipitation generally increases (Kane and Yang, 2004) due to the increasing dominance of the underlying permafrost. The exception is observed in glacierized basins of the NWB region. Because glacial regimes originate in mountainous regions with high annual precipitation amounts, they also tend to have the highest annual streamflow volumes. Conversely, minimum winter discharge decreases in northerly watersheds, where more permafrost reduces groundwater contributions. In Arctic regions, only large rivers have any appreciable winter discharge, whereas in discontinuous permafrost regions, even small streams may display extended streamflow throughout the winter due to continued groundwater contributions (Hinzman and others, 2003).

CLIMATE CHANGE

Annual, winter, and summer air temperatures generally have increased across the entire NWB region during the last few decades, with greater increases in central and northern regions of the study area (Arctic Monitoring and Assessment Programme, 2011). Annual precipitation trends are not consistent. Winter precipitation has increased in northern regions and decreased in southern regions. Summer precipitation has increased slightly throughout, with greater increases observed in southeast and central regions (Arctic Monitoring and Assessment Programme, 2011). Notable increases in extreme events have been documented in 5-day precipitation amounts and in river discharge in several watersheds in Interior Alaska (Bennett and Walsh, 2014; Bennett and others, 2015).

Higher winter and spring air temperatures are producing an earlier onset of spring runoff and more rapid snowmelt events, resulting in a compressed snowmelt runoff period with higher maximum discharge in some regions. The warming climate is also having a significant impact on ice regimes in the NWB region with earlier occurrence of break-up, later freeze-up, and a shorter ice cover period. Mid-winter ice break-ups and associated flooding are becoming more common, especially in transitional areas. Some break-up water-level trends suggest that break-up severity is increasing (Janowicz, 2010). Mid-winter rainfall periods have become more frequent in recent years and are projected to increase in the future (Rennert and others, 2009; McAfee, Walsh, and Rupp, 2013). This has serious implications for almost all aspects of life in northern regions from caribou cratering through ice-crusted snow for lichens to tree damage due to icing, to societal consequences of icy roads and structures. The shorter ice-covered period also yields a longer open-water season with a myriad of biological and social consequences (Arctic Monitoring and Assessment Programme, 2011), including longer periods of growth for migrating fish and waterfowl.

The warming climate seems to be causing a change in the permafrost and glacier distributions of the NWB region. Permafrost warming and the associated thaw results in a thicker active layer and permafrost loss in favorable areas.[3] In the near term, permafrost degradation is expected to be greatest within the discontinuous and sporadic permafrost zones because these permafrost classes are warmer and therefore more susceptible to thawing (Hinzman and others, 2005). The effect of climate warming on glacial regimes has been determined to be significant in the NWB region (Arctic Monitoring and Assessment Programme, 2011; Janowicz, 2011). On a regional basis, glaciers throughout western North America have generally retreated since the Little

3 Favorable areas for permafrost degradation include sites where the underlying permafrost is able to receive heat energy from various sources. Such sites include those with south-facing slopes, lower elevations, thin organic layer, standing water, less snow cover, and lower soil moisture.

Ice Age (Zemp and others, 2007; Moore and others, 2009). The estimated loss varies regionally with the greatest retreats observed in the Coast and St. Elias Mountains, which have the greatest glacier and ice cap coverage (Barrand and Sharp, 2010).

As permafrost properties change with a warming climate, hydrologic responses seem to be changing as well. Degrading permafrost increases the thickness of the active layer, decreases the overall thickness of the permafrost, and in certain areas entirely eliminates the presence of underlying permafrost. These actions place a greater importance on the interaction between surface and subsurface processes. Melting glaciers and ice caps are likewise altering hydrologic response by increasing runoff and altering its spatial and temporal distribution (Janowicz, 2011).

Recorded data of the last few decades indicate that annual mean flows have increased within continuous and discontinuous permafrost zones, with some variability within sporadic permafrost regions (Janowicz, 2008). Annual maximum discharge has decreased within some continuous permafrost regions, and less so in discontinuous regions with variable trends in sporadic permafrost zones. Winter minimum discharge has significantly increased within continuous and discontinuous permafrost regions over the last three decades with variable trends in sporadic permafrost regions (Walvoord and Striegl, 2007).

Streamflow response trends of the glacierized drainage basins in the sporadic and discontinuous permafrost zones of southwestern Yukon indicate an increasing trend of mean and minimum annual discharge throughout the study region (Zhang and others, 2001; Yue and others, 2003). However, trends of annual maximum discharge are more variable, with most streams showing increasing trends and some showing significant decreasing trends (Janowicz, 2011).

Glacierized basins with little permafrost seem to exhibit classic glacial response trend characteristics, whereas basins with substantial amounts of permafrost seem to exhibit more typical permafrost response trends (Janowicz, 2011). Other glacierized basins with variable amounts of permafrost have overlapping signals indicating mixed streamflow trends that are difficult to interpret.

INFORMATION GAPS

- It is unclear how potential increases in precipitation and a warmer climate would effect changes to groundwater processes as permafrost degrades.
- More information is needed on the primary controls, interactions, and responses between surface water balance and dominant vegetation distributions.
- Permafrost is showing the signs of degrading in a number of areas, but there is a lack of fundamental information on how those changes will affect boreal hydrologic regimes and associated impacts to ecosystems.

3.3
PERMAFROST IN THE NORTHWEST BOREAL REGION

By C.R. Burn[1] and M.T. Jorgenson[2]

KEY FINDINGS

- Ground temperatures 15 m (16 ft) or more deep in boreal regions have remained stable or have declined slightly during the last 10 years.
- Active-layer depths vary across the NWB region with interannual changes reflected by variations in summer climate.
- Ground ice amounts and distribution remain poorly quantified across the region.
- Thermokarst terrain is abundant in the NWB region.
- Permafrost degradation has already delivered significant consequences to ecosystems, land use, and infrastructure stability.

Permafrost, ground that remains at or below 0°C (32°F) for two or more years, is a fundamental factor affecting the patterns and processes of boreal ecosystems and the human use of northern lands. Much of the Northwest Boreal (NWB) region has discontinuous to sporadic permafrost, where frozen ground and unfrozen terrain are interspersed. In the northern part of the region and in the mountains, more than one-half of the landscape supports permafrost, but to the south, the proportion is less. To assess how permafrost responds to a warming climate, changes in the thickness of the seasonally thawed active layer above permafrost, the amount of ground ice that affects thaw settlement, and the extent and rates of thermokarst development need to be monitored and evaluated. Permafrost degradation is already extensive and delivers significant consequences for ecosystems, land use, and infrastructure.

1 Department of Geography and Environmental Studies, Carleton University, Ottawa, Ontario, Canada.

2 Alaska Ecoscience, Fairbanks, Alaska, USA.

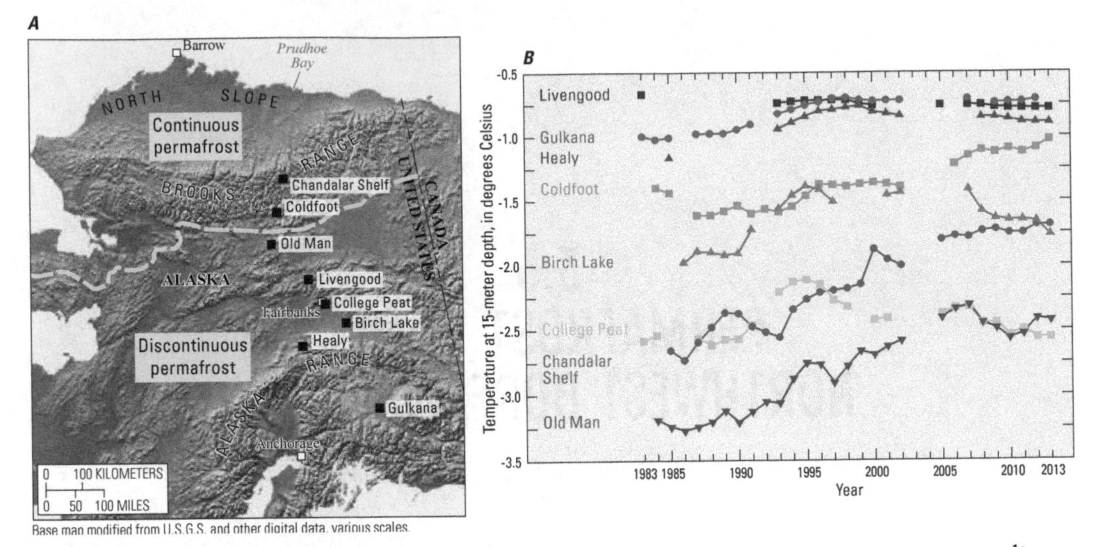

Figure 3.3.1. Permafrost temperature trends at 15 depth-at-monitoring stations across a climate gradient in Alaska showing (*A*) location of monitoring site network and (*B*) data from NWB region sites (provided as an update to Romanovsky and others, 2013).

GROUND TEMPERATURES

Since the 1970s, air temperatures have increased throughout northwest North America, leading to gradual warming of the ground. Annual mean temperatures just below the active layer, near the permafrost surface, show considerable interannual variation, but deeper in the ground the fluctuation occurs more gradually (Smith and others, 2010). Most boreal sites with icy permafrost have slow changes in temperature at depths of 5 m or more because of latent heat effects at temperatures greater than about –1°C (30°F), due to thawing of interstitial ice. In northwestern Canada, such ground temperatures have increased by 0.07–0.2°C (0.13–0.36°F) per decade, although the rate of warming has decreased where the interstitial ice has begun to melt, as at Norman Wells, Northwest Territories (Smith and others, 2012). In boreal Alaska, permafrost temperatures at 15–20 m (50–65 ft) depth have stabilized or decreased at several sites since 2007, reversing the warming trend of the previous decades (Romanovsky and others, 2013; Fig. 3.3.1). Ground temperatures, however, are strongly affected by topography, atmospheric inversions, surface water, and vegetation (Jorgenson and others, 2010). These factors can cause ground temperatures to vary as much as 10°C (50°F) across a landscape, twice the variation of the projected warming of 3–5°C (5.4–9.0°F) by 2100 for boreal regions. Such spatial variation complicates landscape-scale assessment of future permafrost stability.

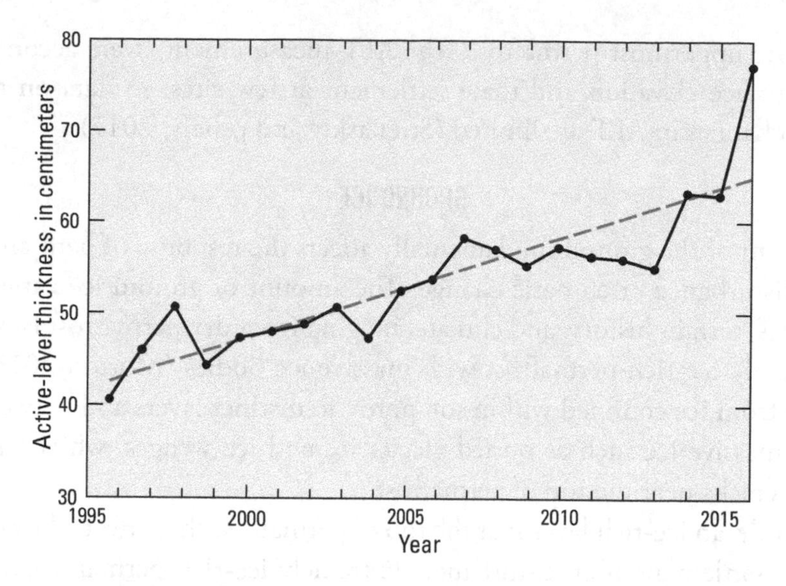

Figure 3.3.2. Average Active-Layer Thickness as measured at three Circumpolar Active Layer Monitoring sites in Interior Alaska. The CALM sites used were Old Man (U16), Wikersham Dome (U17), and Pearl Creek (U19). Site descriptions and active-layer measurements are available at https://www2.gwu.edu/~calm/. (Figure courtesy of Prof. N.I. Shiklomanov, George Washington University.)

ACTIVE LAYER

The active layer is critical to the ecology of permafrost terrain because it is the seasonally thawed layer of soil where plant roots grow, water moves above permafrost, and soil carbon accumulates or decomposes. Freezing and thawing of the active layer often is accompanied by internal soil displacement, leading to the movement and sequestration of organic matter downward through the soil profile and into the permafrost. The organic material stored in permafrost is of great current concern because of the potential for its carbon to be released as carbon dioxide (CO_2) or methane (CH_4) if the active layer deepens and the ground becomes seasonally thawed (McGuire and others, 2009; Grosse and others, 2011).

Long-term monitoring of active-layer thickness (ALT) by a large international collaboration under the Circumpolar Active-Layer Monitoring (CALM) Program has documented trends over the last several decades (Shiklomanov and others, 2012), although there are fewer monitoring sites in boreal regions than at tundra sites. In northwestern Canada, no definitive trend was determined in ALT over the 1998–2005 period, but interannual changes reflected summer climate variation (Smith and others, 2009). In boreal Alaska, average ALT has shown a slight increase since 1996, with a large jump in thickness in 2016 (Fig. 3.3.2). Detecting trends in ALT is complicated by the effects of thick soil organic layers that retard soil warming, and the high ice content of the uppermost permafrost. The ALT measurements were accompanied by surveys of surface elevation and thaw settlement at few sites, so data on the terrain response to changes in ALT are limited (Streletskiy and others, 2017).

GROUND ICE

The ice content of the ground fundamentally affects the response of permafrost terrain to surface disturbance or climatic change. The amount of ground ice varies with soil characteristics, terrain history, and climate, ranging from dry permafrost lacking excess ice to extremely ice-rich permafrost with massive ice bodies (Murton 2013). The ice itself ranges from ice confined within soil pores, to distinct layers and networks within the soil, to massive ice such as buried glacier ice and ice wedges, which fill thermal-contraction cracks near the top of permafrost.

Commonly, an ice-rich layer is at the top of permafrost that affects the amount and rate of thaw settlement after disturbance. Extremely ice-rich permafrost is associated with three distinctive surficial deposits in Canada and Alaska: glacial moraines of late Pleistocene age in Alaska and northwestern Canada (Kokelj and others, 2017); glaciolacustrine deposits with clay-rich sediments that typically have thick horizontal ice layers (Shur and Zheskhova, 2003; French and Shur 2010); and late Pleistocene eolian deposits with large syngenetic (forming at the same time as the surrounding material) ice wedges (Kanevskiy and others, 2011) (Fig. 3.3.3). The ground-ice content and distribution influences the amount of subsidence and its location in flat terrain, and the places where landslides may develop on hill slopes. Such instability is a threat to infrastructure.

THERMOKARST

The thawing and subsidence of ice-rich permafrost, termed thermokarst, is extensive in the boreal region and leads to radical changes in topography, hydrology, soils, vegetation, and wildlife use (Jorgenson and others, 2013). There is a wide range of thermokarst features depending on the amount and type of ground ice, including polygonal troughs and pits associated with degrading ice wedges, thaw slumps on slopes, and thermokarst lakes, bogs, and fens on flat terrain (Kokelj and Jorgenson, 2013) (Fig. 3.3.4). Thermokarst can be initiated by active-layer deepening in response to climate change or by disturbances from fire and human activity.

Permafrost degradation is affected by strong feedbacks, primarily positive effects of standing water restricting freeze back and negative effects of vegetation development during plant succession (Jorgenson and others, 2010). Thawing occurs either from the top by deepening the active layer or laterally with heat flow from surface water and groundwater.

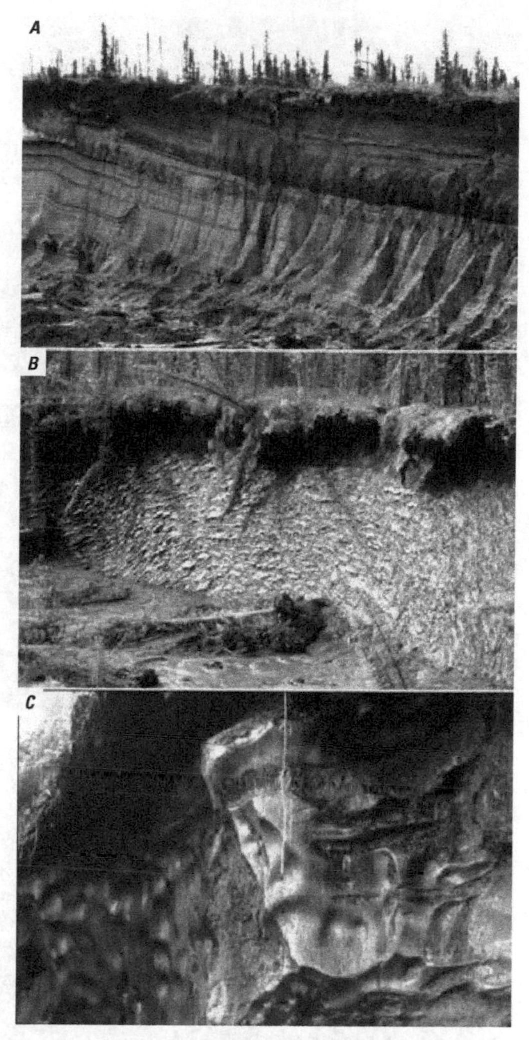

Figure 3.3.3. Surficial deposits with high ground-ice contents in the NWB region. (*A*) Relict Pleistocene buried glacier ice covered by oxidized till, Peel Plateau, Northwest Territories; (*B*) glaciolacustrine sediments with segregated ice, near Mayo, central Yukon; (*C*) large syngenetic ice wedges in Yedoma, Klondike area, western Yukon. Photograph (*A*) by S.V. Kokelj, Government of the Northwest Territories (2011), and photographs (*B* and *C*) taken in 1984 and 1991, respectively by C.R. Burn, Carleton University, Ottawa, Ontario, Canada.

Figure 3.3.4. Common types of thermokarst terrain in Alaska. (*A*) Thermokarst lakes, shore fens, and bogs on the Koyukuk Flats (2008); (*B*) thermokarst fens with drowning forests on the Tanana Flats (2006); and (*C*) thaw slump on glacial moraines in the Noatak drainage (2010). Photographs by M.T. Jorgenson, Alaska Ecoscience.

IMPLICATIONS FOR MANAGEMENT

The societal and ecological consequences of permafrost degradation in the boreal regions include changes to the hydrologic regime and freshwater discharge, changes to habitat for fish and wildlife, increased emissions of methane and changes in soil carbon, impairment of water quality from sediment input from thaw slumping on slopes, and damage to human infrastructure (Arctic Council and the International Arctic Science Committee, 2005; Kokelj and others, 2013). The magnitude of these changes will depend on climate and ground-ice conditions. Habitats affected by thawing permafrost become drier in upland areas and wetter in lowlands as forests are converted to lakes and bogs (Jorgenson and others, 2013). The net effect of permafrost degradation on the carbon balance of boreal soils is still unclear (McGuire and others, 2009).

Risks to infrastructure from permafrost degradation include decreased bearing capacities of pile foundations for buildings and elevated pipelines, impairment of road surfaces, deterioration of mine-tailings impoundments that rely on permafrost for wastewater containment, decreased operation of ice and snow roads in winter, damage to linear infrastructure from landslides, and contamination of runoff or groundwater from failure of industrial and military waste sumps and village sewage lagoons (Larsen and others, 2008).

INFORMATION GAPS

- Long-term rates of permafrost degradation are needed from more field sites.
- Anticipated carbon release from permafrost terrain is under investigation in tundra environments, but few data are available from the boreal forest.
- Wildfire is the most widespread disturbance that leads to degradation of permafrost, which needs to be assessed and monitored.
- Cost estimates for construction and maintenance of linear infrastructure built on thaw-susceptible permafrost are required for effective management of the transportation network.

3.4
GROWING SEASON LENGTH IN THE NORTHWEST BOREAL REGION

By Valerie Barber[1], Glenn P. Juday[2], and Scott N. Williamson[3]

KEY FINDINGS

- The growing season in the NWB region has increased, primarily due to spring snow cover decreasing by 2 and 10 days per decade (period of reference 1972–2008).
- The length of the growing season as measured by leaf out and senescence has increased by approximately 5 days per decade over the last 30 years, but earlier green-up and later senescence are regionally dependent.
- A recent and comprehensive analysis of satellite observations (2000–2010) shows that Interior Alaska and northwestern Canada boreal forests are browning.
- Temperature profiles now resemble those from 7° latitude farther south for the boreal. Climate models indicate that over the next 80 years those southern profiles could shift an additional 20° latitude northward.
- Potential social impacts on rural native populations and subsistence lifestyles include mismatches in timing of the natural system events and the human reliance on those systems, such as hunting and ungulate behavior, unreliable transportation corridors used in winter for traveling and hunting, new species moving north and old species in decline, and increased wildfire occurrence and severity.

---

1 University of Alaska Fairbanks, Adjunct Professor (retired), Fairbanks, Alaska, USA.

2 Professor Emeritus of Forest Ecology, School of Natural Resources and Extension, University of Alaska Fairbanks, Fairbanks, Alaska, USA.

3 Postdoctoral Fellow, Department of Biological Sciences, University of Alberta, Edmonton, Alberta, Canada.

With a potentially warming landscape (Chapter 3.1), it is likely that annual growing season for the Northwest Boreal (NWB) region (defined either as cumulative days above a temperature threshold or as the time between leafout and senescence) is lengthening. Plants, other than specially adapted Arctic and alpine species, make little photosynthetic gain at temperatures less than 5°C (41°F), even if they are not subject to lethal cold. Inherently, the active growing season has been defined as the number of snow-free days that display a daily average temperature greater than 5°C (41°F). Growing season length is the number of consecutive days with greater than freezing temperatures. However, in the case of cold-hardy plants, which includes most boreal and Arctic native species, growing season length and active growth season can be defined by temperature thresholds a few degrees below freezing. The length of time between leafout and senescence is increasing on a continental scale in Europe (Menzel and others, 2006), affecting species similar to those in Alaska particularly because of earlier spring leaf development. If tissues are not damaged by low temperatures, plants can resume growth when temperatures increase to favorable levels. Available evidence indicates that the annual growing season of the NWB region is lengthening, whether defined as cumulative days above a temperature threshold or as the time between leafout and senescence.

DECREASING SNOW COVER

Shorter seasonal snow cover is consistent with a lengthening of the growing season (Jia and others, 2009). Therefore, seasonal snow cover is an important aspect in determining growing season. The extent of pan-Arctic snow cover has decreased by 14 percent in May and 46 percent in June between 1967 and 2008 (Brown and others, 2010). Climate models underestimated recent Northern Hemisphere spring snow cover reduction in 2008–2012 (Derksen and Brown, 2012), which suggests that projections of future snow cover reductions are similarly underestimated.

WARMING TEMPERATURES

The frost-free growing season in Fairbanks (Fig. 3.4.1) has increased in alignment with regional temperature trends, but also seems to have crossed a threshold of much lower risk of occasional extreme short growing seasons. Temperature profiles now resemble those from 7° latitude farther south for the boreal and 4° farther south for the Arctic during the 1980s, and climate models indicate that during the next 80 years the profiles could shift an additional 20° latitude northward (Xu and others, 2013). In Interior Alaska the growing season has increased 45 percent (range: 85–120+ days) in the last century (Wendler and Shulski, 2009).

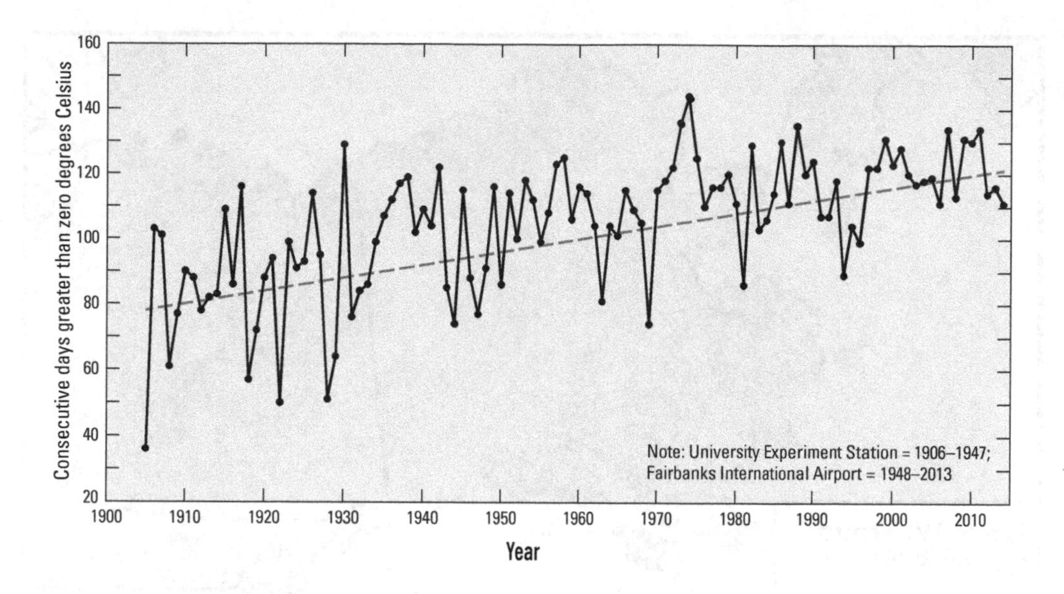

Figure 3.4.1. Frost-free season length based on climate data collected from the University Experiment Station, University of Alaska Fairbanks campus (1906–1947) and Fairbanks Airport, Fairbanks, Alaska (1948–2013). This is one of the longest continuous instrument-based weather records (112 years) in the region.

VEGETATION GROWTH

With the exception of several intensively studied areas within the NWB region, the main source of information related to growing-season length comes from surface-cover changes tracked by satellite observation, which track landscape-wide changes in both snow cover and vegetation productivity. The most common measure of vegetation productivity by satellite is the normalized difference vegetation index (NDVI). The NDVI is a proxy for biomass of green vegetation cover, which can be indexed to the amount of photosynthesis with repeat coverage through a season. Trends from NDVI indicate decreasing greenness (browning) and increasing greenness (greening) in different parts of the NWB region (Potter, 2014). A recent and comprehensive analysis in growing season (June–August) NDVI trends (Fig. 3.4.2) shows that Interior Alaska and northwestern Canada boreal forest is browning and the north slope of Alaska (largely tundra ground cover) is showing considerable greening over the period of record (year- to-year; Guay and others, 2014). Goetz and others (2005) showed that for Canada and Alaska tundra and conifer forests, the NDVI is a strong proxy for gross photosynthesis, which is positively correlated to growing season length. The browning likely is the result of temperature-induced drought stress on upland Interior Alaska boreal species (Barber and others, 2000; Juday and others, 2003; McGuire and others, 2010; Beck and others, 2011). This temperature-induced drought stress (negative

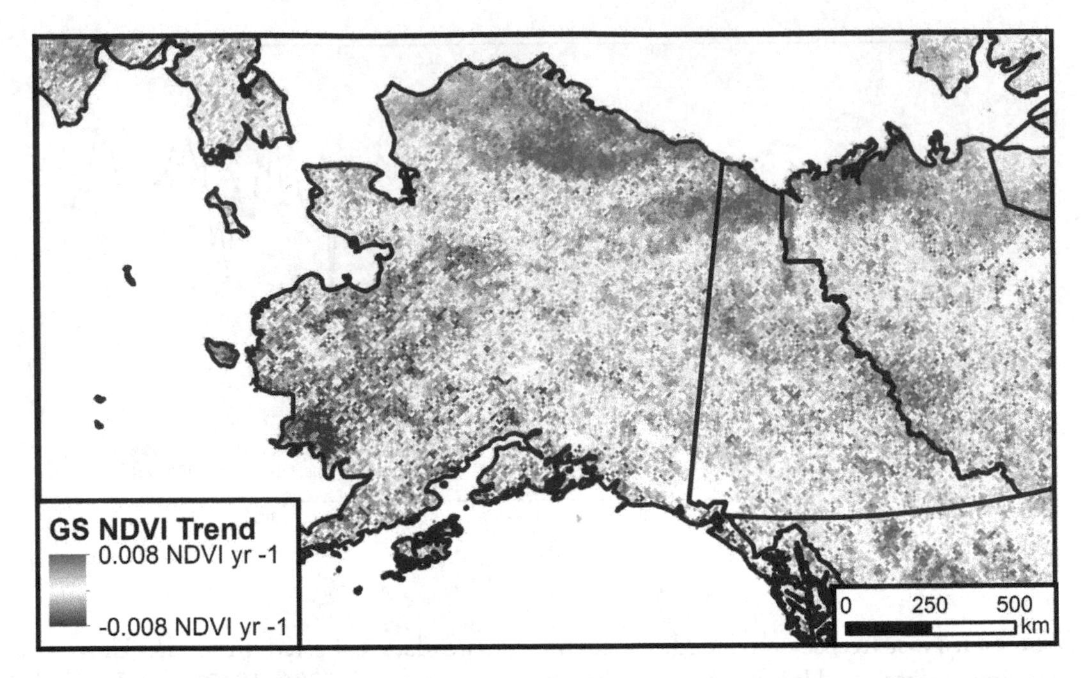

Figure 3.4.2. Trends in growing season from normalized difference vegetation index (NDVI; June to August) using GIMMS 3g data were calculated with the Theil-Sen approach during 1982–2012. Positive and negative trends indicate increases and decreases, respectively, in vegetation productivity over time. Areas classified as cropland, grassland, urban, and permanent snow and ice are not color coded. Data courtesy of C.J. Tucker and J.E. Pinzon, NASA Goddard Space Flight Center.

sensitivity to growing season temperature and positive sensitivity to precipitation) is seen on floodplain spruce as well as upland interior spruce (Juday and Alix, 2012). Further, Gamon and others (2013) show for wet sedge near Barrow, Alaska, that growing season productivity is more related to soil moisture (precipitation) and growing degree days than it is to length of snow cover period with reduced growth as the length of the snow-free season increases.

EFFECTS OF GROWING SEASON LENGTH ON ECOSYSTEMS, HABITATS, SPECIES, AND COMMUNITIES

Increasing growing season length provides the opportunity for the survival of a whole set of plant and animal species previously excluded because they could not complete their life cycles during the short seasons typical of the region 30 years ago before the onset of regional warming. Some novel species are likely to be ultimately beneficial and some may be detrimental, but the incorporation of new species often is disruptive. Highly mobile species, such as birds and flying insects, often are able to respond nearly immediately to the new opportunities. Forest insect outbreaks detected by aerial survey in 2013 increased by 42 percent from 2012, with much of it because of increases in spruce, alder, and birch defoliation in Alaska due to warming and drying conditions

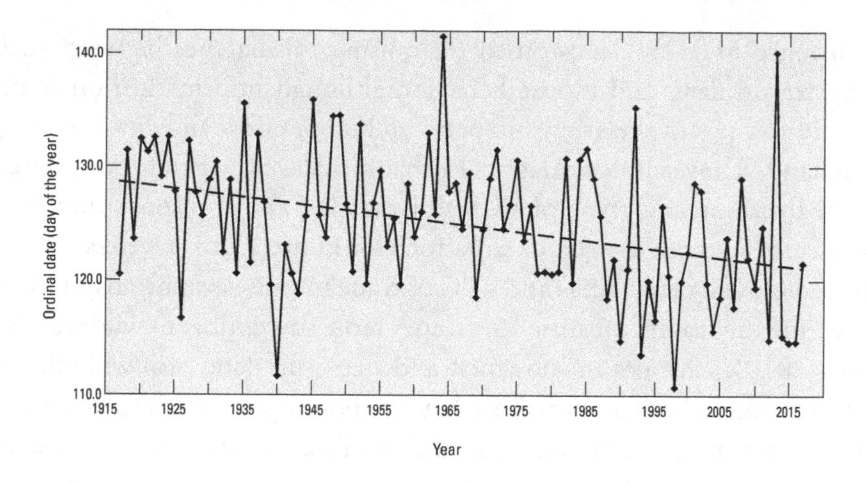

Figure 3.4.3. Spring ice breakup dates for the (roughly) 100 years from the start of the Nenana tripod competition (1917–2017), Tanana River at Nenana, Alaska (http://www.nenanaakiceclassic.com/2018%20Brochure.pdf).

(U.S. Forest Service, 2014, Chapter 2.2). Various geometrid moth species like leaf rollers (Tortricidae) and leaf beetles (Chrysomelidae) were contributors to defoliation, but the major contributor was the birch leaf roller (*Epinotia solandriana*). In 2013, aspen leaf miners (*Phyllocnistis populiella*) increased activity from 2012, but activity was lower than 2010 levels. Spruce defoliation decreased from 2012 levels, but a large outbreak of western black-headed budworm (*Acleris gloverana*) damaged a large population of spruce in southwest Alaska. In the boreal forest of western Canada, tree mortality rates have increased by 4.9 percent per year from 1963 to 2008 (Peng and others, 2011) due to regional drought, warmer temperatures, and subsequent increases in beetle outbreak and insect infestation.

Most of the boreal tree species, and especially the conifers, depend disproportionately on snowmelt for their annual water budget (Callaghan and others, 2011a). Decreased cold season length reduces the period of snow accumulation and increases the length of the above-freezing season, increasing water loss from evaporation and plant transpiration (Callaghan and others, 2011b). The net result is an increased probability of the onset of moisture limitations to growth of trees and other forest vegetation earlier in summer, and a greater risk of critical to lethal shortage of soil moisture in hot, dry summers.

One of the biggest effects that increased growing season will have is on timing of seasonal events such as the blooming of flowers, emergence of insects, migration of birds and range animals, and planning and timing of hunting seasons. This is extremely important to the subsistence lifestyle within the NWB region. These changes are pushing apart the timing of events that occur in the natural system and those activities that the local subsistence users are accustomed to, causing mismatches between such events as hunting season dates and moose behavior, although information on hunting and climate indicate a positive relationship between hunter effort and autumn temperature (Knut Kielland, University of Alaska, Fairbanks, Alaska, written commun., 2016). Changes in freeze-up and thaw of rivers for crossing and transportation are another mismatch, creating more dangerous situations for hunters and travelers.

Increased growing season has and will continue to cause evapotranspiration deficits during the growing season, creating landscapes more susceptible to wildfires (Kasischke and others, 2010). Changes in snowmelt and freeze-up dates also will affect seasonal migration of birds and other animals such as caribou (upon which subsistence hunters depend), the rate of northerly range expansion of native and non-native species, and the habitats of both ecologically important and endangered species (Hezel and others, 2012).

Decreased length of the cold season (or winter) also delays the start of safe winter travel on ice, accelerates the date of spring ice breakup (Fig. 3.4.3), and introduces more risk to the ice travel season. Further, rapid spring breakup triggers more ice jams and flooding of villages located next to rivers.

INFORMATION GAPS

- Information about future changes in effective moisture availability to plants as temperature and precipitation increase or stay the same is needed to indicate whether browning or greening can be expected.
- Information about snow cover changes as it relates to soil moisture and storage is needed to determine its effect on growing season productivity.
- Future scenarios of effects of increased growing season length on the timing and migration of birds and mammals and their use for subsistence lifestyles and potential adaptation if wildlife patterns change.

3.5
ENDURING FEATURES IN THE NORTHWEST BOREAL REGION

By Donald Reid[1], Jim Pojar[2], Gregory Kehm[3], and Dawn R. Magness[4]

KEY FINDINGS

- Abiotic components of landscapes, including latitude, elevation, topography, bedrock, hydrologic regime, soils, and ice, are frequently factors influencing species distributions. These components have been termed enduring features or geodiversity.
- Species are shifting their distributions to cope with changing climate; enduring features will influence future distributions and colonization pathways for some species in some landscapes.
- In the context of conservation planning, mapping a full suite of enduring features can provide both a coarse-filter approach to habitat representation and a fine-filter approach to inclusion of unique geodiversity patches or habitats.
- Considering conservation in the context of climate change, there is significant value in integrating models of connectivity with maps of enduring features to assess pathways and corridors for shifting distributions.
- The value of applying an enduring features approach will depend on the prominence of these features in focal species distributions and colonization processes, their influence as special elements, meaningful connectivity models, and how they might complement other approaches.

"All the world's a stage, and all the men and women merely players" (Shakespeare, *As You Like It*, Act II, Scene VII). Taking this metaphor to the drama of ecosystems,

1 Wildlife Conservation Society Canada, Whitehorse, Yukon, Canada.
2 Smithers, British Columbia, Canada.
3 Gregory Kehm Associates, Vancouver, British Columbia, Canada.
4 U.S. Fish and Wildlife Service, Soldotna, Alaska, USA.

the players also include all life forms as they interact with one another. Yet the stage remains remarkably stable, comprised of a set of physical spaces and structures within which the play unfolds. These are the abiotic "enduring features," "geodiversity" (Beier and others, 2015), "arenas" (Hunter and others, 1988), or "land facets" (Wessels and others, 1999) of the natural drama, and include the geologies, landforms, and drainage patterns that shape the actors' movements but change little through the drama.

CONSERVING NATURE'S STAGE

Conserving nature's stage refers to a conservation strategy that explicitly uses enduring features or geodiversity as a surrogate for conserving biodiversity (Beier and others 2015). In contrast to the suite of drivers of change discussed elsewhere in this volume, enduring features are stabilizers, providing a strong degree of permanence to the habitat template that underpins species distributions. As species redistribute and communities reorganize, enduring features can conceptually provide stable units for conservation planning. Conserving geodiversity, and its attendant range of abiotic conditions, will conserve a diverse stage able to support many actors and the consequent ecological and evolutionary processes (Lawler and others, 2015).

Maintaining connections among enduring features with similar abiotic conditions should help individual species redistribute themselves. The most fundamental way organisms responded to the substantial changes in Holocene climate was by shifting their distributions (Huntley, 2005), so assessing factors affecting distribution is central to understanding effects of current global change. Climate is one of these factors, and often an indirect one. Patterns of topography, drainage, and soils directly influence, and often control, species movements and distributions (Hunter and others, 1988).

Numerous organisms, especially plants and their associated herbivores, depend on rock or soil substrates of restricted chemistry, acidity and (or) structure. Uplands are barriers to movement of lowland species, and lowlands and rivers define the distributions of many others. When responding to declining local habitat quality, aquatic organisms will completely depend on the connectivity of water bodies because their routes will be limited to the enduring drainage patterns.

Although many species in the northern hemisphere are shifting their ranges northward, or upslope (Parmesan and Yohe, 2003), each species must respond locally depending on how its colonization abilities fit within the array of suitable habitats and movement barriers presented by enduring features and other habitat attributes. These features may play a particularly influential role in the NWB LCC because plate tectonics have created a more complex and diverse array of bedrock types and topographies than in other boreal regions (Roots and Hart, 2004). Straightforward northward shifts may frequently be blocked or deflected by plateaus and mountains.

ENDURING FEATURES CATALOGUE

The catalogue of enduring features (Table 3.5.1) is large, and spans diverse degrees of temporal permanence. Climate itself could have been considered an enduring feature until a few decades ago, but it is now a fundamental driver of directional change (Chapter 3.1). The most permanent features are latitude, physiography, bedrock, and landforms. The distributions of these on the landscape generally result from processes with millennial time scales, or episodic events deep in geological time. Some have the potential to change within the human lifetime (for example, alluvial, colluvial, or aeolian landforms), but their dynamism is often regenerative without changing their distribution. Consequently, this broad array of most permanent types can be considered stable or static for conservation purposes. Soil is intermediate in permanence, showing gradual but often directional change, in response to shifting inputs of organic material and moisture. Some aspects of water and associated cryogenic (ice) features are the least permanent because they rapidly absorb increased heat, changing the timing of their density and phase shifts. Thawing permafrost can allow some lakes to drain rapidly. Warming air results in warming waters in lakes and rivers with direct effects on freeze-up and in faster evaporation with effects on lake levels. However, the geographical locations of rivers and their gradient shifts, as well as lake basins, are fairly stable. The underlying distribution of water in terms of drainage patterns, gradient shifts, and lotic and lentic water bodies, will generally remain stable.

IMPLICATIONS FOR MANAGEMENT

The goal of coarse-filter conservation is to protect representation of the full suite of habitats (Groves, 2003). Plant communities often have been used as habitat units in such planning, but are not likely to move as coherent units in a changing climate (Hunter and others, 1988). Mapping of enduring features (one classification of habitats), and their connectivity, may be a more robust approach to coarse-filter conservation planning compared to the multi-model climate and climate envelope projections of future suitable climates which have significant irresolvable error or uncertainty (see Box 3.5.1, "Landscape Connectivity in the Central Yukon Planning Region of Alaska," and Box 3.5.2, "Kenai Mountains to Sea"; Anderson and Ferree, 2010; Brost and Beier, 2012; but see Schneider and Bayne, 2015).

By mapping a large suite of enduring features (Table 3.5.1 and "Enduring Features of the Muskwa-Kechika Management Area, British Columbia"), a geographic representation of habitat richness is developed and zones with higher representation of habitats ("eco-portfolio diversification") are identified. A complementary approach is to highlight the mapped distributions of rare or unusual enduring features (fine-filter elements such as hotsprings, caves, and dunes). These are habitats deserving specific conservation attention such as local protection or the application of spatial buffers and timing windows to human activities. These two approaches will not be sufficient for

Table 3.5.1. Catalog of classes of enduring features, with geographic manifestations and influences on factors potentially limiting species distributions.

Class	Geographic manifestation	Influence on limiting factors
Latitude	Duration of daylight Annual pattern of insolation Synoptic pressure and wind patterns	Seasonality Net primary productivity Seasonal precipitation potential
Physiography	Land–ocean interfaces	Regional precipitation potential Seasonal temperature variation
	Mountain ranges—elevation, orientation, and aspect	Orographic precipitation potential and precipitation shadows Barriers to dispersal of low-elevation species Regional insolation potential (combined with temperature to produce growing degree days and potential evapotranspiration) Regional permafrost potential Wind exposure
	Plateaus—elevation	Seasonal temperature range potential Wind exposure
	Basins—elevation range and orientation	Temperature inversion potential Wind exposure
	Hydrologic drainage pattern—orientation and gradient	Seasonal flow regime—timing and quantity Waterfalls and impassable rapids Barriers to dispersal of high-elevation species
Bedrock	Chemical composition—elements, acid/base, and toxics	Substrate and soil pH Nutrient availability (substrates, soils) Growth-limiting and toxic chemical distribution
	Solubility and ease of weathering	Texture/particle size; nutrient availability and dispersion (substrates, soils) Karst features and subterranean drainage (caves, sinks)
	Physical structure—erodibility	Cliffs, outcrops, tors Mass wasting potential—talus, scree, landslides
Landforms (e.g, fluvial, glaciofluvial, alluvial, morainal, colluvial, aeolian)	Soil parent materials	Soil structure (particle sizes), hardness, nutrient distribution, nutrient replenishment
Water	Flowing (lotic)	Seasonal temperature regime Seasonal flow regime Gradient, rapids, waterfalls Spatial distribution of reaches Point source springs and seeps (mineral licks, hotsprings)
	Static (lentic)	Seasonal temperature regime Seasonal depth changes Spatial distribution of water bodies, and islands within water bodies Point source springs and seeps (mineral licks, hotsprings, tufa pools)
Ice	Permafrost	Active layer phenology and depth Spatial distribution and extent
	Seasonal ice cover	Winter ice-free water (including groundwater effusions along streams)
	Glaciers and ice caps	Seasonal lotic flow regimes and temperatures

conservation planning in the face of climate change because they do not address the diverse scales at which species operate, and they do not address the processes of range shift that warming may necessitate (Groves, 2003). However, greater regional diversity and interspersion of enduring features likely will provide a greater selection of habitat choices for whatever species manage to colonize in the future, and therefore higher biodiversity values (Anderson and Ferree, 2010).

Beier and Brost (2010) advocated combining a subset of enduring features ("land facets"), known to influence distributions of influential species, with models of interspersion and connectivity among facets, to map zones of relative richness and map corridors or linkages between similar facets. This approach, detailed in Box 3.5.2, "Kenai Mountains to Sea," offers real conservation value by combining meaningful mapping of topographic niches with least-cost pathways for distribution shifts in response to a changing climate. However, some enduring features will act as strong or absolute barriers to movement for some species (for example, major waterfalls and canyons for some fish and land mammals; wide lowlands to many alpine species; Table 3.5.1). Algorithms that model connectivity need to realistically deal with these possibilities.

Even so, conservation planning will require more than just the assessment of distribution and connectivity of enduring features, because diverse other factors (climate, human footprint, other organisms) can affect distributions of selected organisms, and climates will not change uniformly across a region. An enduring features strategy is one of a set of tools in the conservation planning toolbox, requiring judicious selection of focal features, species, and connectivity models (Beier and Brost, 2010).

BOX 3.5.1: LANDSCAPE CONNECTIVITY IN THE CENTRAL YUKON PLANNING REGION OF ALASKA

The Central Yukon Planning Area is a large, multijurisdictional landscape in Alaska delineated by Bureau of Land Management (BLM) for planning purposes. The area is more than 239,670 km^2 (59 million acres). Approximately 222,500 km^2 (55 million acres) of lands in the federal conservation estate occur within or directly adjacent to the planning area. This includes Gates of the Arctic National Park and Denali National Park, managed by the National Park Service, and seven National Wildlife Refuges, managed by the U.S. Fish and Wildlife Service. Additionally, Noatak National Preserve and Kobuk Valley National Park are contiguous with this conservation estate and thus provide another 32,360 km^2 (8 million acres) that would benefit from landscape connectivity in the planning area.

A geodiversity approach was used to suggest landscape linkages that would be robust in a changing climate (for detailed methodology see Brost, 2010) across the federal conservation lands. Land facet units were empirically clustered using topography (slope, elevation, solar insolation, and landscape position; Fig. 3.5.1). Other geodiversity elements such as surficial and bedrock geologies could be used. Large contiguous areas of each land facet (for example, gentle, low-elevation, warm slopes) were identified within the conservation estate (defined as National Park Service and U.S. Fish and Wildlife Service lands). Then the most efficient but permeable paths for moving between blocks of the same land facet were identified by simulating the path of least resistance (Fig. 3.5.2, Magness and others 2018).

Figure 3.5.1 (top right). Central area of the Central Yukon Planning Area showing the distribution of land facets in the different classes of canyon, slope, and ridge topographic position. A qualitative search for consistent distribution of similar colors (that is, facets) within and between separate conservation units provides a first approximation of potential corridors of connection across the region.

Figure 3.5.2 (bottom right). Central Yukon Planning Area showing the distribution of conservation units and simulated corridors connecting the units by paths of least resistance between locations of the same land facet.

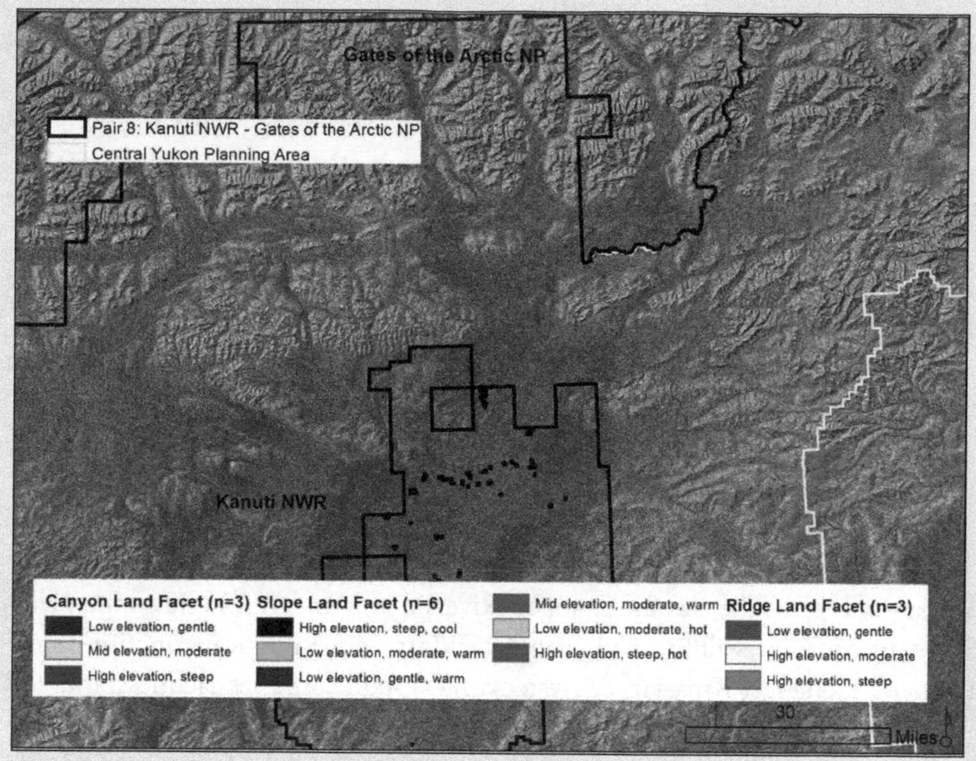

Figure 3.5.1.

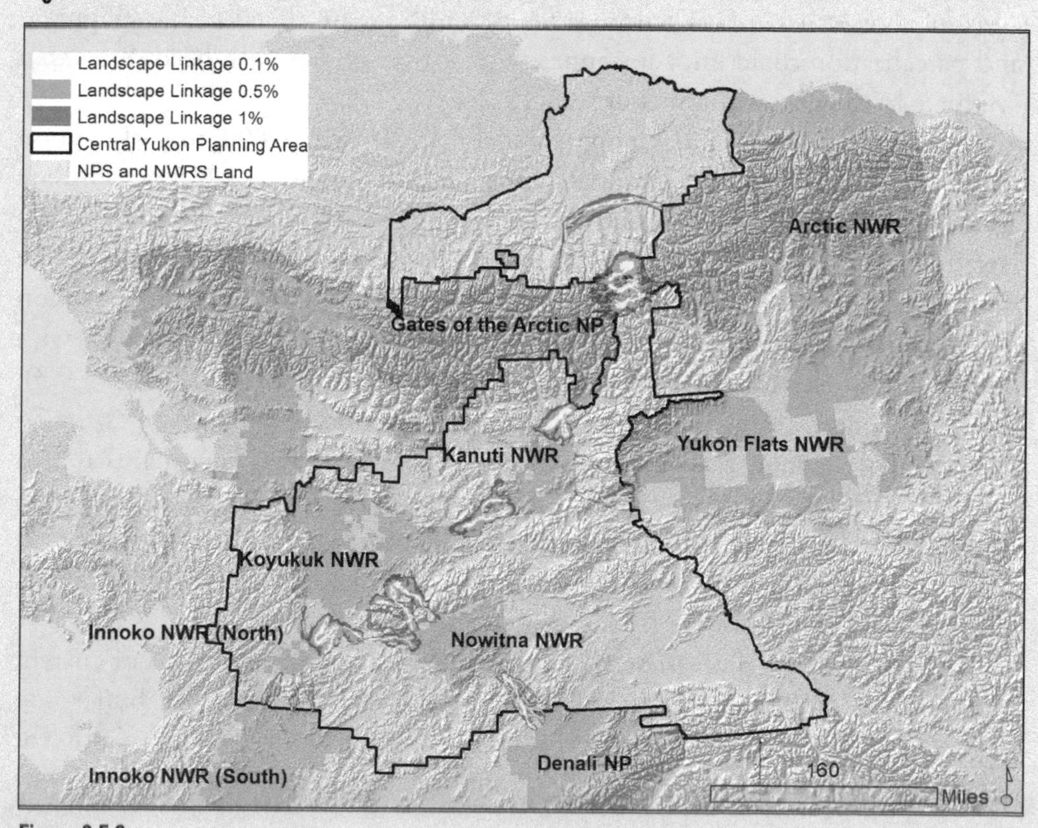

Figure 3.5.2.

BOX 3.5.2: KENAI MOUNTAINS TO SEA

"Kenai Mountains to Sea" is a multipartner conservation strategy that uses rivers as an enduring feature to plan for landscape connectivity (Morton and others, 2015). The 24,270 km^2 (6 million acres) Kenai Peninsula in south-central Alaska sits at the southwestern extent of the boreal forest and the northwestern extent of the Pacific Northwest rain forest. The biome interface and topographic diversity result in rich biological and habitat diversity. Climate change has and will continue to shape ecosystems; documented changes include treelines rising in elevation, wetlands drying, changes in available water, and nonglacial rivers warming (Berg and others, 2009; Klein and others, 2005; Dial and others, 2007, 2016; Mauger and others, 2016).

Habitats and species are not expected to remain static on the landscape. Rivers with ocean outlets can be used as an enduring feature to maintain landscape connectivity even as the climate changes. Rivers maintain hydrologic connectivity and can provide for wildlife and plant corridors and ecological functions such as nutrient exchange. Riparian ecosystems have higher species richness and soil productivity due to the nutrient flows from upstream and from the ocean. The land-use patterns on the Kenai Peninsula increase the vulnerability of river corridors to a loss of connectivity. Most private lands are on the coast, therefore, development and parcelization could impede connectivity of river outlets with headwaters that originate in the conservation estate.

The Kenai Peninsula has approximately 2,900 km (1,800 mi) of salmon rivers draining into 386 ocean outlets. Of these 386 outlets, 212 (55 percent) are already conserved within the federal Kenai National Wildlife Refuge, Kenai Fjords National Park, and Chugach National Forest. However, many of these streams are short, glacial, steep-gradient streams representing only 15 percent of the salmon rivers extent. The Kenai Mountains to Sea strategy identifies 22 salmon rivers with headwaters in the conservation estate that flow into private lands (Fig. 3.5.3). These 22 salmon rivers represent 1,558 river kilometers (968 river miles) or 54 percent of all salmon river miles peninsula-wide. The Kenai Mountains to Sea strategy targets these river corridors (the strategy groups some river outlets due to proximity and shared parcels; therefore, detailed information is provided for 20 rivers) as they provide the most conservation value, leveraging existing but only partial protection within the federal conservation units.

Connectivity goals on these rivers may differ depending on levels of current urbanization, which could make restoration to a predetermined buffer too expensive and controversial to realistically consider. A minimum width for ensuring hydrological function, such as minimized bank erosion, is a 15 m (50 ft) setback for future development, already defined by the Kenai Peninsula Borough. For wildlife movement, 400 m (1,310 ft) on one or both banks would ensure the movement of even the largest mammals. In more urbanized settings, 400 m (1,310 ft) could also provide greenways for bicycle and non-motorized transportation. The largest corridor width would ensure all ecosystem services associated with a riparian corridor are maintained.

The diversity of ownership and landscape context will require different voluntary partnerships and actions to conserve and enhance fish and wildlife habitats for continuing economic, recreational, and cultural benefits to residents and visitors. Parcel acquisition by land trusts is one of many tools considered for conservation and development of these corridors. Less-than-fee-simple approaches include voluntary compliance, perpetual easements, short-term agreements on private parcels as part of publicly funded restoration projects, and land agreement exchanges with tribal and local governments. Different approaches will make sense based on the ownership, wildlife requirements (for example, caribou migration), and other ecological factors. For example, the Swanson River is a 174 km (108 mi) salmon river with only 2.3 km (1.4 mi) outside federal lands requiring agreements on only two parcels owned by the State of Alaska. The strategy for the Swanson River would be quite different than the Kenai River, which has 129 km (80 mi) outside the federal lands requiring agreements on more than 2,000 parcels valued at more than US$735 million.

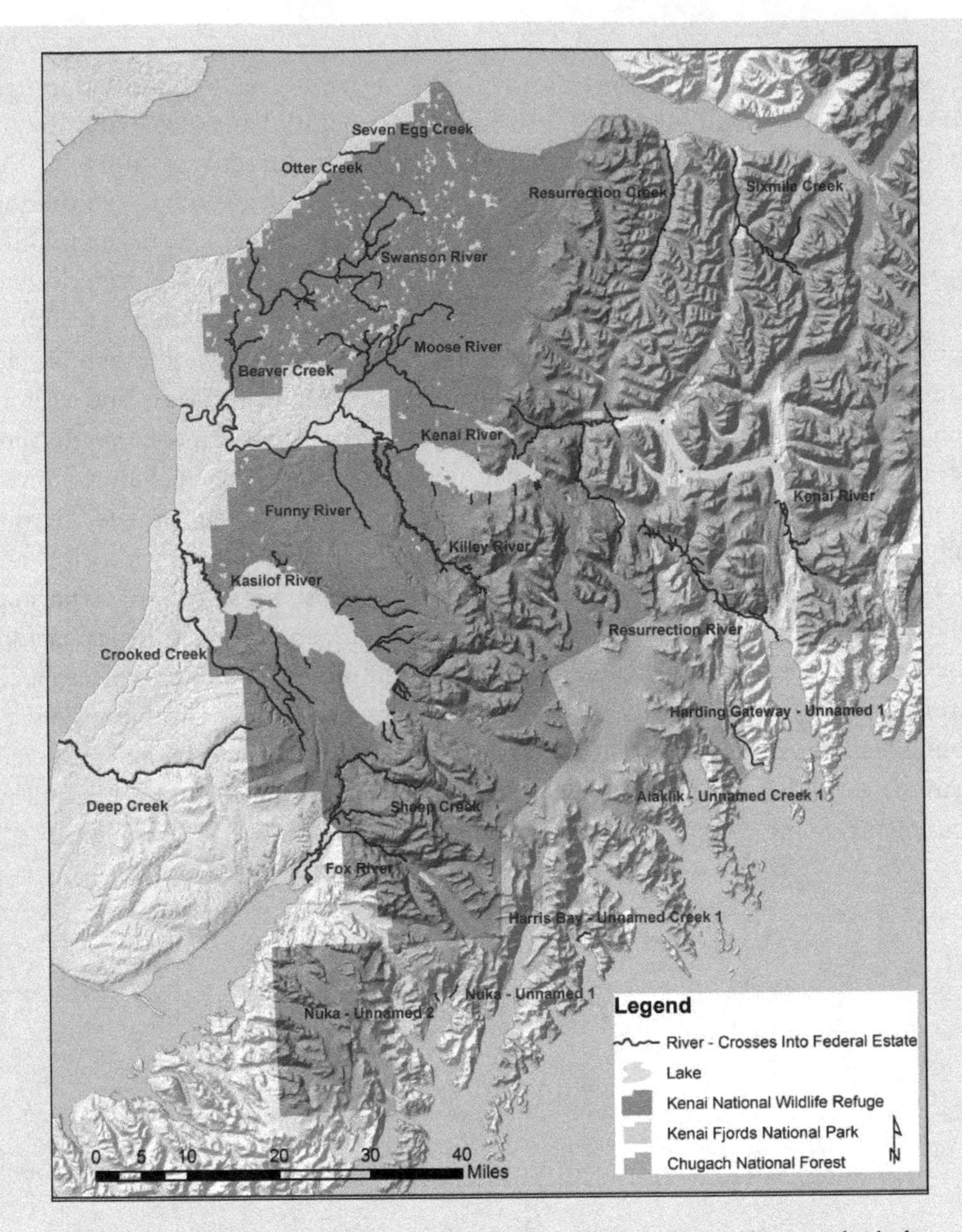

Figure 3.5.3. Focal salmon rivers on the Kenai Peninsula in south-central Alaska that can be used as enduring features to ensure landscape connectivity.

BOX 3.5.3: ENDURING FEATURES OF THE MUSKWA-KECHIKA MANAGEMENT AREA, BRITISH COLUMBIA

The Yellowstone to Yukon Conservation Initiative (2012) conceived a conservation assessment and outreach project in 2010 to address challenges in the future management of the Muskwa-Kechika Management Area (M-KMA) in northern British Columbia, Canada. These challenges include stewardship of the existing network of conservation lands, strengthening conservation measures in light of climate change, planning and evaluating land and resource use proposals within this shifting environment, and enhancing further public and community understanding of and appreciation for the biodiversity values of the M-KMA.

The M-KMA region covers 63,800 km^2 (15.7 million acres) and includes the northern part of the Rocky Mountains in British Columbia and the eastern part of the Cassiar Mountains (Fig. 3.5.4). The region supports fully functioning predator-prey systems remaining largely undisturbed by industrial development. Large mammal species relying on extensive conservation lands in these montane boreal ecosystems include grey wolf (*Canis lupus*), wolverine (*Gulo gulo*), black bear (*Ursus americanus*), grizzly bear (*Ursus arctos* ssp.), moose (*Alces alces*), caribou (*Rangifer tarandus*), elk (*Cervus canadensis*), mountain goat (*Oreamnos americanus*), and Stone's sheep (*Ovis dalli stonei*).

For the purposes of this assessment of enduring features, the study area was expanded to 132,560 km^2 (37.2 million acres) to cover physiographic subprovinces (Holland, 1976), and thereby facilitate a representational analysis of enduring features (full variety and coverage) within the M-KMA proper.

Enduring features were modeled to inform a precautionary approach to stewardship and planning within the M-KMA. Enduring features were analyzed within physiographic provinces to identify areas with a high variety and many rare enduring features. A synthesis product was generated by combining areas of both high variety and concentrations of rarity, and was analyzed for representation within existing conservation lands. Additional analysis included quantifying the common enduring features within a physiographic province, defined as those features representing 50 percent or more of the land area. The methods for modeling enduring features were based on those developed and applied by the Nature Conservancy (Groves and others, 2000). Each 90 by 90 m cell contains a unique physical signature composed of elevation groupings, substrate (bedrock and quaternary geology) groupings, and macro landform types (summit, slope, valley, aspect; Fig. 3.5.5).

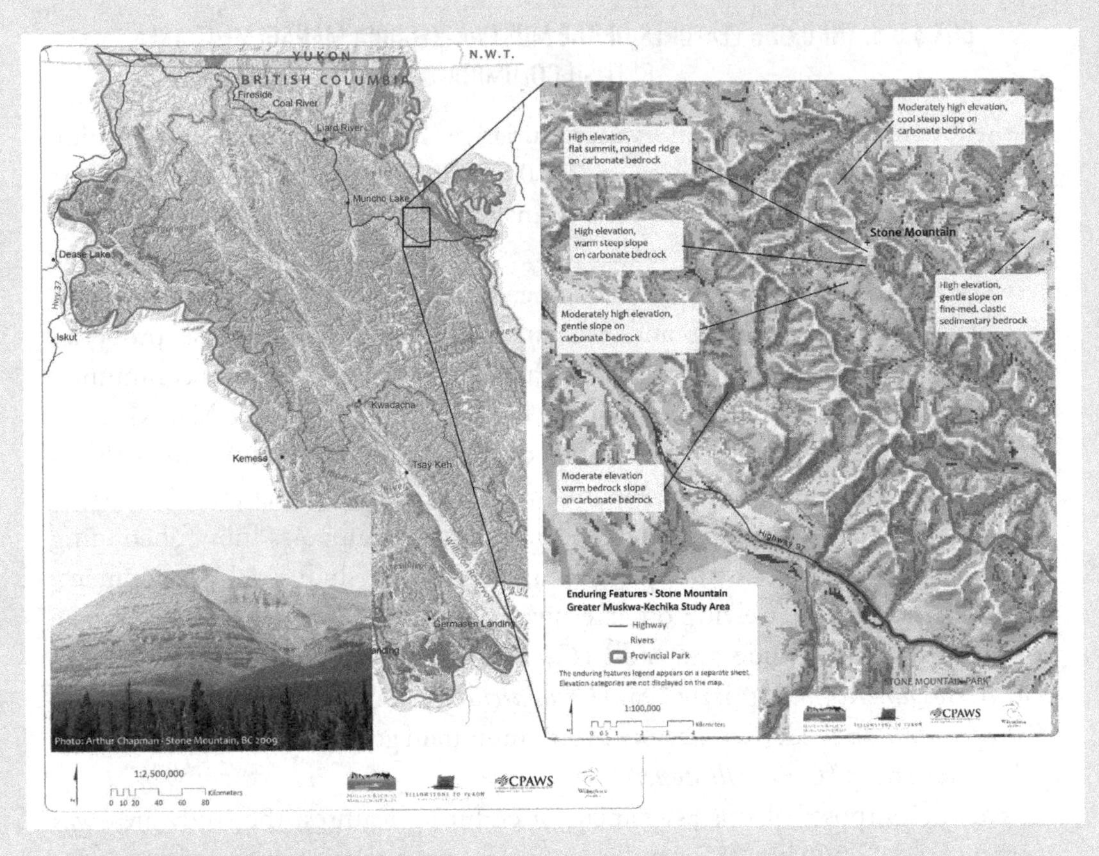

Figure 3.5.4. Composite map of enduring features in the full Muskwa-Kechika Management Area study area (simplified view due to cartographic limits), a detailed map of enduring features in the Stone Mountain area of region, and a photograph of Stone Mountain. The mapping includes more than 1,600 distinct and, in some cases, rare combinations of elevation, geology, and macro landform. Photograph by Arthur Chapman, 2009.

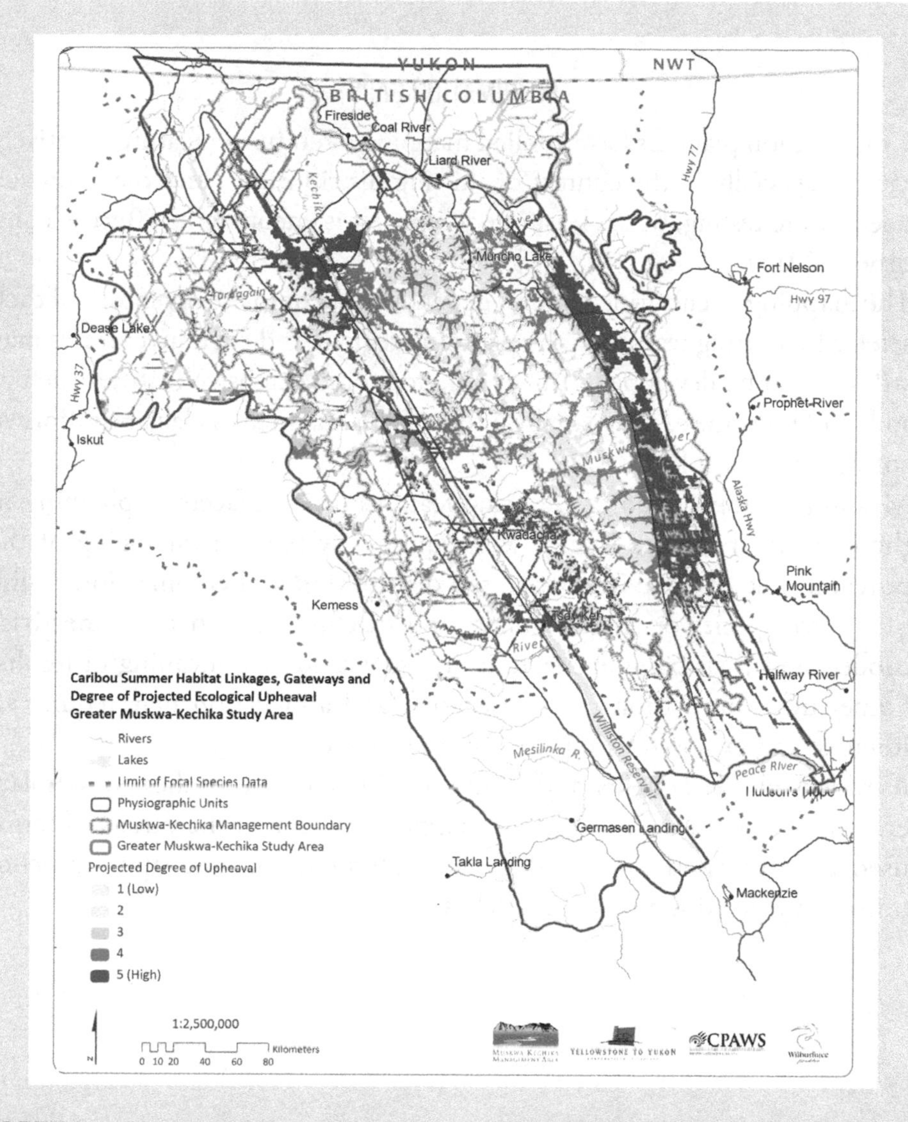

Figure 3.5.5. Map showing the Caribou Summer Habitat Linkage Model Overlain on Degree of Projected Ecological Upheaval, which depicts the relative degree or intensity of ecological upheaval projected over the study area to 2050. The darker areas are projected to change the most, and the lighter areas the least. Areas with a projected high degree of upheaval call for management strategies that maintain wildlife connectivity between areas of suitable habitat, to allow species to move or adapt to the new conditions. Areas with low projected upheaval can be viewed as potential climate change refugia for species that use this habitat today.

INFORMATION GAPS

- Conservation planners have applied mapping of enduring features relatively rarely, especially in the context of shifting distributions, and the approach needs more testing, but at least one recent test is encouraging (Buttrick and others, 2015).
- The mapping of enduring features depends on a complete inventory of each selected enduring feature in the planning region, and such inventories may still have to be developed. This is true for some extensive features such as soils and for some of the spatially limited elements such as glacial landforms, springs, and karst features.
- Meaningful application of enduring features or land facets to planning in the context of climate change will require a systematic cataloging of the potential role of these features in the distribution and colonization ability of diverse species or species guilds, and careful selection of connectivity models and their parameterization, in order to derive meaningful results. Some such information may be lacking, and application may require an iterative and exploratory approach in most regions.
- A review of the benefits and limits of different approaches to simulate enduring features, their applicable mapping scales, and the variety of terms used to describe them would benefit interpretation and application of outputs by practitioners and modelers.

4
BIOLOGICAL DRIVERS OF LANDSCAPE CHANGE

By John DeLapp[1]

INTRODUCTION

The biological components of the Northwest Boreal (NWB) ecosystems comprise a dynamic and ever-changing web of life. Although the extent of the boreal forest has remained relatively stable since the last ice age, current climate change is driving profound changes in the distribution of plants and animals and their habitats within the NWB region. Novel ecosystems may evolve in response to our changing environment. New and existing infectious diseases could emerge to affect important food species in the north. Long-established interactions of keystone species such as salmon, wolves, and brown bears may change, resulting in different future biological and cultural landscapes. This chapter describes the current scientific understanding of how the biological components of the environment are driving these aspects of landscape-scale change in our region

This chapter consists of four subchapters:

- Vegetation Composition Change in the Northwest Boreal Region
- Novel Communities and Trophic Assemblages in the Northwest Boreal Region
- Wildlife Parasite and Pathogen Life Cycles in the Northwest Boreal Region
- Marine-Derived Nutrients in the Northwest Boreal Region

These four subchapters offer an introduction to the concepts of biological drivers of landscape change in the NWB region. For each topic area, resource managers and the interested public are provided with a summary of the latest relevant research that will serve to inform management activities and natural resource policies.

1 U.S. Fish and Wildlife Service, Anchorage, Alaska, USA, Retired.

4.1
VEGETATION COMPOSITION CHANGE IN THE NORTHWEST BOREAL REGION

By T.N. Hollingsworth[1], R.E. Hewitt[2], and J.F. Johnstone[3]

KEY FINDINGS

- Modern vegetation in the NWB forest has remained fairly stable for the last 5,000 years.
- Due to physical isolation and harsh environmental conditions, plant species diversity is extremely low in comparison to other biomes.
- Due to low species diversity, there is low redundancy in the functional roles of species, contributing to high vulnerability to the effects of climate change.
- Vegetation is the basis for all terrestrial subsistence resources, and therefore understanding drivers of vegetation change is critical for management.

HISTORICAL VEGETATION

In the Northwest Boreal (NWB) forest, paleoecological reconstructions of vegetation suggest that there have been substantial shifts in dominant vegetation since the Last Glacial Maximum (LGM, about 21–18 thousand years before present [Ka BP]) related to climate and fire disturbance regime. During the repeated glaciations of the Quaternary Period, most of Interior Alaska and much of north-central Yukon escaped glaciation (Pewe and others, 1965; Hopkins, 1982). Most of this region was not hospitable for boreal taxa during the Pleistocene due to the extreme aridity and cold air temperature and was characterized by a mosaic of steppe and tundra (Cwynar and Ritchie, 1980; Schweger and others, 2011). During the LGM, boreal taxa, including the *Picea* species, are hypothesized to have either persisted on the Bering Land Bridge, which connected Asia with North America (Brubaker and others, 2004), or been

1 U.S. Forest Service, Pacific Northwest Research Station, Boreal Ecology Cooperative Research Unit, Fairbanks, Alaska, USA.

2 Institute of Arctic Biology, University of Alaska Fairbanks, Fairbanks, Alaska, USA.

3 Department of Biology, University of Saskatchewan, Saskatoon, Saskatchewan, Canada.

driven to the south or east of the ice sheets (Ritchie and MacDonald, 1986), including localized glacial refugia in Canada (Gamache and others, 2003; Jaramillo-Correa and others, 2004). With warm and dry conditions during the early Holocene (about 13–10 Ka BP) the mosaic of shrub, graminoid, and forb tundra transitioned to deciduous woodland (Keenan and Cwynar, 1992; Bigelow and others, 2003; Edwards and others, 2005). The later development of the modern coniferous boreal forest occurred when conditions shifted to a cooler, moister climate beginning in the mid-Holocene (about 10–6 Ka BP). White spruce (*Picea glauca* [Moench] Voss) spread back into Alaska through multiple migration routes (Edwards and Brubaker, 1986; Ritchie and MacDonald, 1986; Keenan and Cwynar, 1992), whereas in the Yukon area there is less understanding of the migration pathways (Rainville and Gajewski, 2013). White spruce dominated the landscape from 10–6 Ka BP, followed by a transition to black spruce–dominated (*Picea mariana* [Mill.] Britton, Sterns, and Poggenburg) forests in the late Holocene (about 5 Ka BP to present). These changes in canopy dominance were strongly linked to fire activity on the landscape. In particular, the prevalence of fires in the last 5 Ka BP correlates to black spruce dominance (Lloyd and others, 2005). The paleoecological perspective on the correlation between vegetation cover and fire can aid in hypothesis development and projections of future climate-fire-vegetation dynamics.

CURRENT VEGETATION

Species composition has remained relatively stable in the NWB region during the last 5,000 years, with only scattered evidence of species migration from the south or west. The stability of vegetation is a result of the severe environmental conditions and the relative isolation of the area, with the Bering Sea to the west and the Alaska Range and St. Elias Range to the south (Chapin, Robards, and others, 2006). This stability has created a vegetation mosaic that reflects both environmental variables and disturbance history, without the influence of species migrations and extinctions. The circumboreal forest has fewer vascular plant species than any other forested biome (Pastor and Mladenoff, 1992), and therefore dominant species have broad distributions.

There are 5–8 dominant tree species in the NWB forest. The predominant conifers across Interior Alaska and Yukon are white and black spruce (*Picea glauca* and *P. mariana*, respectively), with local populations of Alaskan larch or tamarack (*Larix laricina*). The predominant deciduous species are quaking aspen (*Populus tremuloides*), balsam poplar (*Populus balsamifera*), and paper birch (*Betula papyrifera*) and black birch (*B. neoalaskana*). One noticeable difference in dominant tree species between the Yukon area and Interior Alaska is the presence of lodgepole pine (*Pinus contorta* var. *latifolia*), which is a recent colonist of south-central Yukon during the past 1–2 millennia (MacDonald and Cwynar, 1986).

There is evidence that the northern spread of lodgepole pine is unusual within the context of Pleistocene interglacials (Schweger and others, 2011), and it may still be migrating north and west into Interior Alaska as a function of fire disturbance (Johnstone and Chapin, 2003). In contrast to vascular species, the NWB region is richer in lichen and bryophyte diversity than southern forests (Hollingsworth and others, 2006). Ecosystems of high vascular and nonvascular plant diversity often are non-forested, such as bogs, fens, alpine tundra, and shrub tundra or older black spruce stands that have not burned at high severity in more than 100 years.

DRIVERS OF VEGETATION CHANGE

The direct and indirect effects of climate warming are strong drivers of vegetation change in Alaska and northwest Canada. The direct effects of unprecedented warming (Clegg and Hu, 2010) are correlated with decreased forest productivity in the interior boreal forest (Goetz and others, 2005; Parent and Verbyla, 2010; Baird and others, 2012) that may be caused by temperature-induced declines in tree growth associated with water stress (Barber and others, 2000; D'Arrigo and others, 2004; Wilmking and Myers-Smith, 2008). In contrast, increased productivity at the forest-tundra ecotone (Chapin and others, 2005; Beck and others, 2011) suggests the initiation of a northward biome shift. Together this suggests that a complex suite of abiotic factors may influence boreal forest productivity and composition within the current range limits of the boreal biome and the potential for state changes particularly at range limits. To date, the ecotypic differentiation, or the genetic variation, of most NWB tree species is not known across their landscape distributions, and therefore the ability to accurately project the capacity of boreal tree species to adapt to changing climate conditions is uncertain.

Simultaneously, the indirect effects of climate change have strong effects on vegetation. Changes in the fire regime related to climate warming are expected to disrupt stabilizing feedbacks that have maintained the composition and geographic distribution of NWB forests during recent millennia (Chapter 2). For example, wildfire not only directly affects seed availability and establishment site suitability (Johnstone and Chapin, 2006b), but also affects the influence of biotic interactions, such as plant-fungal symbioses, on seedling establishment and forest migration (Hewitt and others, 2016). Furthermore, changes in the hydrological regime, including shifts in permafrost stability, can have profound effects on species composition, especially in wetlands.

For example, lake drying can create new colonizable landscapes (Winterstein and others, 2016) and shifts in flooding and drought rapidly change vegetation composition and productivity (Chivers and others, 2009; Churchill and others, 2015). In uplands and lowlands, short-term fluctuations in animal populations can have long-lasting effects on vegetation composition. For instance, moose preferentially browse on willows and deciduous species other than alder, altering the dominance of early successional

BOX 4.1.1: ESTABLISHING A NEW BONANZA CREEK LONG-TERM ECOLOGICAL RESEARCH PROGRAM SITE NETWORK

A Case Study for How to Create a Network of Regionally Extensive Sites to Be Monitored in a Changing Climate

Long-term vegetation monitoring enables a scientist to link changes in vegetation, disturbance regime, and climate. The monitoring of vegetation across a large spatial scale allows for the evaluation of landscape-level patterns and processes that can inform modeling and mapping efforts. Long-term vegetation monitoring is a key focus of the Bonanza Creek Long-Term Ecological Research (BNZ LTER) program in interior Alaska. Since 2010, the BNZ LTER has focused on the regional effects of climate change and climate-disturbance interactions on boreal ecosystems. To address regional variability in climate-fire-vegetation interactions, monitoring sites were selected to specifically address variations in site conditions that drive divergence of successional pathways. Ecoregions, defined by geographical zones in surficial geology, parent material, climate, and local flora were used to delineate major regions of interior Alaska. A large proportion of the NWB forest is situated between the Yukon River and the Alaska Range within three ecoregions: the Ray Mountains, the Tanana-Kuskokwim Lowlands, and the Yukon-Tanana Uplands (Fig. 4.1.1).

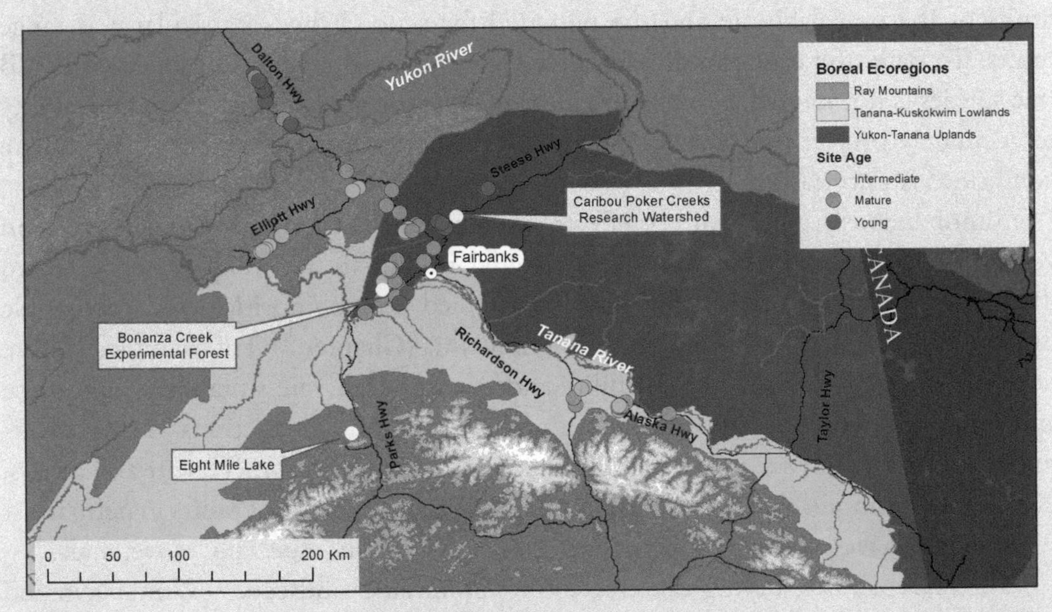

Figure 4.1.1. Extensive Bonanza Creek Long-Term Ecological Research (LTER) program site network encompassing three ecoregions, multiple fire scars, and various post-fire successional pathways. Caribou Poker Creek, Bonanza Creek, and Eight Mile Lake are core research sites. Modified from figure created by Jamie Hollingsworth, Site Manager for the Bonanza Creek LTER.

To capture the variability in community composition and ecosystem processes across the three ecoregions, multiple factors are monitored at each site: vegetation composition and productivity; permafrost status (soil temperature and moisture); climate (air temperature, precipitation, and radiation) in stands of various ages (time-since-fire); and site drainage regimes. Sites were selected based on accessibility, age, and site drainage status. Across ecoregions, approximately 90 young sites (less than 20 years post-fire), approximately 90 intermediate aged sites (40–70 years post-fire), and approximately 90 mature sites (more than 80 years post-fire) were selected for monitoring. Monitoring networks such as this are essential to capture current and project future vegetation patterns. BNZ LTER is well poised to link vegetation change to the direct and indirect effects of climate change across boreal Alaska.

species on the floodplains with landscape-level consequences for soil nitrogen (Butler and others, 2007; Nossov and others, 2011). Significant browsing of snowshoe hares on conifer seedlings during population peaks can delay or alter the development of a conifer forest canopy (Angell and Kielland, 2009). Across the NWB region, despite historically low levels of industrial, anthropogenic land-use change, the potential for intensified oil and gas production may have unforeseen ecosystem consequences and be a future driver of vegetation change.

IMPLICATIONS FOR MANAGEMENT

Vegetation provides habitat for all terrestrial wildlife subsistence resources. Certain plant species also are extremely valuable as subsistence resources in their own right; therefore, changes in composition, cover, and extent of the NWB forest can have large implications for humans. Management could focus on preservation of optimal habitat on a local scale, while being cognizant that landscape-level shifts on the vegetation mosaic are occurring outside the control of management policies at the regional scale. Using a framework of adaptive management will likely enable local and regional managers and communities to help meet conservation goals within a changing landscape.

INFORMATION GAPS

- Increase in the spatial resolution and extent of paleorecords in the NWB forest will improve the ability to reconstruct the influence of past climate change on vegetation patterns and lead to a better understanding of important mechanisms in relation to future climate change.

- Vegetation mapping based on species composition as opposed to dominant cover in the NWB region is limited. Mapping that links the interior of Alaska and western Yukon will contribute to a synthetic view of the region.
- Future experimental vegetation studies should focus on the manipulation of both the direct and indirect effects of climate, to better disentangle the importance of temperature, moisture, and disturbance independently and in combination.
- Increased resolution of the genetic diversity, local adaptation, and gene flow across species' ranges is needed.
- Understanding the drivers of vegetation change at both the northern extent and southern boundary will increase our ability to accurately project vegetation change into the future.

4.2
NOVEL COMMUNITY AND TROPHIC ASSEMBLAGES IN THE NORTHWEST BOREAL REGION

By John M. Morton[1], Donald Reid[2], Dylan Beach[3], and Douglas Clark[4]

KEY FINDINGS

- Novel ecosystems have repeatedly emerged in the NWB region since the last glacial maximum, and current climate change will result in rapid establishment of novel communities.
- Changing ecosystems pass through thresholds at which dominant functions change; sometimes these thresholds are irreversible.
- Understanding the most influential ecological functions and processes characterizing an ecosystem is the most robust means of assessing and dealing with change.
- Shifts in species composition will dominate our view of change, but ecological functions can be maintained by novel and emerging assemblages of species.
- Both old and new management approaches can be utilized to steward ecological processes and novel assemblages as ecosystems change.

Since the maximum extent of the last continental glaciation, ecosystems in the Northwest Boreal (NWB) region have been in flux as the climate changed and species dispersed at different rates. Some plant communities that existed in that period are not found today (Edwards and others, 2005). Current climate change is particularly rapid in the NWB region (Stocker and others, 2013), and is perhaps faster than during the Holocene (Loarie and others, 2009). What is considered the "historical" or "natural" set of ecosystems is the result of biome-scale changes that are ongoing. Species, however,

1 U.S. Fish and Wildlife Service, Soldotna, Alaska, USA.
2 Wildlife Conservation Society Canada, Whitehorse, Yukon, Canada.
3 Case Western University, Cleveland, Ohio, USA.
4 University of Saskatchewan, Saskatoon, Saskatchewan, Canada.

are changing their distributions, as habitats are reassembling themselves, and human actions and land-use changes add to the emerging novelty.

NOVEL ECOSYSTEMS, THRESHOLDS, AND ECOLOGICAL REORGANIZATION

Novel ecosystems, thresholds, and ecological reorganization are well-established concepts in both restoration (Hobbs and others, 2013) and resilience (Scheffer and others, 2001; Biggs and others, 2012) literature. Novel ecosystems arise when "historical" ecosystems become "hybrid" ecosystems following loss or gain of some species and sufficient time for processes to adjust. Likewise, hybrid ecosystems can become truly "novel" ecosystems. In both instances, a threshold is crossed, but novel ecosystems cannot return to the historical condition, whereas hybrid ecosystems may revert (Hallett and others, 2013). Changes in human land use, natural disturbance regimes, and climate can, together or alone, force an ecosystem across a threshold. The source of novelty driving the change can be abiotic (for example, temperature, snow conditions) and (or) biotic (novel species and interactions) (Mascaro and others, 2013). Change needs to be measured in intensity (number of shifts in species composition and their functional relationships) and time. Thresholds are not reached instantaneously, because processes take time to respond to shifts in abiotic or biotic conditions. Furthermore, some sort of disturbance often is the trigger for ecological reorganization after crossing a critical threshold (Scheffer and others, 2001; Biggs and others, 2012). For example, plants may survive some time in gradually changing growing conditions (that is, inertia) until an acute disturbance kills them (Johnstone, McIntire, and others, 2010); existing or immigrant species may adapt, behaviorally or genetically, to novel conditions.

Ecologists have tried to derive rules that explain the patterns of community composition and how species assemble in these communities (for example, Weiher and Keddy, 1999). Some species-specific traits (for example, dispersal method and pattern; food specificity) may predispose species to successful integration in an ecological community at certain times. However, the complexity of trophic interactions, along with stochastic events, still reduce the ability to predict which of a pool of species can fit into a community under changing abiotic or biotic conditions. Ecologists and adaptation practitioners need to move beyond individual species, to assess ecological processes or functions, because thresholds are best characterized in terms of shifts in processes that underpin functions of ecosystems (Harris and others, 2006; Hallett and others, 2013; Starzomski, 2013).

Hallett and others (2013) suggested three classes of ecosystem functioning: (1) within community (primarily trophic), (2) whole ecosystem (energy flow and nutrient cycling), and (3) ecosystem services (functions of high human interest). Functions derive from processes (Harris and others, 2006), and processes can be classed by the component in flux: water (precipitation, temperature, evapotranspiration); carbon (photosynthesis, sequestration, decomposition, permafrost dynamics); minerals (erosion, decomposition, permafrost, nitrification, soil formation); energy (temperature, insolation, wind, fire); primary producers (distribution, dispersal, primary production, succession, competition, pathogens, consumption); and consumers (distribution, dispersal, competition, predation, parasitism). Fluxes are interactive, so efforts to understand novelty require thoroughly assessing linkages among all fluxes in the target ecosystem (for example, see Resilience Alliance, 2010). Management focus often is on the distribution and abundance of producers and consumers, which are the most derivative of all the fluxes.

Climate envelope models, which correlate recent species distribution with climate parameters, are the dominant tools for assessing possible changes in species distribution from projected climate parameters (Starzomski, 2013). This approach has considerable value when climate parameters (fluxes of water and energy) actually limit distributions, so it holds most value for plants and invertebrates. For example, the climate projection models provide a useful assessment of potential shifts in habitats in the NWB region (Scenarios Network for Arctic Planning and Ecological Wildlife Habitat Data Analysis for the Land and Seascape Laboratory, 2012). This approach can be augmented by better integrating related and contingent fluxes (for example, carbon and minerals on soil formation; consumption [that is, herbivory and predation] effects on energy flow) and taking account of extreme events, phenological thresholds, and spatial scale in climate shifts (Pearson and Dawson, 2003; Garcia and others, 2014).

IMPLICATIONS FOR MANAGEMENT

With the acknowledgment that carbon emissions must be significantly reduced, preparations in the short-term for pragmatically responding to the cascading effects of rapid climate change include (1) establishing and maintaining a suite of appropriately scaled and connected protected areas (Danby and Slocombe, 2005) to allow for dispersal and migration, to serve as reference areas (see Chapter 3.5) for monitoring and experimentation regarding hypothesized climate effects, and perhaps as climate refugia (for example, Murphy and others, 2010); (2) developing robust, landscape-scale interjurisdictional monitoring programs to detect changes in species distributions and in regimes of disturbance and land use; (3) designing long-term research to better understand dominant ecological processes and functional interrelationships; and (4) improving projection models (Starzomski, 2013). Novelty must be engaged to avoid the most costly outcomes in terms of losses of biodiversity and ecosystem function, and to provide more rapid transfer of information to policy makers and the public (Seastedt and others, 2008). Policy (Shaffer, 2014) and legislation (Garmestani and Allen, 2014) may need amending to allow consideration of a more comprehensive suite of adaptation tools in the management of transitions toward future conditions.

In the longer term, ecologists and climate adaptation practitioners need to be cognizant that scientific hubris can be a double-edged sword: just as there is neither a

BOX 4.2.1: ARE NOVEL ECOSYSTEMS AND ASSEMBLAGES BAD?

In a world of directional change, novel futures should be expected (Williams and Jackson, 2007). Intrinsically, change is neither bad nor good but in this context is simply the reshuffling of species distributions and ecological processes in response to directional perturbations. However, the difference between the late Holocene and the unfolding Anthropocene (sensu Steffen and others, 2007) is the rapid rate of change (Loarie and others 2009) and therefore potentially high rate of species extinction and loss of genetic diversity as boreal ecosystems respond to rapid climate warming or potentially increasingly fragmented landscapes.

As Mascaro (2013, p. 155) wrote, "while extinction may not initially destabilize ecosystem function, continued extinction will do so inevitably." Even as managers struggle to define time-dependent management goals—restore to the past or future?—the absence of clarity in how to define biodiversity in a dynamic and human-driven landscape muddies the waters. Leading-edge species (Starzomski, 2013) and even some exotic invasive species (Richardson and Gaertner, 2013) are envisioned as propagules for novel ecosystems, yet the current management paradigm is to eschew deliberately translocating species to places they have not occurred before while combating the accidental spread of many nonnative species. Why hesitate to deliberately transplant white spruce (*Picea glauca*) to a warming high arctic, implicitly willing to accept a fledging boreal forest dominated by wind-dispersed hardwoods (*Salix* spp. and *Populus* spp.), but depauperate of cone-bearing softwoods, even though the latter will eventually migrate over the Brooks Range with enough passage of time (Rupp and others, 2001)? Yet the accidental introduction of spruce is implicitly welcomed to the arctic coastal plain by an unknown vehicle traveling up the Dalton Highway (Elsner and Jorgenson, 2009), whereas farther south on this same road, efforts are underway to control the spread of sweetclover (*Melilotus officinalis*), an Old World species that began colonizing North America more than three centuries ago. The basis of these contradictions is as much societal as ecological and is rooted in the beliefs and values of the practitioner (Clark, 2002; Kennedy, 2013). This tension is most often (and ultimately ineffectively) resolved at local scales rather than with strategies that are coherent over large geopolitical landscapes (for example, NWB region boreal biome).

complete understanding of current ecological processes nor the prescience to forecast ecological "surprises" and no-analog futures (sensu Williams and Jackson, 2009), social and economic drivers that will not wait for scientific certainty also must be appreciated (see Box 4.2.2, "Novel Assemblages: Nexus of Rapid Climate Change and Exotic Species on the Kenai Peninsula"). Consequently, practicing adaptive management (Nichols and others, 2011; Seastedt and others, 2008) and networking adaptation efforts (see Box 4.2.3, "Participatory Scenario Planning for 'New' Wildlife in the Southwest Yukon") will help ensure that smart approaches are used while minimizing ecological risk and unintended consequences. Ecologists and climate adaptation practitioners should be prepared to work with directional change to steward outcomes (Chapin, Robards, and others, 2006), evolving from retrospective to prospective adaptation approaches (Magness and others, 2011). As discussions about facilitating climate adaptation proceed, perhaps with the explicit goal of designing novel community and trophic assemblages, recognize that there is not a single interjurisdictional or transboundary entity with the authority to implement such action.

Lastly, even as the initial responses may be experimental and exploratory (for example, common gardens), the collective aim of our management responses among NWB LCC partners must ultimately be coherent and strategic at the biome scale to ensure we are not at cross-purposes, particularly in our approach to managing exotic species. Serving as a bridging organization (Folke and others, 2005), then, may be one of the most important roles of the NWB LCC.

INFORMATION GAPS

- Fine-scale spatially explicit species richness data are lacking; that is, extant species assemblages.
- Well-estimated historical rates of ecosystem drivers are lacking (for example, fire, spruce bark beetle), as is knowledge about potential thresholds or tipping points (for example, extreme weather events, available water).
- New measures, tools, and management approaches are needed; for example, assessing the need to intervene; measuring ecological difference in unknown metrics to determine what is novel; capacity to detect ecological surprises; articulating management goals in a directional world (Hobbs and others, 2013).
- A better understanding is needed of the underlying values that shape social, ethical, and policy dimensions of embracing and even designing novel ecosystems (Chapin, Robards, and others, 2006).
- The potential implications of adding (deliberately or otherwise) novel species to boreal ecosystem functions and services is needed.

BOX 4.2.2: NOVEL ASSEMBLAGES: NEXUS OF RAPID CLIMATE CHANGE AND EXOTIC SPECIES ON THE KENAI PENINSULA

The 24,000 km^2 (9.266 mi^2) Kenai Peninsula is connected by a 16 km^2 (6 mi^2) wide isthmus to the adjacent mainland in south-central Alaska. Converging lines of evidence strongly suggest that boreal ecosystems on the western peninsula are responding to rapid climate warming. Consecutive summers of above-average temperatures enabled unprecedented mortality of 400,000 ha (988,000 acres) of white, Lutz, and Sitka spruce (*Picea glauca* × *sitchensis* and *Picea sitchensis*, respectively) by spruce bark beetles in the 1990s (Berg and others, 2006). Over the last half century, alpine treeline rose 50 m (164 ft) (Dial and others, 2007), lowland wetlands decreased 6–11 percent per decade (Klein and others, 2005, 2011; Berg and others, 2009), the Harding Icefield shrank 5 percent in surface area and 21 m (69 ft) in elevation (Rice, 1987; Adageirsdottir and others, 1998), and available water decreased by 55 percent (Berg and others, 2009). Forests previously burned by late-summer canopy fires are being replaced by spring fires in bluejoint (*Calamagrostis canadensis*) grasslands, and a 2005 wildfire in mountain hemlock (*Tsuga mertensiana*) was outside historical fire regimes (Morton and others, 2006). These empirical observations support climate envelope models that forecast 30–80 percent of the peninsula will experience ecosystem transition (not simply succession), including deforestation of Caribou Hills and afforestation of tundra in

Figure 4.2.1 (right). Conversion of spruce beetle-killed forest to grass-dominated savannas is the most dramatic change on the Kenai Peninsula in response to rapid climate warming, perhaps being sustained by human-caused spring fires. Photo by John Morton, U.S. Fish and Wildlife Service.

the Kenai Mountains through 2100 (Magness and Morton, 2018). Under extreme scenarios, more than 400 terrestrial species could be extirpated if certain land-cover types disappear. Conversely, *e*Bird (www.ebird.org, accessed November 9, 2017) records show 13 new bird species in the last 5 years, and earlier arrival and later departure dates, respectively, for 33 and 38 species of migratory birds.

"Winners and losers" can be expected in almost any climate change scenario (O'Brien and Leichenko, 2003), so for some native species to be extirpated or replaced by new species is not inherently problematic. However, the isolated geography of the peninsula may restrict migration of native biota to and from the adjacent mainland. In disturbed and highly transitional areas, it seems likely that novel assemblages will be at least partially composed of accidentally introduced species from extant populations of exotic earthworms, arthropods, slugs, feral pheasants (*Phasianus colchicus*) and turkey (*Meleagris gallopavo*), more than 110 exotic vascular plants and chytrid fungus (*Batrachochytrium dendrobatidis*; Reeves and Green, 2006), as well as deliberately introduced non-native Sitka black-tailed deer (*Odocoileus hemionus sitkensis*), ruffed grouse (*Bonasa umbellus*), and northern pike (*Esox lucius*). In response to expanding grasslands in the aftermath of a spruce bark beetle outbreak, interagency fire management programs promote reforestation with non-native lodgepole pine; indeed, of which some mature trees have produced seedlings. Novel assemblages on the Kenai Peninsula may be based not on a simple reshuffling of native boreal species in response to climate change, but on human-dispersed species from the Old World and temperate North America (Harris and others, 2006).

BOX 4.2.4: PARTICIPATORY SCENARIO PLANNING FOR 'NEW' WILDLIFE IN THE SOUTHWEST YUKON

The southwest Yukon is marked by a climate-induced directionally changing landscape, an increasing shift away from traditional subsistence lifestyles, considerable governance innovation through land-claim settlement, and changing species composition. These changes may result in a novel ecosystem state, although its precise nature is uncertain. Recent introduction of elk (*Cervus canadensis*), reintroduction of wood bison (*Bos bison athabascae*), and immigration of mule deer (*Odocoileus hemionus*) have created a complex suite of changes only beginning to be understood, and raise hard questions for wildlife managers. Working with management groups, Beach (2014) experimentally applied and evaluated a participatory scenario planning approach that used scientific and traditional knowledge to explore alternative future scenarios and management goals for these three ungulates. Three workshops were held in 2012–13 with the Alsek Renewable Resource Council, Yukon Wood Bison Technical Team, and the Yukon Elk Management Planning Team.

Participants completed surveys during and after the workshops. Through a set of structured exercises, participants identified drivers of social-ecological system change and articulated how they could interact to shape the system by 2032. Four scenarios developed warn of plausible future events such as novel disease, pest outbreaks, and increasing land development. The most significant contribution of scenario planning was to create a forum to share perspectives and develop mutual trust and understanding which facilitated the quality and extent of cross-cultural dialogue.

Two types of lessons, content and process, were learned from the workshops that have implications for other wildlife managers in the NWB LCC. *Content* lessons include: (1) disease status, introductory status, future context, and food security implications will influence the value society attributes to new species; (2) management actions taken today will have serious implications for the population levels of new ungulate species; (3) hunting and education opportunities may influence society's value of new species; (4) the desire of Yukon residents to maintain a harvest represents an opportunity to facilitate the valuing of new species; and (5) increased disease monitoring is needed because novel diseases are likely to be future drivers of ungulate populations. *Process* lessons include: (1) improved knowledge sharing as a way to increase First Nations engagement; (2) longer-term wildlife management planning processes can accompany conventional planning time horizons; (3) planning can include holistic thought processes that consider how driving forces interact into the future; and (4) scenario planning can help practitioners examine causal links between driving forces and, therefore, uncertainty when planning.

Figure 4.2.2. Scenario planning workshop participants group individual drivers of change into axes of change, or themes, depicting the dominant forces causing change in the southwest Yukon Territory social-ecological system. Photo by Doug Clark, University of Saskatchewan, Saskatoon, Saskatchewan, Canada.

4.3
WILDLIFE PARASITE AND PATHOGEN LIFE CYCLES IN THE NORTHWEST BOREAL REGION

By Jane Harms[1] and Caroline Van Hemert[2]

KEY FINDINGS

- Climate warming and anthropogenic activity are linked to emerging infectious diseases in wildlife. The rapid rate of ecological change in the NWB region may lead to emergence or establishment of diseases that affect animal and human health.
- Parasites with a free-living stage in the environment and vector-borne pathogens are especially responsive to higher temperatures and a longer growing season.
- Changes to ecological conditions induced by a warming climate may influence the susceptibility of hosts to disease, alter rates of disease transmission, and facilitate invasion of novel pathogens and parasites. Increasing anthropogenic activities also could contribute to changing patterns of disease in animals and humans.
- Infectious diseases can have profound effects on wildlife population dynamics through direct morbidity or mortality or through influences on physiological parameters such as reproduction, energy budgets, and immune function.
- Diseases in wildlife influence human health through transmission of zoonotic pathogens and through demographic effects of disease on wildlife populations. Changes in subsistence harvest and food security may have direct cultural, health, and economic impacts on northern communities.

1 Environment Yukon, Whitehorse, Yukon, Canada.
2 U.S. Geological Survey, Anchorage, Alaska, USA.

BACKGROUND AND CURRENT STATE OF KNOWLEDGE

Infectious diseases in wildlife populations are receiving increasing attention, partly due to the high number of emerging pathogens in wildlife that have the potential to cause disease in humans (Daszak and others, 2001) or domestic animals (Miller and others, 2013), and because diseases can be transmitted from domestic animals to wildlife (Dobson and Foufopoulos, 2001; Hughes and Macdonald, 2013). Infectious diseases also have been implicated as causes of marked declines or local extinctions of wildlife species (Frick and others, 2010; Rosa and others, 2013; Hollings and others, 2014), with associated ecosystem-scale consequences. Climate warming has been linked to the emergence or expansion of many infectious diseases in many different regions of the world (Harvell and others, 2002; Altizer and others, 2013). In the Arctic and subarctic, warming temperatures and other ecological effects such as increasing human activity may lead to establishment of diseases that have important effects on animal and human health (Fig. 4.3.1; Bradley and others, 2005; Kutz and others, 2009; Hueffer and others, 2011).

Historically, cold temperatures, geographic isolation, and lack of appropriate vectors in northern regions have limited the diversity of wildlife pathogens (Hueffer and others, 2011; Hoberg and others, 2012). Climate warming may release or alter some of these constraints, providing opportunities for introduction or expansion of pathogens into areas that were previously unsuitable. In particular, shifts in temperature, precipitation, and season length are likely to affect the distribution and abundance of wildlife pathogens (Harvell and others, 2002; Hoberg and others, 2008). Although baseline information about wildlife disease in many parts of northern Canada and the United States is relatively scarce, the Northwest Boreal (NWB) region and areas farther north provide an important arena for the study of the effects of climate warming and other landscape changes on wildlife health and disease.

DRIVERS OF WILDLIFE PARASITES AND PATHOGENS

The emergence and establishment of pathogens, their vectors, or both depends on many factors, but it is clear that climate warming and anthropogenic alteration of ecosystems often are important contributors (Fig. 4.3.1; Ogden and others, 2013). Parasites with a free-living stage in the environment and vector-borne pathogens are particularly responsive to warming (Altizer and others, 2006). For example, climate-related changes in parasite range and life cycle have been well documented for several nematode parasites of Arctic ungulates (Kutz and others, 2009; Kutz and others, 2013) (see Box 4.3.1, "Parasites and Climate Change"). Climate warming is likely to promote conditions that are more favorable for arthropods that serve as parasites and vectors of disease (Harvell and others, 2002; Hoberg and others, 2008). Recently, winter ticks (*Demacentor albipictus*) were documented on moose (*Alces alces*) in the Sahtu region of the Northwest Territories, which represents a new northern limit for this parasite

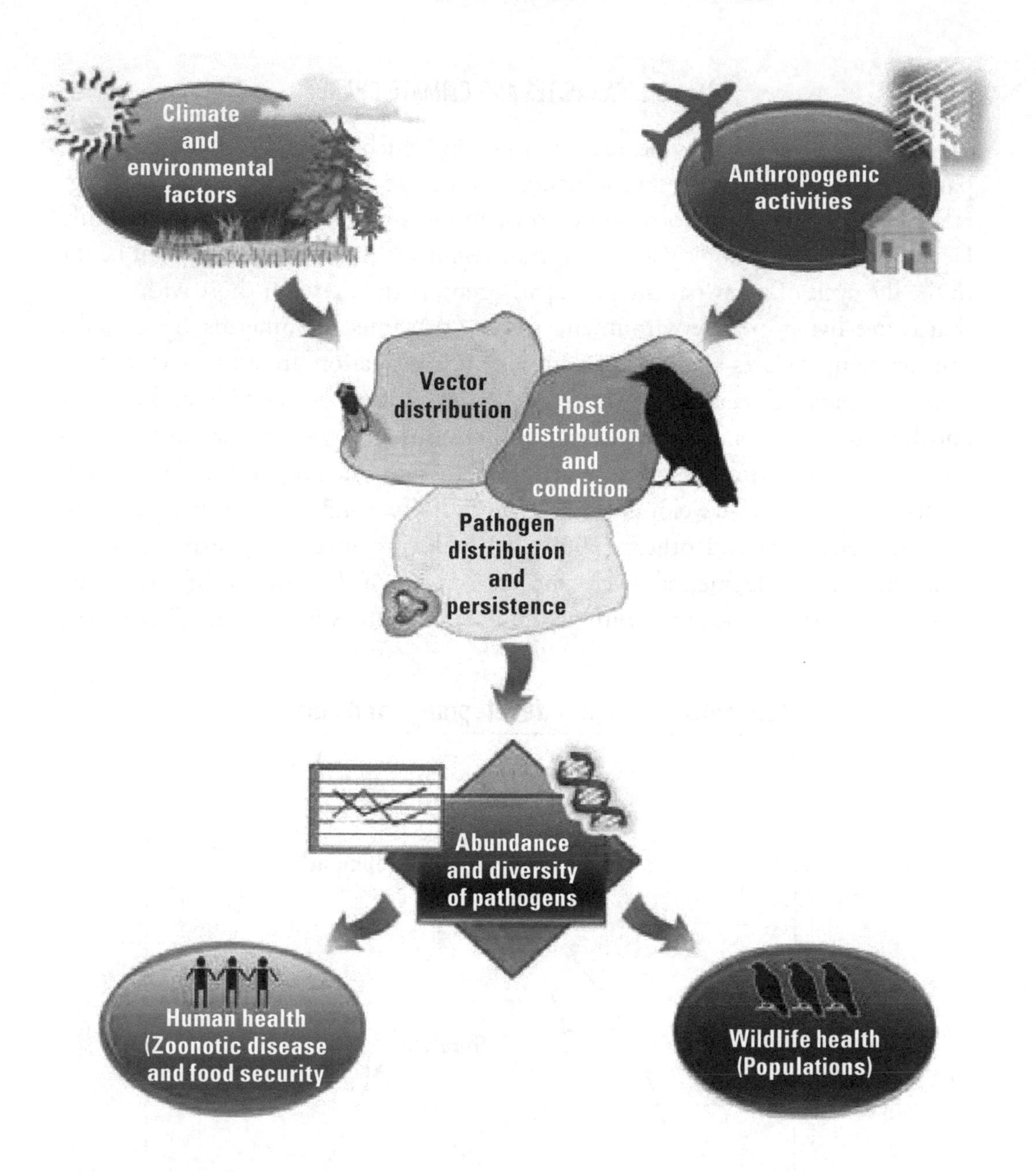

Figure 4.3.1. Conceptual diagram showing factors that may affect wildlife disease dynamics in the Northwest Boreal region. Rapid environmental changes in this region are currently underway due to climate warming and increased anthropogenic activities. As a result, shifts in the distribution and persistence of hosts, pathogens, and vectors are expected. These changes will influence both wildlife and human health through exposure to disease and availability of subsistence food resources. Based on diagram from Van Hemert and others (2014). Reprinted with permission from the Ecological Society of America. John Wiley and Sons.

BOX 4.3.1: PARASITES AND CLIMATE CHANGE

Northern ecosystems provide ideal settings in which to study the response of parasites and pathogens to environmental change due to the rapid rate of warming, relatively low biodiversity of wildlife hosts, and fewer confounding anthropogenic factors compared to tropical and temperate regions. Climate is an important factor in the life cycle of many parasites and pathogens, particularly for those with a stage that is free-living in the environment or those that infect ectotherms. Specifically, warmer temperatures and altered patterns of precipitation are estimated to have effects on parasite transmission rates, virulence, patterns of infection and disease, and host-parasite dynamics (Polley and Thompson, 2009). Higher temperatures and a longer growing season also may lead to shifts in geographic distribution or accelerated larval development and survival (Fig. 4.3.2) for some parasites. For example, Kutz and others (2009; 2013) documented range expansion and a shortened developmental cycle for an important lungworm of muskoxen (*Ovibos moschatus*) associated with warming temperatures in the Canadian Arctic.

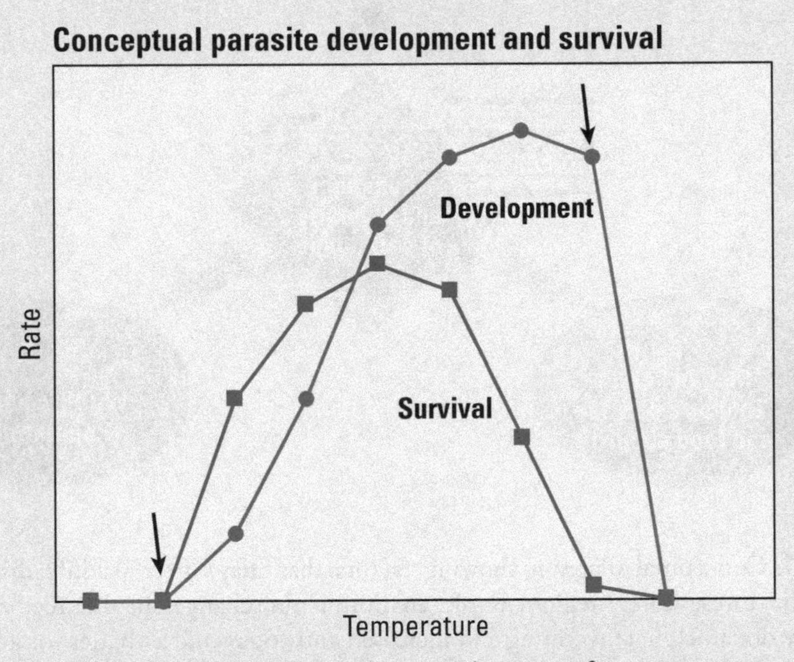

Figure 4.3.2. Hypothetical development and survival curves for parasitic nematodes. Arrows indicate lower and upper developmental thresholds. Parasites with egg and larval stages in the environment (free-living) are exposed to the entire range of conditions. In contrast, parasites with invertebrate intermediate hosts or vectors may be exposed only to the development regions of the lines because part of their development relies on an intermediate host. Note the offset between development and survival rates as temperatures increase. Reprinted from Kutz and others (2009), with permission from Elsevier.

Similarly, temperature and precipitation strongly influenced the rate of development and timing of transmission of parasitic nematodes of red grouse (*Lagopus lagopus scotia*) in Scotland (Cattadori and others, 2005). In subarctic Fennoscandia, the recent emergence of epidemic disease due to mosquito-borne filaroid nematodes in cervids has been associated with higher temperatures (Laaksonen and others, 2010). Ongoing work to fill knowledge gaps and develop conceptual and predictive models will be key to understanding the role of parasitism in wildlife health and in implementing feasible and appropriate mitigation measures (Kutz and others, 2009).

(Kashivakura 2013; see Box 4.3.2, "Changing Distributions of Vectors and Vector-Borne Diseases"). Although timing of insect emergence is important because insects are a key food source for many wildlife species, it also may result in exposure to vector-borne diseases such as avian malaria (Bradley and others, 2005; Garamszegi, 2011).

Climate-induced changes in ecological communities are likely to influence the susceptibility of hosts to disease, alter host exposure and pathogen transmission rates, permit invasion and establishment of novel pathogens or vectors, and support invasion of new hosts and their associated pathogens, thereby creating new opportunities for intra- and interspecific disease transmission (Burek and others, 2008; Kutz and others, 2013). Changes in phenology, including seasonal advances in green-up, will affect the timing of reproduction and migration, with subsequent impacts on animal distribution (Visser, 2008).

Anthropogenic activity, particularly environmental degradation caused by human activities such as pollution and habitat fragmentation, is an important contributor to disease emergence (Dobson and Foufopoulos, 2001). Encroachment of humans and domestic animals into natural habitats through resource exploration, industrial development, and tourism provides opportunity for exposure of humans to new zoonotic agents or transfer of pathogens between animal species (Cutler and others, 2010). Habitat degradation or loss may limit food availability and restrict movement of animals, and increase the opportunity for contact among wildlife, domestic animals, and humans (Acevedo-Whitehouse and Duffus, 2009). Development of industrial infrastructure and transportation corridors can alter patterns of wildlife distribution or movement, support increasing numbers of species that benefit from anthropogenic resources and introduce invasive species or pathogens (Kutz and others, 2009). Some zoonotic diseases, such as giardiasis, are likely to be introduced directly from humans into the environment (for example, by runoff into freshwater habitats; Burek and others, 2008; Jenkins and others, 2013). Exposure to environmental contaminants introduced as a result of industrial activities has the potential to influence susceptibility to disease through mechanisms such as alteration of immune function (Acevedo-Whitehouse and Duffus, 2009).

BOX 4.3.2: CHANGING DISTRIBUTIONS OF VECTORS AND VECTOR-BORNE DISEASES

Climate warming may result in conditions that permit the northward range expansion of arthropods, including mosquitoes, ticks, and biting flies, which may be vectors for wildlife and zoonotic diseases (Harvell and others, 2002; Hoberg and others, 2008). Recent increases in the range and abundance of ticks in Sweden have been attributed to milder winters and longer growing seasons (Jaenson and others, 2012), and several tick species appear to have expanded northward in parts of Russia and northern Canada (Kutz and others, 2009; Revich and others, 2012). Range expansion of arthropods such as ticks may be facilitated by the northern dispersal of tick hosts (including small mammals and birds) that are responsive to changing climatic conditions (Ogden and others, 2013). Increasing temperatures also improve the survival and replication rates of insect vectors, allowing vectors and the diseases they carry to become established in new locations (Revich and others, 2012). For example, northward expansion of Lyme disease and its vector, deer tick (*Ixodes scapularis*), recently has been documented in North America (Ogden and others, 2013). Additional studies of host dispersion and establishment of vectors are needed for making estimates about the effects of climate change on the invasion and diversity of vector-borne diseases (Ogden and others, 2013).

Warming temperatures may encourage agricultural activities in northern regions, creating opportunities for disease transmission from domestic animals to wildlife. Currently, there is substantial concern about the potential for transmission of respiratory pathogens from domestic sheep and goats to wild thinhorn sheep, which could result in disease outbreaks similar to those seen in bighorn sheep in southern North America (Jenkins and others, 2000). Translocation of wildlife to northern regions also can facilitate spread or introduction of pathogens. For example, translocation of elk (*Cervus canadensis*) in the 1950s and 1990s likely increased the northern range of the winter tick in Yukon, Canada (Leo and others, 2014).

EFFECTS ON WILDLIFE AND HUMANS

The NWB region plays a critical role in maintaining resilient populations of migratory and resident birds, mammals, and fish (Blancher, 2003; Pastor and others, 1996). Increasing numbers of new pathogens and parasites in the ecosystem may have significant conservation implications for local wildlife. Infectious diseases can have profound effects on population dynamics through direct mortality or other physiological parameters such as effects on reproductive success or immune function.

Infectious diseases also can influence wildlife health in subsequent stages of the annual cycle through carryover effects (Harrison and others, 2011).

Wildlife species are an important source of zoonotic diseases; therefore, changes in distribution or abundance of pathogens and parasites could have significant impacts on human health (Cutler and others, 2010). Wildlife often serve as pathogen reservoirs and contribute to the movement of pathogens over large distances through migration. Consumption or handling of tissue from infected or diseased animals can transmit pathogens from wildlife to humans. Partly because of substantial wildlife harvest activities, numerous zoonotic diseases from wildlife have been documented in the NWB region, including tularemia, echinococcosis, and trichinellosis (Hueffer and others, 2013).

In addition to direct effects on human health through zoonotic diseases, changes in wildlife health and resulting population demographics have important implications for food security in the NWB region. Many residents rely heavily on wildlife resources, including locally harvested fish, birds, and mammals. In particular, First Nation and Alaska Native residents have a long history of subsistence harvest in this region. Threats to the viability of wildlife populations would have direct cultural, health, and economic effects on communities.

INFORMATION GAPS

- Baseline information about the current distribution of pathogens and parasites in the NWB region is lacking, which limits the ability to identify and track emergence or expansion of infectious diseases.
- More research is needed to understand relationships between environmental variables (such as temperature and precipitation) and life cycles of pathogens and parasites that infect northern wildlife. Models that simulate how the relationships between climate and disease may change over time will provide important information that can be used to protect human and animal health.
- Key aspects of the ecology of many wildlife diseases, such as transmission routes, reservoirs, and carryover effects, have not been established in northern ecosystems.
- Demographic and fitness consequences of disease on wildlife populations are not well known, making it difficult to project the outcomes of changing pathogen and parasite communities.

4.4 MARINE-DERIVED NUTRIENTS IN THE NORTHWEST BOREAL REGION

By Ramona Maraj[1] and Mark Wipfli[2]

KEY FINDINGS

- Freshwater food webs often are greatly stimulated by marine nutrients, such as delivered by returning salmon, and in some places, insect densities and growth rates can increase several-fold.
- Naturally occurring stable isotopes show riparian vegetation near spawning streams acquire up to 70 percent of the nitrogen from salmon in their foliage.
- Bears and other vertebrates distribute the nutrients from salmon through waste excretion and by leaving salmon carcasses scattered along stream banks.
- Wolves also redistribute salmon to the forest floor. In some inland places, the contribution of salmon to wolf diets is similar to estimates reported for coastal wolves.

Salmon are a good example of fish species that deliver marine subsidies to inland boreal ecosystems and are a primary source of marine nutrients in the region. The return of salmon to natal freshwater systems completes a nutrient cycle via their spawning grounds and decomposition in the streams. In the Northwest Boreal (NWB) region, the salmon, wolves, and bears interact as keystone species, and removing any of these species could have trophic consequences that reduce other species. There are broad ecosystem productivity and species abundance benefits from maintaining high salmon numbers or maintaining "nutrient capital" provided by the salmon within watersheds.

1 Fish and Wildlife Branch, Department of Environment, Whitehorse, Yukon, Canada.
2 U.S. Geological Survey, Alaska Cooperative Fish and Wildlife Research Unit, University of Alaska Fairbanks, Fairbanks, Alaska, USA.

ROLE OF SALMON

There are five species of Pacific salmon (*Oncorhynchus* spp.) in the NWB region: Chinook (*O. tshawytscha*), coho (*O. kisutch*), chum (*O. keta*), pink (*O. gorbuscha*), and sockeye (*O. nerka*), all having similar life cycles. Salmon begin life as an egg embedded in stream or lake bottoms, emerge in spring as swim-up fry, and generally spend as much as two years rearing in fresh water. Pink and chum salmon migrate to salt water soon after hatching, whereas the other three species spend at least one winter in fresh water. Salmon make their way to the ocean as smolts, where they grow to adulthood, generally taking 2–5 years (Groot and Margolis, 1991). The lifecycle is completed when mature fish return to fresh water to spawn and die. Salmon are a classic example of fish that deliver marine subsidies to land-based ecosystems. Other examples include smelt (*Osmerus eperlanus*), stickleback (*Stylephorus chordates*), and shad (Clupeidae), although their contribution is significantly lower at a broad landscape scale (Gende and others, 2002; Naiman and others, 2002; Hocking and Reynolds, 2011).

The return of salmon to natal freshwater systems completes a nutrient cycle (Fig. 4.4.1; Naiman and others, 2009). Over millennia, nutrients within the stream have been carried downstream from higher elevations, eventually reaching the ocean. Salmon, while in the ocean, may assimilate some of these nutrients when they feed. When salmon return to their spawning grounds and die, they provide these nutrients to freshwater and terrestrial ecosystems. Salmon can deposit thousands of tons of nutrients and energy (carbon). An adult male chum salmon may average 130 g (4.59 oz) of nitrogen, 20 g (0.71 oz) of phosphorous and more than 20,000 kilojoules (kj) (4,780 kilocalories [kcal]) of energy as protein and fat. In a 250 m (820 ft) reach of a stream in southeastern Alaska within a one-month window, 80 kg (176 lb) of nitrogen and 11 kg (24 lb) of phosphorus were deposited into the stream when salmon died (Gende and others, 2004).

The nutrients encourage primary productivity, providing food for insects, which then feed salmon and other animals. Freshwater food webs supporting fishes often are greatly stimulated by marine nutrients, and in some places, insect densities and growth rates can increase several-fold (Wipfli and others, 1998; Wipfli and others, 2003; Scheuerell and others, 2007).

Beyond the aquatic ecosystem, plant growth in northern forests often is limited by either nitrogen or phosphorous, so nutrients from the redistribution of fish carcasses can have a strong influence on the rates of growth of many plant species (Ben-David and others, 1998). Naturally occurring stable isotopes indicate riparian vegetation near spawning streams acquire as much as 70 percent of the nitrogen from salmon in their foliage (Ben-David and others, 1998; Helfield and Naiman, 2002; Wilkinson and others, 2005).

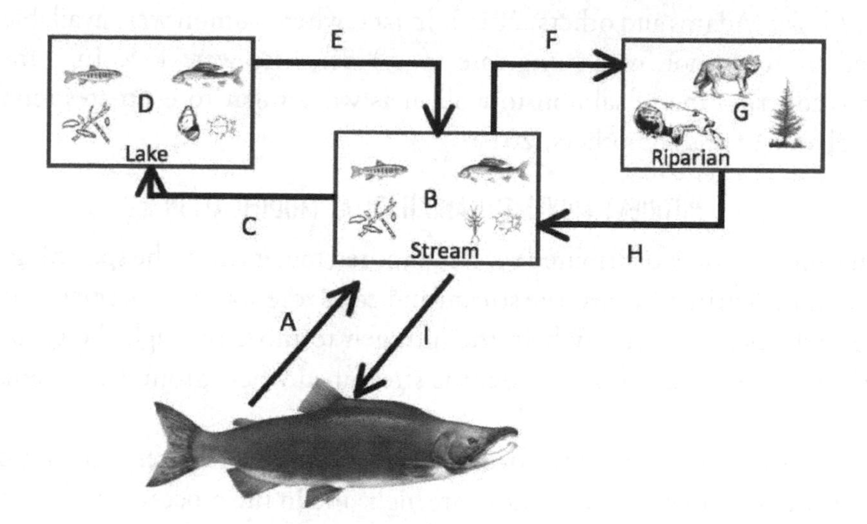

Figure 4.4.1. Diagram showing cycling of marine-derived nutrients. (A) Spawning salmon transport nutrients upstream; (B) nutrients are incorporated into stream food web; (C) nutrients are delivered to lakes by beach spawners and discharge from spawning tributaries; (D) nutrients are incorporated into lake food webs; (E) nutrients are washed from lakes into receiving streams; (F) nutrients are delivered to riparian zones by bears, wolves, and other vertebrates; (G) nutrients are taken up by riparian vegetation and incorporated into the food web; (H) riparian trees enhance the quality of instream habitat for salmonids by providing shade, bank stabilization, organic matter, and large woody debris; (I) nutrients enhance smolt production and potentially lead to increased return of adult salmon spawners. Adapted from Naiman and others (2009).

ROLE OF BEARS AND WOLVES

Salmon are a critical food source for some bear (*Ursus* spp.) populations, helping them to deposit fat quickly for hibernation and reproduction. Bears can kill and consume a large amount of the total spawner biomass (Ruggerone and others, 2000; Gende and others, 2001; Quinn and others, 2001). Bears distribute nutrients through waste excretion. An adult female bear can deliver about 46 g (.66 oz) of salmon-obtained nitrogen per hectare (acre), primarily as urine (Hilderbrand and others, 1999). Urine is converted quickly to ammonium and becomes an available nutrient source for plants. If salmon are abundant and easy to catch, the bears often selectively eat only the parts of the fish with the highest fat content (for example, eggs [roe]). The remaining discarded carcass is then carried away, broken down by scavengers and decomposers, or both, which further distributes the nutrients.

Wolves (*Canis lupus*) also redistribute salmon to the forest. Although wolves are considered to be obligate predators of ungulates, spawning salmon provide a marine subsidy affecting inland wolf–ungulate food webs. Stable-isotope analyses showed that salmon made significant contributions to wolf diets in Denali National Park and

Preserve, Alaska (Adams and others, 2010). In fact, where salmon were available, wolves consumed more salmon when ungulate (prey) densities were low. In some inland places, the contribution of salmon to wolf diets was similar to estimates reported for coastal wolves (Adams and others, 2010).

NUTRIENT MOVEMENT AND HABITAT MODIFICATION

Nitrogen from salmon is distributed several hundred meters from the spawning streams, with the carcasses primarily near the stream and carnivore waste more prevalent distally. Leaching and microbial activity help the nitrogen to move through the ground, even into the hyporheic zone, the area under the streambed where groundwater and surface water interact.

The vegetation near spawning streams has high nitrogen content; trees grow faster and denser when nutrient concentrations are high, and in the process, stabilize the banks, filter sediment, shade the water, and deposit debris (Naiman and others, 2009). Aquatic invertebrate biomass and density increase (Wipfli and others, 1999). Cederholm and others (1989) reported that more than 100 species benefit from salmon biomass.

MAINTAINING THE NUTRIENT FEEDBACK

A keystone species is one on which other species in an ecosystem largely depend, such that if it were removed the ecosystem would change drastically, and may have a disproportionately larger effect on its environment than its numbers would suggest (Paine, 1995). Removing a keystone species can have profound ecosystem implications. In some cases, two keystone species—such as bear or wolf and salmon—interact synergistically.

Anything that lowers the number of bears, wolves, or salmon in these systems can have trophic consequences that decrease numbers of other species. Broader ecosystem productivity and species abundance benefit from maintaining high salmon numbers, or maintaining "nutrient capital" within watersheds. To maintain the nutrient feedback, forests and their riparian ecosystems can be managed to provide woody debris and habitat complexity for carcass and nutrient retention, and therefore broader ecosystem productivity.

NUTRIENT SUBSIDIES FROM OTHER SOURCES

Marine nutrients are one example of resource subsidies. Other sources (Fig. 4.4.2; Wipfli and Baxter, 2010) of nutrients (and prey) for fishes include terrestrial invertebrates from surrounding riparian ecosystems (Wipfli, 1997; Kawaguchi and Nakano, 2001; Baxter and others, 2005). Small headwater channels also are a source of prey for stream fishes (Wipfli and Gregovich, 2002) at certain times of the year (Fig. 4.4.3; Wipfli and Baxter, 2010). Stream salmonids can rely more on terrestrial- than

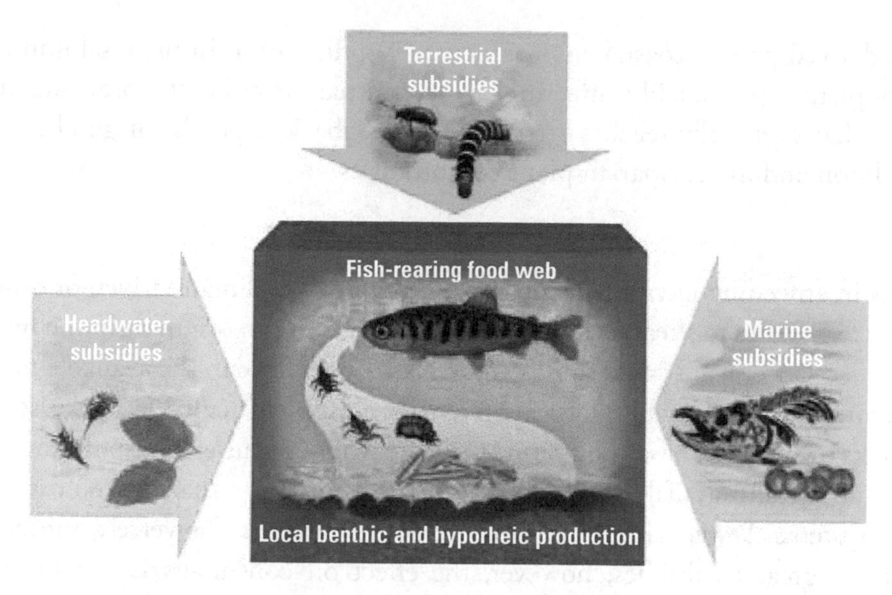

Figure 4.4.2. Subsidies important to freshwater food webs and fishes. Diagram from Wipfli and Baxter (2010).

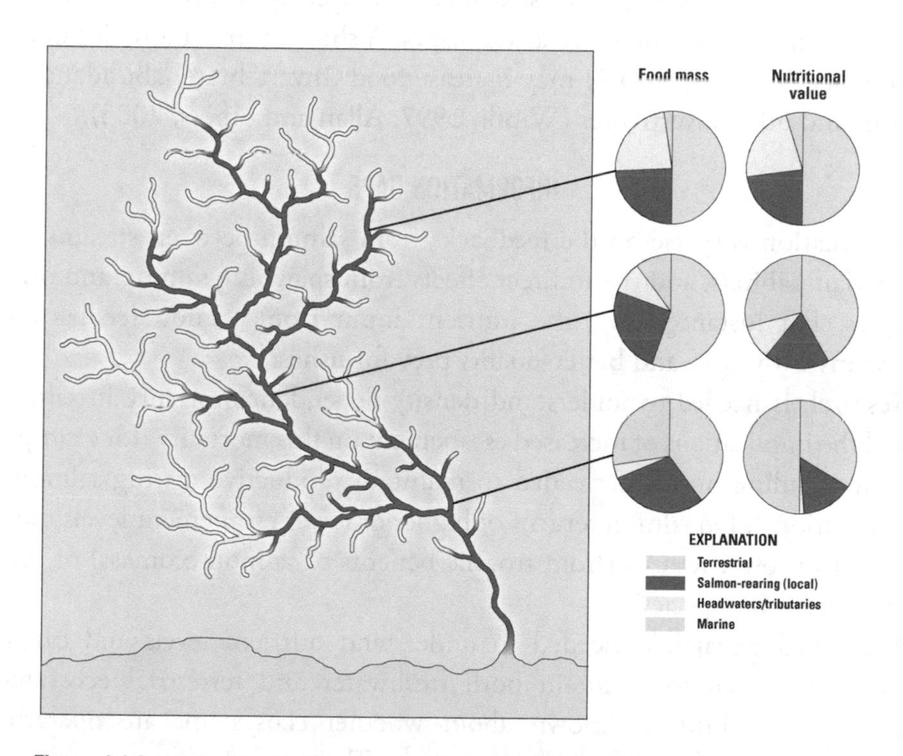

Figure 4.4.3. Relative importance of subsidies through time and space. Diagram from Wipfli and Baxter (2010).

aquatic-derived prey in coastal environments (Wipfli, 1997). In turn, salmon fertilize riparian plants, presumably affecting the terrestrial invertebrate prey quality and quantity that eventually reaches streams. This feedback depends on good returns of adult salmon and intact riparian plant communities.

CLIMATE AND LANDSCAPE CHANGE

Changes in any component of the marine nutrient cycle, climate-driven or otherwise, may affect subsidies to stream ecosystems. These changes can originate in the marine or terrestrial ecosystem and can result in an increase or decrease in the available nutrients. In the marine system, any biophysical driver (for example, Pacific Decadal Oscillation) that affects the number of salmon returning to spawn may cause corresponding changes in the amount and availability of marine-derived nutrients that can be carried into riparian systems. Fewer salmon result in reduced subsidies. Conversely, more salmon translate to greater subsidies; however, the effect on consumers is not linear, with incremental effects likely tailing off at high densities.

Shifts in the terrestrial system, such as those resulting in changes to riparian plant cover or species composition, will alter terrestrial invertebrate subsidies to streams. This influences prey abundance for fishes. Climate change, forest fires, or other causes of successional changes may alter tree assemblages. A shift towards more deciduous from coniferous species (Chapter 4.1) may increase food (invertebrate) abundance for fish, songbirds, and other invertivores (Wipfli, 1997; Allan and others, 2003).

INFORMATION GAPS

- Information is sparse on the feedbacks from salmon between streams and riparian habitats, and the indirect effects from spawning salmon and other fishes. For instance, does the nutrient input from salmon feedback to streams with more and better-quality prey for fishes?
- Research is needed to understand density-dependent mortality in salmon and the implications of increased escapement on this mortality (for example, corresponding nutrients needed to maintain productive rearing salmonid populations). Do salmon returns of higher or lower escapement levels mean higher or lower returns (from trophic benefits of salmon biomass) of their own and other species?
- Additional research is needed to understand nutrient levels and carcass biomass needed to maintain both freshwater and terrestrial ecosystem components. Little is known about whether ecosystems are operating on the legacy effects of past salmon runs. These riparian systems may be functioning at a higher level of inherent productivity from hundreds of years of salmon returns.

5
SOCIOECONOMIC DRIVERS

Philip J. Burton[1]

Landscape dynamics in the Northwest Boreal (NWB) region are driven by human as well as natural factors. Many of the triggers to environmental impacts and landscape change in boreal Alaska and Canada are a result of complex interactions among regional demographics and traditions, local resource concentrations, and global economic and political incentives. These drivers unfold in resource extraction initiatives and land use change (Chapter 5.1), an uncertain future for subsistence resource use and a limited documented understanding of its ecological effects (Chapter 5.2), and the distribution and persistence of chemical pollutants (Chapter 5.3). Less well appreciated are the roles of laws, regulations, policies, and treaties among the many political entities of the region and around the world (Chapter 5.4), and the typically unexplored value systems of different managers and policy makers (Chapter 5.5).

Transparent documentation of these drivers and their effects is identified as a common need for understanding these pressures and agents of change. Improved information on the socioeconomic drivers of landscape change is needed to capture beneficial opportunities and manage or mitigate negative consequences on ecosystems and society in the NWB region.

1 University of Northern British Columbia, Terrace, British Columbia, Canada.

5.1
LAND USE AND RESOURCE EXTRACTION IN THE NORTHWEST BOREAL REGION

By Philip J. Burton[1], Andrew W. Balser[2], Daniel J. Hayes[3], and Marcus Geist[4]

KEY FINDINGS

- The NWB region remains largely wilderness, dominated by natural ecological processes and relatively low human impact.
- Mining and transportation infrastructure are the dominant land uses in the region, with military installations and related activities important in Alaska.
- Although timber harvesting has been a minor ecosystem-modifying activity in the region, initiatives for community-based bioenergy systems are likely to increase its impacts.
- Land use changes and industrial development are greatly influenced by external factors, including world petroleum and metal prices, and various governmental initiatives.

Compromised ecological integrity has historically occurred because of deforestation and other forms of land conversion. Such threats can be of greater concern than those stemming directly from climate change because they can completely destroy an ecosystem and can occur rapidly (Harte, 2001). In a global context, landscapes of the Northwest Boreal (NWB) region still consist of mostly natural and comparatively intact and unsettled woodlands, wetlands, and mountains (Ellis and others, 2010). Large areas are designated for conservation and recreation, much of which is unallocated public land; there is little private land or urbanization, and even industrial and infrastructure uses (for example, mining, roads) occupy a small part of the region (Table 5.1.1). Nevertheless, demand for natural resources from within and outside

1 University of Northern British Columbia, Terrace, British Columbia, Canada.
2 U.S. Army Corps of Engineers, Fairbanks, Alaska, USA.
3 University of Maine, Orono, Maine, USA.
4 University of Alaska, Anchorage, Alaska, USA.

Table 5.1.1. Land uses and estimated human footprint across the Northwest Boreal region.

					Northwest Boreal Landscape Conservation Cooperative	
Land use	Alaska	Yukon	Northwest Territories	British Columbia	Total area	Percentage of total
Areal footprint (ha)						
Protected areas (no extractive uses)	22,862,264	4,256,066	1,600,151	4,775,007	33,493,488	24.49
Forest harvesting (past records, current openings)	28,753	7,511	0	65,273	101,537	0.07
Mining (claims, leases, applications)[1]	836,539	3,805,230	64,170	1,552,739	6,258,678	4.58
Agriculture (pasture/hay, cultivated crops)	33,200	1,700	0	< ~500	35,400	0.03
Urban (built up, and other developed/cleared land)	102,400	28,728	786	7,331	139,245	0.10
Linear infrastructure (if average 50 m rights-of-way)	13,692	3,117	234	996	18,038	0.01
					Total length (km)	Density (km/km^2)
Linear Infrastructure (km)						
Streets and roads	23,172	4,651	187	1,535	29,545	0.022
Railroads (active and abandoned grade)	744	103	0	227	1,074	0.001
Pipelines (petroleum and natural gas)	1,653	532	280	161	2,626	0.002
Electrical transmission lines	1,814	947	0	69	2,830	0.002

[1]Surface land-cover disturbance is lower than these registered tenure areas.

the region, changing demographics, and increased human footprints will continue to affect the landscape. This chapter briefly reviews some historical and projected land uses across the region, their ecological impacts, and the socioeconomic factors that influence land use and resource development.

LAND USE OVERVIEW AND SCALING

In the northwest boreal region, as elsewhere, important land uses and their associated effects span broad gradients of size and duration. Management implications for activities related to (1) resource extraction (particularly petroleum development, mining, and forestry); (2) agriculture; (3) national security; (4) tourism and recreation; (5) transportation infrastructure; and (6) habitation and commerce are linked with these spatiotemporal characteristics. Figure 5.1.1 depicts critical NWB land uses

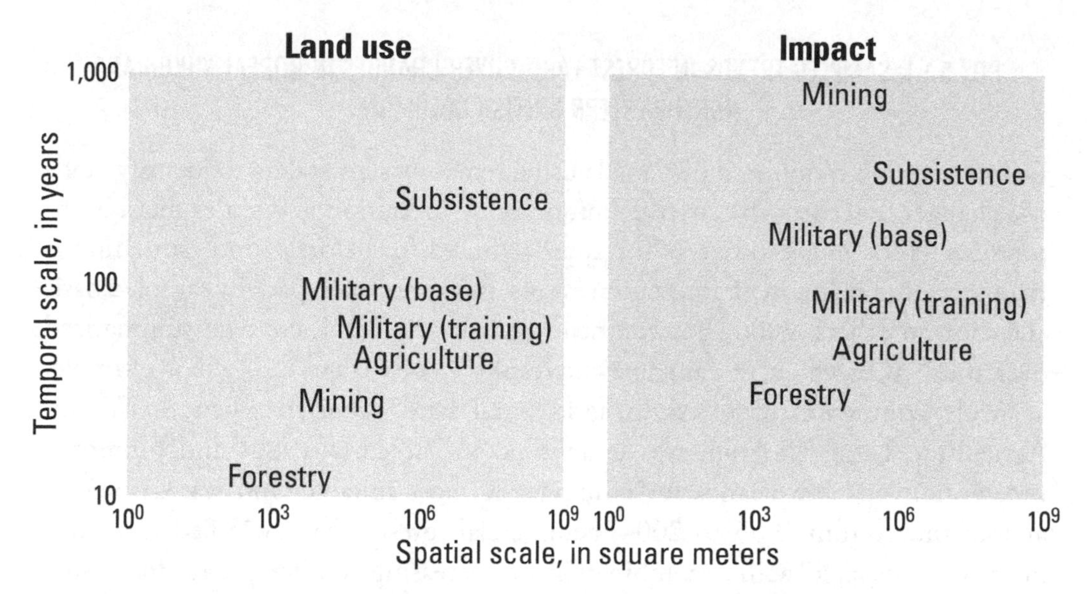

Figure 5.1.1. Land uses and perceived footprints scaled in space and time.

in this context and offers an initial schema for consideration of integrated, regional management. In this hypothetical comparison, it is logical to conclude that the potential scale of impacts often can surpass the physical footprint of human activities on the land. This conceptual figure is not authoritative but can provide a useful framework for evaluating the multiplier effect of different land uses, while recognizing the variability among individual examples of each land use. Mechanisms of impact and the opportunities for mitigation vary by sector; for example, the off-site effects of base metal mines often are associated with the acidification of surface waters (see Chapter 5.3), whereas the roads associated with mining and forestry operations contribute to wildlife mortality and sediment generation (Forman and Alexander, 1998).

RESOURCE EXTRACTION—PETROLEUM

With the vast majority of petroleum production in Alaska focused around Prudhoe Bay and the National Petroleum Reserve on the Alaska North Slope, oil and gas development in the NWB region is minor. Oil and gas has been produced in and around Cook Inlet since the 1960s, but is now at production levels only 10 percent of what it was 40–45 years ago (Alaska Oil and Gas Association, 2015). There is non-producing oil and gas potential in southeastern and north-central Yukon, particularly in the Liard and Eagle Plain geologic basins (Yukon Government, 2017). The natural gas industry in the Liard Basin of northeastern British Columbia is well developed, but is largely east of the NWB region. It can be argued that there is greater disruption to landscapes and wildlife associated with exploration surveying and drilling (typically consisting of considerable helicopter traffic, crisscrossed seismic lines cut in the forest,

BOX 5.1.1: SAMPLED TRENDS IN FOREST LAND-COVER CHANGES IN BOREAL YUKON AND NORTHWESTERN BRITISH COLUMBIA

Deforestation is monitored in Canada using remotely sensed data to identify land-use changes that meet the strict definition of deforestation and scales them up to produce nationwide estimates. The approach used for deforestation estimation is to observe and delineate deforestation events over time, on a spatially sampled basis (Leckie and others, 2006; Environment Canada, 2013). These data summarized over the NWB region reveal some interesting trends (Fig. 5.1.2). For example, a steady conversion of forest to agricultural land occurred when the Yukon Agriculture Lands Program was established in the early 1980s and promoted preemption of Crown lands for farming; the most suitable land was deforested at that time. From 1995 to 2004, commercial forestry has dwindled to almost nothing, associated with a combination of decreasing lumber prices, increasing fuel costs, and the closure of (out-of-region) sawmills and pulp mills. The peak in transportation infrastructure development from 1995 to 1999 was associated with improvements to the Alaska and Cassiar-Stewart highways. A surge in mining activity that commenced in 1997 was primarily associated with an expansion of the Kemess Mine in north-central British Columbia, but several additional mines in Yukon and British Columbia were in various stages of development through 2010 (and beyond) under conditions of favorable metal prices and supportive government policies.

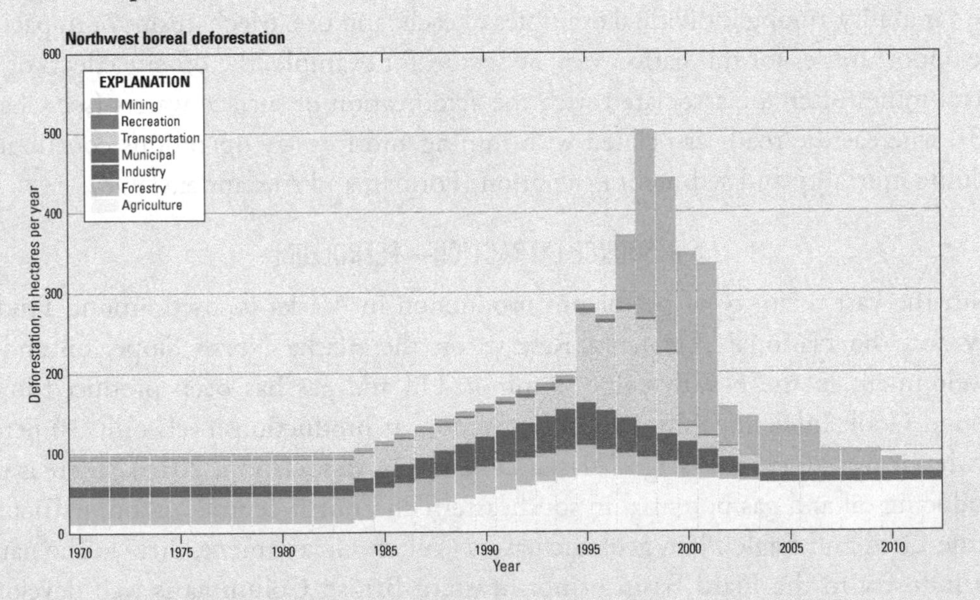

Figure 5.1.2. Trends in forest land-cover changes in boreal Yukon and northwestern British Columbia.

and many failed test wells) than is associated with oil and gas extraction itself (Borthwick, 1997). Additionally, crude oil spills and pollution impacts (see Chapter 5.3) are an ever-present risk associated with oil pipelines, whereas natural gas pipeline ruptures release high quantities of methane (a potent greenhouse gas) into the atmosphere and sometimes trigger fires.

RESOURCE EXTRACTION—MINING

Mining is arguably the single most important resource use in the NWB from a land-management perspective. Gold and coal have been produced in significant commercial quantities since the early 20th century, but zinc, gold, lead, and silver production are economic mainstays today. Tens of thousands of gold claims are scattered throughout the region, ranging from small, individual placer and dredging claims to large, industrial operations displacing more than 1 million m^3 (1.3 million yd^3) of surface material. Hard rock mines have a relatively small footprint, whether associated with underground shafts or open pits, but also have environmental effects associated with waste rock disposal, sedimentation ponds, milling and processing infrastructure, and transportation. Ecosystem displacement, sediment generation, and potential chemical impacts to soils (for example, from fuel spills and abandoned equipment) and adjacent aquatic ecosystems (for example, from acid rock drainage) require intensive, long-term management that considers use and recovery trajectories at decadal time scales (Akcil and Koldas, 2006; Lottermoser, 2010).

RESOURCE EXTRACTION—FORESTRY

With slow tree growth and long regeneration times, forest harvesting in many boreal regions can be considered more exploitative than sustainable (Burton and others, 2003). Commercial forestry is the dominant land use in much of coastal Alaska and in southern British Columbia, but plays a relatively minor role in the lower-productivity forests of the NWB region. Small logging operations and sawmills have partially met local needs through much of the 20th century, but the cost of shipping timber or pulp products to southern or international markets is prohibitive. Consequently, forest fires regularly affect much larger areas of land than is affected by logging and associated management activities (see Chapter 2). However, land-use changes and resource development such as forestry, agriculture, and mining result in complex interactions with the fire regime, altering patterns of fuel distribution, ignition, spread, and suppression (Csiszar and others, 2004; also see Chapter 2). In recent years, initiatives to create wood-fueled bioenergy plants have been undertaken to help meet electricity needs, especially for isolated communities throughout boreal Alaska and Yukon (Warren and Maynard, 2009; Yukon Government, 2009; Brackley and others, 2010). The scale of harvesting operations required to feed those energy demands sustainably has not yet been determined, but could be significant. Although individual boreal tree species such as trembling aspen (*Populus tremuloides*) may show improved growth under a warmer, drier climate, it is expected that the elevated occurrence of forest fires and insect pests may offset any forest productivity gains anticipated under climate change (Gauthier and others, 2015).

AGRICULTURE

Commercial agriculture typically requires a long-term conversion of natural vegetation to open ground for cultivated crop production or to grow forage for livestock. Although the NWB region is considered marginal for agriculture in a global sense, its isolation from major food-producing areas in North America and a pride in local self-sufficiency has been expressed in a long history of strong-willed food production and agricultural experimentation. Only 2 percent of Yukon land area is suitable for agriculture, with about 10,000 ha (24,700 acres) currently in production (Bell and others, 2012), and the NWB region of British Columbia has even less agricultural land. Less than 0.3 percent of the land area in Alaska is used for agriculture, most of it (>300,000 ha, >741,000 acres) in the boreal region (Dinkel and Czapla, 2012). The conversion of forestland to forage or crop production continues sporadically throughout the region, averaging about 445 ha per year (ha/yr^{-1}) (1,100 acres per year [ac/yr^{-1}]) in Yukon (Bell and others, 2012). This land use conversion can be expected to accelerate under a warming climate, resulting in the loss of 51 percent of carbon stocks or 11 kg/m^{-2} (98,140 lb/acre^{-1}) at the ecosystem level (Grünzweig and others, 2004). Most agricultural land use in the NWB region consists of forage production, on both native range (including woodlands) and in seeded pasture and hay fields.

Cattle, swine, and poultry are raised and marketed locally. Field crops include oats, potatoes, and carrots, but greenhouse and nursery production close to the cities are more important economically (Dinkel and Czapla, 2012). With such a small footprint and little intensive industrial production, environmental impacts such as waterbody contamination from pesticides or fertilizers are rare. Although agricultural lands have disrupted the matrix of natural forest habitat, they often provide food for wildlife (ungulates and bears) and result in human-wildlife conflicts that do not usually end well for the wildlife.

MILITARY ACTIVITIES

Major military installations located near Fairbanks and Delta Junction in Alaska exert both intensive and extensive demands on land resources. The military bases house personnel, operating units, equipment, and other facilities. The bases themselves have clearly defined intensive-use footprints, whereas associated large tracts of undeveloped land are designated as training areas or management units under Department of Defense control. In contrast, the Canadian side of the 141st meridian supports insignificant levels of military presence and activity in this region. Effects include trampling and compaction from off-road vehicles and occasional disturbances from incendiary and explosive devices. Especially in the conterminous United States, military lands and their disturbance regime often are important in maintaining open and fire-maintained habitats and species that otherwise may have become regionally rare (Stein and others, 2008).

TRANSPORTATION AND OTHER INFRASTRUCTURE

Linear disturbances and land use changes associated with roads, railroads, and pipelines dissect forested landscapes and generate edge effects into the forest. From the perspective of wildlife biology and landscape functioning, however, those linear corridors can either deter normal wildlife movement and the spread of disturbances such as wildfire, or (if traffic is light) channel accelerated movement of wildlife along such routes (Forman and Alexander, 1998). With some 20,000 km (12,400 mi) of public roads and 770 km (480 mi) of active railroad (U.S. Department of Transportation, 2001), Alaska has a transportation network ten times as extensive as that in Yukon or northwestern British Columbia (Table 5.1.1). A framework of oil and gas pipelines and high-voltage transmission lines also have rights-of-way (typically 30–50 m or 33–66 yd wide) that are kept free of trees and often are traveled by hunters and recreationists. Many remote communities still hope to be connected to the road networks, and power generation facilities of all scales continue to be built and linked to the electrical grid. Proposals for linking North Shore oil and gas production to the rest of the continental North American market, plus aspirations for overseas export of natural gas (which is primarily methane) through coastal liquefied natural gas facilities, collectively indicate that more pipelines are likely to be built in the future.

POPULATION GROWTH AND URBANIZATION

In many ways, most human effects on the environment are a function of population size and resource consumption patterns. As reflected in its transportation infrastructure, Alaska has a population more than 15 times that of Yukon and northwestern British Columbia. If considering the city of Anchorage and Alaska Interior regions as indicative of the boreal portion of the state, projections call for the population to increase from 508,000 to 680,000 people over the next 30 years (Robinson and others, 2014). Conversely, the Yukon population has been growing at 18 percent per decade (Yukon Bureau of Statistics, 2017), compared to 9.9 percent per decade in Alaska. If recent tendencies continue with respect to preferences for large suburban homes, exurban acreages, consumer products, and mechanized outdoor recreation, any population increase can be expected to proportionally increase pressures on the natural landscape—particularly around urbanized areas such as the cities of Anchorage, Fairbanks, and Whitehorse. Those impacts will include ongoing forest loss and fragmentation, off-road vehicle use, hunting pressure, and the introduction of more invasive species.

DIFFUSE LAND USES

Some land uses go unrecognized and unreported because they are diffuse and do not necessarily result in detectable land-cover conversion. For example, subsistence uses of the land—including hunting, fishing, berry picking, and firewood gathering—remain poorly documented but can be expected to continue into the future while meeting a decreasing proportion of average household needs (see Chapter 5.2). Even with increasing population levels, subsistence uses can be sustainable and are assumed (perhaps erroneously) to have little effect on overall ecosystem integrity, and therefore often are considered compatible with the designation of protected areas (parks and conservancies) that exclude extractive industries. Many land tenures—such as forest licenses, mining claims, and military lands—have the potential for great landscape modification, whereas others, such as registered traplines, guided hunting territories, and commercial recreation, likely influence only selected ecosystem components, such as the populations of targeted mammals or particularly sensitive wildlife species. It is also worth noting that trends in resource development also interact with a changing climate in complex ways, as exemplified by the implications of lengthened forest fire seasons, the loss of permafrost and ice roads, the gradual expansion of plant hardiness zones, and the geographic range of pests and diseases (Prowse and others, 2009).

GLOBAL DRIVERS

Many trends in land use and resource extraction in the NWB region reflect the influence of fluctuating commodity markets and government initiatives. For example, the relation of mining activity to international metal prices is well documented (McDowell Group, 2014; Fig. 5.1.3). Like forestry, agricultural land use (primarily grazing but also some local vegetable production) sometimes has assisted in meeting local demand, but land conversion to agriculture has historically reflected the existence of government programs and policies. For example, Yukon is one of the few remaining places in North America where homesteaders can apply to buy public land for agricultural purposes (Bell and others, 2012). Likewise, the development of transportation infrastructure typically depends on government initiatives (see Box 5.1.1, "Sampled Trends in Forest Land-Cover Changes in Boreal Yukon and Northwestern British Columbia). For example, the Dalton Highway and the Trans-Alaska Pipeline System were constructed in the mid-1970s through combined government and corporate investment in response to the Organization of the Petroleum Exporting Countries (OPEC)-induced energy crisis (Tremont, 1987). The role of these global market factors, plus the effects and opportunities associated with institutional initiatives, are now recognized as global drivers of land use change (Lambin and others, 2001). Other important thresholds in land management, such as the transition to widespread comanagement arrangements between governments and aboriginal groups, have emerged more organically through regional negotiations (Christensen and Krogman, 2012).

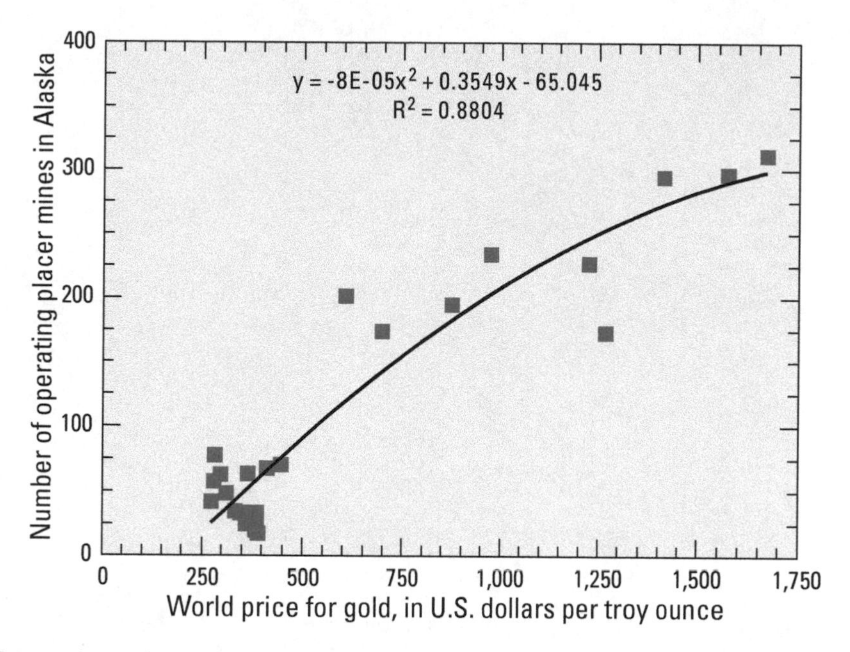

Figure 5.1.3. Number of operating placer mines in Alaska as a function of the world price for gold. Source is Alaska Division of Geological and Geophysical Surveys (1989–2015). Note the apparent importance of a $500 per ounce threshold in stimulating more development and production.

INFORMATION GAPS

- The relative sensitivity of resource extraction industries to market forces, regulatory requirements, and government incentives requires additional documentation and analysis to help ascertain influences on land use and resource extraction.
- Some land uses can be inferred from remotely sensed land-cover data but need to be supplemented and cross checked with permits and other records compiled by government agencies.
- The ecological impacts of diffuse land uses (those that do not result in land-cover change) are poorly documented.
- More information is needed as to when (or how much) resource development and ecosystem modification will actually proceed as a result of the many resource tenures and rights, development proposals, and applications currently registered with authorities.

5.2
RURAL AND INDIGENOUS LIVELIHOODS IN THE NORTHWEST BOREAL REGION

By Philip Loring[1] *and Brenda Parlee*[2]

KEY FINDINGS

- Many Aboriginal peoples of the NWB region continue to depend on the harvesting of wild mammals, birds, fish, berries, and fuelwood to fulfill their nutritional, cultural, and spiritual needs.
- In Canada, the rights of Indigenous peoples to harvest boreal resources is protected through treaties, Section 35.2 of the Canadian Constitution, and under Supreme Court law; in Alaska, access is protected in part under the state constitution and in part by federal law, the Alaska National Interest Lands Conservation Act (Public Law 96–487, 94 Stat. 2371).
- Local subsistence and recreational fishing and hunting also are important influences on boreal ecosystems in Alaska and Yukon; recreational fishing in Yukon is understood to have the largest effect on freshwater fisheries.
- Consistent harvest data are lacking for the NWB region, which limits the understanding of the impacts of hunting and fishing on ecosystems.
- Many Indigenous and multigenerational users of boreal ecosystems have well-developed knowledge, practices, and norms for managing boreal resources sustainably.
- Fish and game are managed in Alaska for enhancement, for example, with predator control, though landscape or ecosystem level effects of these policies are understudied.
- Although specific impacts of hunting and foraging on broader ecosystem processes are largely unknown, work elsewhere suggests they are greatest near villages.

1 University of Saskatchewan, Saskatoon, Saskatchewan, Canada.
2 University of Alberta, Edmonton, Alberta, Canada.

In many communities in the Northwest Boreal forest, the sustainability of local livelihoods is closely coupled to the productivity and sustainability of local ecosystems. Those subsistence-based livelihoods and foodways refer collectively to the use of hunting, fishing, small-scale agriculture, and gathering of forest products (for example, botanical resources, fuelwood) to meet various needs for food and for spiritual and cultural fulfillment. Although boreal livelihoods in northwestern Canada and Alaska have changed greatly over the past 100 years, Indigenous and other rural communities continue to rely on many kinds of wildlife, including terrestrial mammals, fish, aquatic birds, timber, and non-timber forest products (for example, fuelwood, berries, and edible plants).

HARVEST LEVELS

The total harvest and significance of using renewable resources in the Northwest Boreal (NWB) forest is difficult to determine. Average non-commercial harvests by Alaskans are estimated to be 28.3 kg (62.3 lb) of wild fish and game per capita per year, or a total of about 20.8 million kg (45.9 million lb) per year (Alaska Department of Fish and Game, 2016). Similar data are not known to be available for communities in the Canadian part of the NWB region, although hunting and fishing are important in these areas to First Nations and other residents and tourists. Across the Canadian boreal region as a whole, it is estimated that 52,000 Aboriginal households depend on forest ecosystems for livelihood and well-being (Anielski and Wilson, 2009). Although total harvests and economic value of subsistence use of boreal resources has not been calculated, the value of non-timber forest products harvested in boreal Canada has been estimated at $79 million CAD per year (Anielski and Wilson, 2009).

TRADITIONAL MANAGEMENT PRACTICES

Most residents in rural areas practice livelihoods that are adaptable and well-tuned to local ecosystems and the seasons of the North. As such, there is a wealth of local and traditional knowledge about ecosystems, knowledge that can provide a long-term and synoptic perspective on many of the new challenges that are being faced in the region (Turner and others, 2000; Berkes 2012). However, new challenges such as climatic and environmental change are creating myriad ecological stresses for people, their livelihoods, and food security in the region (Stevenson and Webb, 2003; Gerlach and others, 2011). Many local people say that "the world is faster now," by which they mean that the environmental cues or "sense-makers" that they use to read the environment have changed or are changing (Krupnik and Jolly, 2002). Nevertheless, local people continue to develop and test their ecological knowledge about the correlation of seasonal phenological events, knowledge that can provide an important window into how global changes will interact at the ecosystem level.

Just as it is important to acknowledge the important role of northern landscapes in the lives of local people, it is likewise important to recognize the role of humans in these systems as predators, stewards, and ecosystem engineers. Many people who hunt and fish do so within a mindset of reciprocity and stewardship of the resource. Inherent in this stewardship role is the informal monitoring of resource levels and harvesting effects, with depletion signaling the need for restraint or shifts in geographic focus. Practices such as selective burning to create or maintain habitats for valued species (for example, moose), cyclical shifting of harvest locations and species, and biodiversity conservation are all common in many areas of the boreal Northwest Territories, Yukon, Alberta, and British Columbia (Gottesfeld, 1994a; Turner and others, 2000; Sherry and Myers, 2002; Parlee and others, 2005; Parlee and others, 2014). Among many families and cultures, rules or norms for respecting nature and the sharing of harvests are common (Padilla, 2010; Wray and Parlee, 2013). Embedded here, too, among Indigenous families and cultures, is often the belief in the connectedness of the flora and fauna of boreal ecosystems to the spirit world and the responsibility of people to protect the lands and resources for future generations.

ECOSYSTEM EFFECTS OF GAME MANAGEMENT REGIMES

The active and intensive management of wild fish and game for use by rural and small-town residents also is an important driver in the structure and function of boreal forest ecosystems, although the effects of these management regimes are not fully understood. In Alaska, fish and game are managed by the state with an enhancement mandate, meaning that the goal is to achieve the maximum sustainable yield each year, whether for moose, caribou, salmon, or other game or fish (Huntington, 1992). In the context of ungulate management, targeted enhancement has entailed controversial predator control programs designed to inflate ungulate populations for human hunting (Decker and others, 2006; Boertje and others, 2010). In British Columbia and Yukon, the regulation of sport and commercial (guided) hunting is controversial in many environmental and First Nations circles; there are concerns that hunting regulations and quotas frequently are not sufficiently responsive to subsistence harvesting needs and variability in the size and distribution of game populations. Apart from the social effects of these strategies, effects on the ecosystem by enhancement are poorly understood, although marine examples from elsewhere in the world of single-species stock enhancement have been shown to cause degradation of the surrounding food web (Steneck and others, 2011).

ECOSYSTEM EFFECTS OF HARVEST

For the Canadian part of the NWB region, data on harvests and their effects are more limited than for Alaska. Harvested fish stocks in Yukon are healthy and the recreational, First Nations subsistence and commercial fisheries in the region are all considered

BOX 5.2.1: BERRY PICKING IN THE BOREAL REGION

Non-timber forest products, including berries and edible plants, are an important dimension of the livelihoods of many northern Indigenous peoples (Turner and Davis, 1993; Gottesfeld, 1994b; Marles and others, 2000). Dene (see http://www.thecanadianencyclopedia.ca/en/article/dene/) use of berries and medicinal plants was documented as early as the 1800s by European explorers in the region, although it is clear that these practices have been critical to Indigenous boreal livelihoods for countless generations. Numerous species of berries and medicinal plants continue to be harvested across the boreal for economic, cultural, social, and spiritual reasons (Parlee and Berkes, 2005). Most boreal berry-producing species, such as blueberries (*Vaccinium* spp.), normally spread and re-sprout vegetatively, facilitating sustainable berry production even under intensive harvesting, although berry yields can be susceptible to year-to-year variation in weather conditions (Wallenius, 1999). The health of berry-dependent wildlife species (for example, bears, migrating songbirds) can be susceptible to food scarcities, but the effects of berry harvesting on animal populations do not seem to have been studied.

sustainable (Environment Yukon, 2010). Of these, recreational fisheries account for the greatest impacts on freshwater species; an ongoing management challenge is that this pressure is not distributed evenly across the territory, with bodies of water that are closest to roads and population centers being the most heavily fished (Environment Yukon, 2010). The consequences of harvesting pressures are not limited to the Yukon or Alaska; as in many parts of the world (for example, the Amazon River Basin; Redford, 1992; Peres, 2000), there is some evidence of reduced populations of medium- and large-sized mammals due to subsistence hunting near human settlements. Moose hunting also is common in Yukon; many populations are of mammals harvested at levels less than maximum sustainable yield, although a few populations of concern exist (Environment Yukon, 2014). Habitats also tend to be more open around human settlements, reflecting a history of fuelwood collection and prescribed burning (Lewis and Ferguson, 1988; Gottesfeld, 1994b).

As elsewhere in the world, there is evidence of prey switching among Indigenous peoples of the NWB, whether as the result of changes in regulations or changes in the distribution and abundance of specific prey species. In the Yukon Flats of Alaska, for example, people report increasing their hunting for moose because of salmon fishery closures (Loring and Gerlach, 2010). The effects of prey switching strategies have proven difficult to monitor with current data collection practices in Alaska (Hansen and others, 2013), but the effects may be problematic if these hunts are illicit, because some moose populations in the Yukon Flats already have low population densities (Loring and Gerlach, 2010).

CUMULATIVE EFFECTS OF CHANGE ON SUBSISTENCE USERS

The fur trade and related colonization of western Canada and Alaska initiated many changes to boreal societies and cultures (Carter, 1999), including, for example, the forced settlement of people into fixed villages and extinguishment of Aboriginal title in Alaska (Mitchell, 2003). Dramatic changes have continued in the last half-century with commercial forestry, the building of the Alaska Highway and the Alaska Pipeline, and hydroelectric and petroleum operations all creating opportunities as well as challenges (Hall and others, 1985; Parlee and others, 2012). Although some Indigenous peoples have been able to take advantage of the economic opportunities associated with such development, many others have been marginalized—socially, politically, and economically (Kendall, 2001). Household poverty rates for Indigenous communities across the region are high by United States and Canadian standards (Indigenous and Northern Affairs Canada, 2010), as are rates of food insecurity and environmental health challenges such as diabetes, obesity, and colorectal cancer (Council of Canadian Academies, 2014). The meaning of these trends to the level of human activities on the land is not clear. Due to cultural disruption, less opportunity to earn cash through employment likely means less ability to participate in subsistence activities rather than prompting a return to traditional practices. It also may mean that local people grow increasingly supportive of extractive natural resource development for the purposes of creating jobs.

In concert with these societal challenges, warming temperatures, droughts, and extreme weather also challenge the viability of rural communities across the region (Loring and Gerlach, 2009; Markon and others, 2012). Policies for natural resource management also are struggling to keep up with the fast pace of these environmental changes. Arguably, change has been the norm for the Arctic and subarctic regions as long they have been peopled, and historically, northern peoples were able to respond to changes in the land and seascapes and to the distribution of fish and game (Binford, 1978). Recent research highlights how governance and management structures can limit the options and flexibility for people to respond to change. Two examples include restrictive land tenure regimes and hunting/fishing seasons that increasingly do not match with changing seasonality and phenology of fish and game (Loring and Gerlach, 2010; McNeeley and Shulski, 2011).

Although access to traditional/country foods is confounded in myriad ways, alternative options from the store tend to be limited and low in quality, especially in remote bush communities not typically connected to urban centers by roads or other infrastructure (Gerlach and others, 2011). Nevertheless, finding that their food needs cannot be met with locally available wild food resources, many northern peoples now

BOX 5.2.2: THE NUTRITION TRANSITION

The trajectory of change away from traditional foods and toward industrially produced market foods of lower quality has been described as a "nutrition transition" by some (Kuhnlein and others, 2004) and "new world syndrome" by others (Stephenson, 1995). Multiple economic, biomedical, psychological, and sociocultural effects result, including increases in Type II diabetes, obesity, coronary heart disease, and cancer, as well as depression, substance abuse, alcoholism, and violence (Graves, 2003; Alaska Department of Health and Social Services, 2008). The extent and manner to which these health trends are directly and indirectly linked to changes in community food systems, climate-driven or otherwise, still needs extensive research and quantification; however, much research has shown the numerous benefits of traditional foods for the health and well-being of people in northern Alaska and Canada (Bersamin and others, 2007). Comparatively sparse human populations in the NWB tend to obscure evidence of any shift to greater or lesser dependence on the harvesting of wild resources and for any subsequent effects on ecosystems.

fill their cupboards with foods of far lesser quality and cultural relevance (see Box 5.2.2, "The Nutrition Transition"). This food is purchased either from the meager selections available at village stores (for those communities with a store) or from costly periodic provisioning trips to urban supply centers (Gerlach and others, 2011).

PROSPECTS FOR THE FUTURE

The lands and resources of the NWB region have long been fundamental to the subsistence of many Indigenous and rural people in Canada and the United States. Many Indigenous livelihoods developed in areas of productive biodiversity and in areas where people could easily adapt to seasonal and year-to-year variability in the availability of forest resources, fisheries, migratory birds, and wildlife. According to many oral histories and ethnographies, this capacity to adapt ensured the sustainability of local natural resources as well as the health and well-being of Indigenous communities. Fluctuations in historical and current levels of harvested wildlife and botanical resources probably represent a complex conflation of yearly weather events and human influences. Over the last two centuries, and most especially in the last 50 years, however, the added and cumulative effects of natural resource exploitation and climate change have infringed on the adaptive capacity of many communities. Both science and policy are now emphasizing adaptability as a strategy for coping with the impacts of change, but some scholars have raised important questions about the environmental justice implications of adaptation policy for climate change (Loring and Fazzino, 2014). Although some may perceive that these traditional livelihoods and foodways are anachronous to the modern development of western Canada and the United States, they are arguably an important "canary in the coal mine" of how the human relationship with the boreal forest has been adversely transformed and are indicative of future challenges of boreal sustainability.

It is important to note that although there are many similarities in livelihood practices across Indigenous cultures of North America, Canadian First Nations and Alaska Native[3] communities are socioeconomically, culturally, and politically diverse and distinct (Natcher and Hickey, 2002). As such, the issues reviewed here should be understood as representative, but they do not provide a comprehensive picture of the issues being faced by people across the region.

INFORMATION GAPS

- The social, economic, and cultural value of subsistence uses of boreal resources is poorly documented, especially in Canada.
- The relationship of subsistence resource use to economic factors (for example, access to market products and the wage economy) and environmental factors (for example, resource abundance and reliability) has been little investigated in boreal regions.
- Effects of modern hunting and foraging, including strategies such as prey switching, on ecosystem processes and the sustainability of wild resources are largely unknown.
- The ecosystem- and landscape-level effects of fish and game enhancement policies (for example, by means of predator control) are under-studied.
- Future research is needed to identify current patterns and distributions of risk and vulnerability.
- Understanding is needed of the many pathways through which fundamental changes to food systems can undermine individual as well as community social, cultural, and ecological health outcomes, so that communities can understand, plan for, and effectively manage these changes.

3 Due to numerous reasons, including treaty and land claims status, but more importantly the right of people for self-determination, there is no standard terminology regarding Indigenous groups that applies to both the United States and Canada. In the United States, the term "Alaska Native" denotes all Indigenous peoples, including those that are under the more general categories of American Indian and Eskimo (both accepted terms by Indigenous peoples in the United States); in Canada, Indigenous peoples often are referred to as First Nations, Metis, and Inuit (with the term "Eskimo" considered to be derogatory).

5.3
POLLUTION AND CONTAMINANTS IN THE NORTHWEST BOREAL REGION

By Mary Beth Leigh[1], Angela Matz[2], and Mary Gamberg[3]

KEY FINDINGS

- There are hundreds to thousands of sites contaminated from military, government, and commercial uses.
- The NWB region is subject to contaminant loading from long-range transport, including atmospheric deposition of persistent organic pollutants (POPs), mercury, and other pollutants.
- There may be increased mercury deposition by long-range transport from industrially expanding areas such as Asia, even though mercury emissions have decreased in North America and Europe.
- Increased oil, gas, and mining development and transport may potentially result in increased pollution risks.
- A warming climate is likely to release previously ice-sequestered contaminants from melting glaciers and thawing permafrost.
- Wild fish, game, and plants are susceptible to the accumulation of contaminants, which is of concern for local subsistence.
- Careful attention to regulations and compliance for ongoing and future resource development and transport projects is critical to protecting the environmental and human health of the NWB region.

CONTAMINATED SITES

The Northwest Boreal (NWB) region has hundreds to thousands of sites contaminated with petroleum, persistent organic pollutants (POPs), solvents, pesticides, herbicides, mercury, other heavy metals, radionuclides, and other contaminants, resulting from

1 University of Alaska Fairbanks, Fairbanks, Alaska, USA.
2 U.S. Fish and Wildlife Service, Fairbanks, Alaska, USA.
3 Gamberg Consulting, Whitehorse, Yukon, Canada.

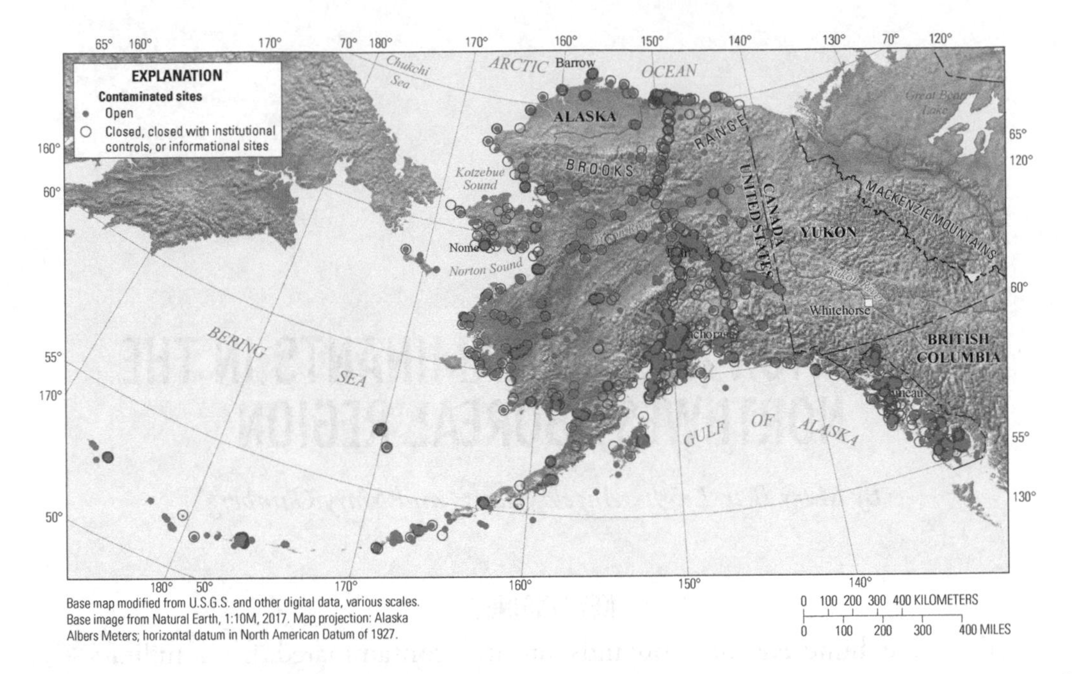

Figure 5.3.1. Distribution of Alaska Department of Environmental Conservation contaminated sites (n=5,181) in Alaska. Courtesy of Alaska Department of Environmental Conservation, (http://dec.alaska.gov/Applications/SPAR/PublicMVC/CSP/Search). There are 1,961 non-remediated sites, considered open, and 3,200 remediated sites, considered closed sites because they have signage, fencing, or deed restrictions.

a wide variety of military, commercial, and governmental activities. Contaminated site inventories are maintained by Canada (Government of Canada, 2014) and the Alaska Department of Environmental Conservation (ADEC; http://dec.alaska.gov/Applications/SPAR/PublicMVC/CSP/Search), including Formerly Used Defense Sites (FUDS) for which the U.S. Army Corps of Engineers has regulatory responsibility (https://dec.alaska.gov/Applications/SPAR/PublicMVC/CSP/SiteReport/3255). Contaminated sites are widely distributed, have affected soils and groundwater both in populated and in remote areas, and tend to cluster around roads, rivers, and pipelines; this also is true for sites that have been remediated, as shown in Figure 5.3.1.

Contaminated sites across the NWB region pose risks to human and environmental health through a number of exposure pathways. Drinking-water contamination is a major issue, because the diffuse population often obtains water from wells or surface water rather than regulated municipal sources. Exposures also can occur through direct contact with contaminated soil, or from volatile organic compounds entering buildings from subsurface contamination. Ingestion in foods is a particular risk for NWB populations, who often rely heavily on subsistence foods (see Chapter 5.2). Some contaminants, like mercury, PCBs, and pesticides, can accumulate in the food

BOX 5.3.1: PHYTOREMEDIATION OF ORGANIC CONTAMINANTS

Organic contaminants such as petroleum and polychlorinated biphenyls (PCBs) can undergo biodegradation by naturally occurring microorganisms in boreal forest soils, and certain plants may facilitate this process. Some plants seem to accelerate the breakdown of organic contaminants by stimulating the activity of contaminant-degrading microbes in the root zone (rhizosphere) in a process known as rhizoremediation, which is a form of phytoremediation (the use of green plants and associated soil process to either contain, remove, or render toxic environmental contaminants harmless). Rhizoremediation may occur through the release of natural plant compounds that serve as cometabolites or inducers, and also provide aeration and other stimulatory conditions in the root zone (Singer and others, 2004; Gerhardt and others, 2009). Phytoremediation is potentially a much more affordable cleanup method for contaminants than conventional methods such as incineration and landfilling, which are especially costly and logistically challenging in remote NWB sites. Native plants are particularly advantageous for phytoremediation applications because of their adaptation to local conditions and stressors, and their use helps prevent the introduction of invasive plant species. Several studies have focused on the potential for native boreal forest plants to promote the biodegradation of organic pollutants.

Alaska feltleaf willow (*Salix alaxensis*) roots promoted microbial biodegradation of PCBs in laboratory studies (Slater and others, 2011). Alaskan soils contaminated with diesel or crude oil were remediated to below regulatory limits in a long-term (16-year) study that involved 2–3 years of treatment with grasses followed by colonization and growth of native boreal forest plants (Leewis and others, 2013). Robson and others (2003) developed a list of cold-adapted plants from western Canada that are adapted to petroleum-contaminated soils as potential phytoremediation candidates. Additional long-term studies are needed to identify the most effective plant species and best practices for rhizoremediation of organic contaminants in the boreal forest region.

chain, presenting risks to fish and game populations, ecosystem function, and humans relying on subsistence foods.

LONG-RANGE CONTAMINANT TRANSPORT

Although there are point sources within the NWB region, many contaminants such as mercury (see Box 5.3.2, "Mercury Pollution in the Boreal Region"), POPs, and black carbon (Law and Stohl, 2007) are transported to the NWB region in air currents (within days) and ocean currents (over years to decades). Persistent organic pollutants from a

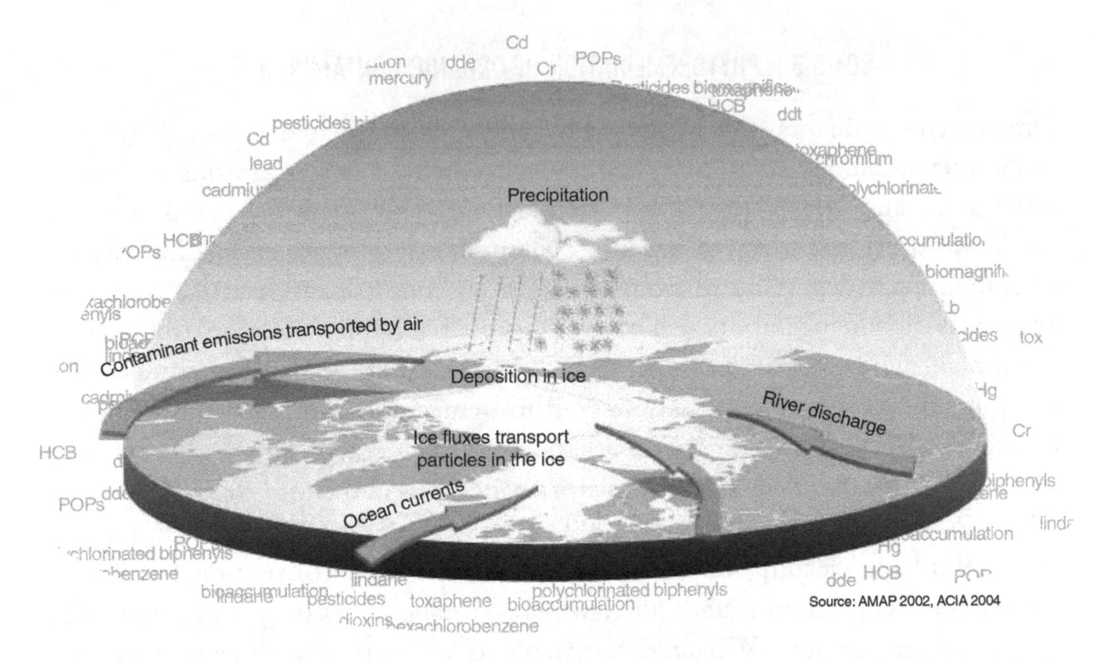

Figure 5.3.2. Mechanisms for long-range transfer of contaminants to northern latitudes including the northwest boreal region. Figure courtesy of Ahlenius (2006), United Nations Environmental Program/GRID-Arendal (http://www.grida.no/resources).

wide range of industries in southern latitudes travel easily to the north on air currents due to their high volatility (Fig. 5.3.2). Many legacy POPs, particularly organochlorine pesticides (for example, dichlorodiphenyltrichloroethane [DDT], hexachlorobenzene, chlordane, dieldrin, toxaphene), have decreased in the north after use restrictions in the south, but some (for example, PCBs) are still high enough to affect the health of some wildlife and humans. Some current and emerging POPs are not yet regulated by international agreements and accumulate in northern food webs, including brominated flame retardants, fluorinated compounds, polychlorinated naphthalenes, and the pesticide endosulfan (Arctic Monitoring and Assessment Programme, 2010a). Smoke from forest fires, which are increasing in both size and severity across the circumpolar north (Arctic Council and the International Arctic Science Committee, 2005), may be blown into the NWB region, as can ash from Pacific Rim volcanoes.

Radionuclides have been deposited in Arctic and subarctic regions from atmospheric fallout, including nuclear weapons tests, the 1986 Chernobyl accident, the 2011 Fukushima accident, and reprocessing plant releases (Arctic Monitoring and Assessment Program, 2016). They can persist for long periods in soils and some plants, and migrate through the terrestrial food web leading to human exposures. Radioactivity from these historical pools is expected to continue to decrease unless mobilized by the effects of climate change, such as the melting of glaciers and permafrost. Climate change may also affect rates of radon gas release (Arctic Monitoring and Assessment Programme, 2010b).

INCREASING RESOURCE EXTRACTION

As the demand for fossil-fuel-based energy continues to increase worldwide, and as some northern regions are becoming more accessible as the climate moderates, there is a growing interest in increasing oil, gas, and mineral development in the NWB region. Oil and gas reserves, such as in the Yukon Flats National Wildlife Refuge in Alaska, are being assessed and proposed for development (Stanley and others, 2004; Resource Development Council, 2010). Access to mineral resources throughout the NWB LCC area is a focus of development interests, including a large-capacity road west from the Dalton Highway to the Ambler Mining District and perhaps beyond (Alaska Industrial Development and Export Authority, 2015). Although it brings some economic benefit, large-scale industrial development brings potential pollution: accidental oil and chemical spills, acid mine drainage, cyanide heap-leach spills, and mobilization of metals and sediments into aquatic systems.

With increased extraction of resources comes increased need for transport of oil, gas, ore, and processing chemicals. Because the NWB region is large and mostly remote, transport by pipelines, truck, or tanker is necessary, and each method poses risks for accidental spills. For example, several new pipelines have been proposed in the NWB region, including liquefied natural gas pipelines from Alaska's North Slope oil fields (for example, federal Energy Regulatory Commission, 2015). The Trans-Alaska Pipeline (TAPS) has had several oil spills in its 37 years (http://dec.alaska.gov/spar/), including an approximately 16,000 barrels (BBL) spill from a gunshot in 2001 (Fig. 5.3.3). Decommissioned pipelines can become contaminated sites, such as the CANOL pipeline in Canada (Indigenous and Northern Affairs Canada, 2010).

Although resource development and extraction activities are increasing in number and density, so are environmental regulations. Careful attention to regulations and compliance for ongoing and future resource development and transport projects will be important to protectt the environmental and human health of the NWB region.

Although local regulations can be effective in limiting point source contamination, contaminants from long-range transport must be regulated on a global basis. The Stockholm Convention on POPs is a global treaty created in 2001 by the United Nations Environmental Program (UNEP) to limit the emissions of specific POPs to the environment. As of November 2015, 179 countries had signed the treaty, with 152 countries (including Canada) ratifying the agreement (meaning it has come into force in that country). Although the United States has signed the agreement, they have not ratified it (Stockholm Convention, 2015). The treaty includes a review process for POPs in addition to the original 12 POPs (http://chm.pops.int/TheConvention/ThePOPs/AllPOPs/tabid/2509/Default.aspx).

BOX 5.3.2: MERCURY POLLUTION IN THE NORTHWEST BOREAL REGION

Mercury enters the environment from natural sources such as forest fires, volcanoes, and weathering of mercury-rich rock as well as from anthropogenic industrial activities—primarily coal burning. Although emissions in North America and Europe have decreased over the last 30 years, emissions from Asia are increasing. China—directly upwind of much of the NWB region—headed the list of countries with highest mercury emissions from anthropogenic activities, contributing about 28 percent of the global emissions of mercury in 2000 (Pacyna and others, 2006). Turetsky and others (2006) determined that mercury formerly sequestered in cold, wet peat soils is released to the environment during fires in Canadian boreal forests and presents a growing threat to sensitive aquatic habitats and northern food chains as the climate warms.

Estimates of circumboreal mercury released from peat by fires were 15-fold greater than estimates without peat. Mercury concentrations in fish have increased following fires in Alberta, by increasing mercury inputs and by restructuring food webs (Kelly and others, 2006). Collectively, these studies suggest that these fire-related mercury emissions to and within the NWB region will continue or increase with a warming climate.

To become bioavailable, elemental mercury must be transformed into methylmercury in anaerobic environments, such as wetlands and impoundments, and these environments are likely to increase with permafrost thaw and other hydrologic changes. Warming temperatures in lakes and wetlands also may increase methylmercury production and concentrations in biota. Food webs that form the basis of traditional and subsistence diets already exhibit efficient mercury biomagnification—a 1,000- to 3,000-fold concentration increase from particulate organic matter to apex predators in some systems (Arctic Monitoring and Assessment Programme, 2003).

The Minimata Convention on Mercury is another global treaty created by UNEP in 2013 to limit global emissions of mercury to the environment. As of November 2015, 128 countries had signed the agreement and 18 (including the United States) had ratified. Canada subsequently ratified the convention in April 2017.

The Canadian Environmental Protection Act (S.C. 1999 c. 33) was created in 1999 and is the cornerstone of Canada's environmental legislation aimed at preventing pollution of the environment. Regulations under that Act are updated regularly (for example, The Toxic Substances List was updated in November 2013; Environment Canada, 2013). The Fisheries Act (R.S.C. 1985, c. F-14) plays a major role in regulating pollution of waters in Canada and the Arctic Waters Pollution Prevention

Figure 5.3.3. Oil spilling from the Trans-Alaska Pipeline due to sabotage by a gunshot in 2001. Photograph courtesy of Federal Bureau of Investigation (https://www.fbi.gov/news/stories/north-to-alaska-part-4).

Act (R.S.C. 1985, c. A-12) regulates waste from natural resource development, specifically in the Canadian Arctic (http://laws-lois.justice.gc.ca/eng/acts/f-14/). The U.S. Environmental Protection Agency (EPA) is the primary agency tasked with protecting the environment in the United States. The EPA does this through regulating discharges of hazardous substances and chemicals under the Clean Water Act (CWA [Public Law 92–500; 86 Stat. 816]) and the Clean Air Act (CAA CWA [Public Law 88-206; 77 Stat. 392]), regulating transport, disposal, and cleanup of hazardous waste under laws such as the Comprehensive Environmental Response, Compensation, and Liability Act (CERCLA, also known as "Superfund" [Public Law 96-510; 94 Stat. 2767]), and regulations promulgated in CFR Title 49B, 1A. The EPA also would be the lead federal agency responding to oil or hazardous waste spills occurring in the freshwater boreal forest ecoregion, under the Oil Pollution Act (Public Law 101-380; 104 Stat. 484), with assistance from the responsible party, other federal agencies with resources at risk, and the State of Alaska.

Alaska environmental regulations are included in the Alaska Administrative Code (AAC), Title 18: Environmental Conservation (indexed at http://dec.alaska.gov/commish/regulations/index.htm) by ADEC. The AAC provides regulatory guidance on topics including oil and other hazardous substances pollution control (AAC Chapter 75) and underground storage tanks (AAC Chapter 78). The ADEC Contaminated Sites Program is responsible for regulating contaminated sites and follows the cleanup guidance listed in Title 18, AAC Chapter 75, Article 3, "Discharge Reporting, Cleanup, and Disposal of Oil and Other Hazardous Substances." Article 3 includes cleanup operations requirements and cleanup levels for contaminants of soil and groundwater. Water Quality Standards are listed in Title 18 AAC 70, which primarily governs the use and protection of surface water. Other sections of Title 18 regulate air quality control (Chapter 50), solid waste management (Chapter 60), and drinking water (Chapter 80), all of which are relevant to contaminant issues within the boreal forest region of Alaska.

CLIMATE CHANGE

The potential effects of climate change on contaminant dynamics within this region are difficult to estimate, because many physical, chemical, and biological factors are involved. Temperature fundamentally influences contaminant fate and transport, because water solubility and atmospheric transport are temperature-mediated (Arctic Council and the International Arctic Science Committee, 2005). Under warmer climatic conditions, chemicals will more readily partition from soil to air, increasing the amount of atmospheric transport and net transport to the north (Arctic Monitoring and Assessment Programme, 2003). Temperature has long been known to affect the uptake and toxicity of pollutants (reviewed by Schiedek and others, 2007). For example, increased fish metabolic and ventilation rates in response to higher temperatures can increase uptake of pollutants, whereas exposure to certain organic contaminants can lower the tolerance to high temperatures in fish (Patra and others, 2007) and in invertebrates (reviewed in Schiedek and others, 2007). Contaminant accumulation may increase in some food chains and decrease in others, depending on food chain structure and the particular responses to environmental change. For example, increased productivity in freshwater food webs may increase mercury or POP concentrations in fish prey, but those concentrations also may be "diluted" by higher fish growth rates.

Erosion rates in coastal and riverine Alaska recently have increased, sometimes dramatically, and thawing permafrost is redrawing aquatic boundaries. These changes threaten release of pollutants from contaminated sites such as landfills and drilling mud pits, and from sites that depend on permafrost stability such as sewage lagoons, dumps, tailings ponds, and fuel delivery, storage, and pipeline sites (Arctic Monitoring and Assessment Programme, 2003).

Increased threats to shoreline fuel delivery, storage, and power generation facilities are expected in the future (Alaska Climate Impact Assessment Commission, 2008; Climate Change in Alaska Adaptation Advisory Group, 2010). Klaminder and others (2008) reported mercury loss from a Swedish palsa mire (an organically rich, permafrost-dominated peat bog) of between 40 and 90 percent as the bog changed with thawing.

The Arctic Council and the International Arctic Science Committee (2005) review suggested that a warming climate will lead to more and larger northern wildfires than in the past, with models suggesting wetter scenarios would produce more large fires compared to drier scenarios that produce more frequent fuel-reducing medium-sized fires. In addition to soot and smoke, boreal fires release mercury (see Box 5.3.2, "Mercury Pollution in the Northwest Boreal Region").

Glacier ice-mass loss and snowmelt due to warming can release contaminants (Blais and others, 1998), and glacial streams contain unusually high proportions of persistent organic pollutants in solution, because of the low organic matter and, therefore, low adsorption potential on suspended sediments. High dissolved-phase organochlorine concentrations combined with low dissolved organic carbon makes these compounds highly bioavailable (Blais and others, 2001). Glacial meltwater was the main cause of DDT peaks observed in biota of downstream lakes in the Italian Alps (Bettinetti and others, 2008); indeed, glacial meltwater may contribute high concentrations of pesticides to cold aquatic ecosystems for decades or centuries (Donald and others, 1999).

INFORMATION GAPS

- Existing contaminated sites and non-point sources of contamination have not yet been fully documented for the NWB region.
- The quantity of contaminants sequestered in the environment and subject to liberation due to climate change and other factors is not yet known.
- The rates and rate-limiting factors for contaminant transformation and biodegradation need to be more fully understood to better estimate contaminant fate and to develop technologies for mitigation in the face of climate change.
- Contaminants in subsistence foods are not well characterized due to the varied diet and vast geography of the NWB region

5.4
LAW AND POLICY DRIVERS IN THE NORTHWEST BOREAL REGION

Julie Lurman Joly[1] *and Scott Slocombe*[2]

KEY FINDINGS

- International agreements create part of the framework in which management decisions are made in the northwest boreal region.
- Changes in the legal structure may have important consequences for the health of the ecosystem and its utility for human inhabitants.
- Major international law drivers in the northwest boreal region focus on wildlife treaties and species-specific joint management plans.
- Land in Alaska largely falls into one of three ownership categories: federal government, state government, and Native corporations.
- Legal drivers in the northwest boreal area stem most significantly from the diverse agencies directing land management; federal lands in particular are under the control of several different agencies, each with its own legal directives.
- Legal conflict among various management groups (particularly between the federal and state governments) over the implementation of laws continues to be an important source of instability in the region.
- In Canadian territories, land and resource management is largely influenced by land and wildlife comanagement boards arising from comprehensive claims negotiated between First Nations and territorial and federal governments. In northern British Columbia, there are no comprehensive claims institutions, but First Nations roles are growing through legal decisions.

---

1 Formerly associated with the School of Natural Resources and Extension, University of Alaska Fairbanks, Fairbanks, Alaska.

2 Wilfrid Laurier University, Waterloo, Ontario, Canada.

- In this legal and comanagement context, the critical issue in effectiveness is intergovernmental relations and implementation of both comanagement and standard provincial, territorial, or state government management regulations and policy.

In the Northwest Boreal (NWB) region, international agreements of relevance to landscape conservation focus largely on cross-border wildlife management. Several important wildlife treaties and management agreements as well as a shared international protected area comprise the world's largest protected wilderness area.

INTERNATIONAL WILDLIFE AGREEMENTS

The major cooperative efforts between the United States and Canada that significantly affect management of the region focus on wildlife management. Primary among these is the Convention for the Protection of Migratory Birds. In 1916, the United States and Great Britain (on behalf of Canada) signed a bilateral bird protection treaty. The Convention contains a list of native birds present in both countries and required management prescriptions. The initial agreement established three categories of migratory birds: migratory game birds, migratory insectivorous birds, and migratory nongame birds. The Convention established closed seasons for birds in each category. For insectivorous and nongame birds, the closed season is year-round, which effectively prohibits any hunting of species in those categories. For migratory game birds, the closed season is between March 10 and September 1.[3]

In 1995, the treaty was amended by a protocol[4] that ensures continued subsistence use of migratory birds by "Indigenous inhabitants" of the two countries. This harvest is allowed outside of the established open seasons, because in the far north migratory birds generally are not present during the designated open season. The term "Indigenous inhabitant" is understood in Alaska to mean "Alaska natives and permanent resident non-natives with legitimate subsistence hunting needs living in designated subsistence harvest areas."[5] The Protocol also requires that each country establish a management body for the development of recommendations for the management of these hunts, and ensures that native groups have substantial representation on that body.

3 The Convention for the Protection of Migratory Birds, August 16, 1916, 39 Stat. 1702, 12 Bevans 375.

4 Protocol Between the Government of Canada and the Government of the United States of America Amending the 1916 Convention Between the United Kingdom and the United States of America for the Protection of Migratory Birds in Canada and the United States, December 14, 1995.

5 Message from the President of the United States Transmitting a Protocol between the United States and Canada Amending the 1916 Convention for the Protection of Migratory Birds in Canada and the United States, August 2, 1996, at 8 (1996).

The United States and Canada also share several species-specific agreements and joint management plans. These target Yukon River Salmon,[6] the Porcupine Caribou Herd,[7] Chisana Caribou Herd (Chisana Caribou Herd Working Group, 2012), and Forty-mile Caribou Herd (Harvest Management Coalition, 2012), and attempt to ensure coordinated cross-boundary management. In the case of Yukon River salmon, the agreement promises that Canada will meet a certain escapement objective and the United States will meet an agreed spawning objective. Most importantly, the parties agree to coordinate research and management efforts in order to meet these goals. The agreement for the protection and management of the Porcupine Caribou Herd requires that both countries conserve the herd and its habitat, cooperate internationally, and ensure long-term opportunities for customary and traditional use of the caribou. The Forty-mile and Chisana Caribou Herds have joint management plans coordinated between the wildlife management agencies on either side of the border. These plans also ensure coordinated management and data sharing. Finally, although endangered species are not currently an important factor in the management of the northwest boreal region, this may change as systems are increasingly impacted by climate change and human development.

INTERNATIONAL PROTECTED AREAS

The Convention Concerning the Protection of the World Cultural and Natural Heritage[8] provides United Nations Educational, Scientific, and Cultural Organization World Heritage Site status for the Wrangell-St. Elias National Park and Preserve, Kluane National Park and Reserve, Glacier Bay National Park and Preserve, and Tatshenshini-Alsek Wilderness Provincial Park complex (the largest international protected wilderness straddling an international border) (National Park Service, 2015). Under this treaty the parties promise to "ensure that effective and active measures are taken for the protection, conservation and presentation of the cultural and natural heritage situated on its territory," including "to take the appropriate legal, scientific, technical, administrative and financial measures necessary for the identification, protection, conservation, presentation and rehabilitation of this heritage."[9] This overlay system of management mandates exist on top of the various domestic land management laws already in place for those areas. A range of collaborative activities has taken place over

6 Yukon River Salmon Agreement of 2001, Amendment to Annex I of the Pacific Salmon Treaty, December 4, 2002 (TIAS 02-1204), available at http://yukonriverpanel.com/salmon/about/yukon-river-salmon-agreement/.

7 Agreement Between the Government of Canada and the Government of the United States of America on the Conservation of the Porcupine Caribou Herd, 1987, available at http://www.treaty-accord.gc.ca/text-texte.aspx?id=100687.

8 Adopted November 16, 1972, available at http://whc.unesco.org/en/conventiontext/.

9 Convention Concerning the Protection of the World Cultural and Natural Heritage, Article 5, available at http://whc.unesco.org/en/conventiontext/.

the years, in some cases with specific, ongoing agreements, such as for managing rafting activities on the transboundary Alsek River (Danby and Slocombe, 2002).

In addition to these agreements, the International Boundary Waters Treaty of 1909, which created the binational International Joint Commission and related bodies, has been invoked in the context of proposed mines. These mine proposals, such as the Windy Craggy mine proposed in far northwest British Columbia in the early 1990s, have been mainly in the Canadian part of the NWB, and could affect transboundary rivers and adjacent areas significant for wildlife (Day and Affum, 1995).

UNITED STATES LEGAL DRIVERS IN THE NORTHWEST BOREAL REGION

In Alaska, the foremost legal drivers in the NWB region are the diversity of federal and state land and wildlife management laws and resulting conflicts. federal and state laws in this area have widely divergent goals and at times can contain conflicting objectives and methodologies. Incompatible legal objectives can lead to management clashes on the ground.

Land ownership patterns are spatially intermingled because of the land allocation history of the state. The intermingled land ownership makes land management complicated by spillover effects from neighboring lands and the need to obtain agreement from many entities in order to take landscape-wide actions.

Land Allocation

Land in Alaska is largely under the control of the federal government, state government, or Alaska Native corporations. This structure stems from a series of land allocation laws. In the Alaska Territorial days, nearly the entire land base of Alaska belonged to the federal government, which purchased it from Russia in 1867 (Treaty of 1867[10]). In 1959, Congress passed the Alaska Statehood Act (Public Law 85–508, 72 Stat. 339, enacted July 7, 1958), which established the State of Alaska. Part of that statute granted to the new state the ability to select more than 41.3 million hectares (102 million acres) of "vacant, unappropriated and unreserved" federal lands. Native groups objected to the characterization of their ancestral lands as vacant and unappropriated (Case and Voluck, 1984). The Alaska Native Claims Settlement Act (Public Law 92-203; 43 U.S. C. 1601 et seq., Ch. 33), passed in 1971, extinguished all native land claims in Alaska, thereby allowing the state land selection process to move forward. That statute created the Alaska Native corporate structure (12 regional corporations and numerous village corporations) and compensated Indigenous groups for the loss of land by allowing the new corporations to select as much as 18.2 million hectares (45 million acres) of unappropriated federal lands (Case and Voluck, 1984). In 1980, the Alaska National

10 Actual title is "Treaty concerning the Cession of the Russian Possessions in North America by his Majesty the Emperor of all the Russias to the United States of America," available at https://www.loc.gov/rr/program/bib/ourdocs/Alaska.html, accessed October 30, 2017.

Interest Lands Conservation Act established 42 million hectares (104 million acres) of federal conservation units (parks, monuments, forests, refuges, rivers, wilderness, and Bureau of Land Management [BLM] lands) and established rural subsistence as a priority activity use on most of those lands. The State of Alaska currently comprises approximately 147.7 million hectares (365 million acres), and most of that land is either in federal, state, or Native corporation hands.

Federal Land Management Laws

In the United States, conservation unit-specific laws have enormous effects on the management of those areas, which in total encompass a large percentage of the NWB region. The boreal region of Alaska hosts several different categories of federal conservation units including BLM lands, National Wildlife Refuges, Wilderness Areas, and National Parks and Preserves (see Box 5.4.1, "United States Land Management Laws (major examples)"). Each of these categories is animated by its own series of laws and regulations, which dictate how the areas are managed, who is able to use them, and for what purposes. Differences in statutory mandates are reflected in fire management strategies, overflight and noise management, sport hunting and subsistence management, timber and minerals management, recreational opportunities, and the opportunity for various commercial enterprises. National Parks are required to "conserve the scenery and the natural and historic objects and the wildlife . . . and to provide for the enjoyment of the same in such manner . . . as will leave them unimpaired" (National Park Service Organic Act of 1916). Wildlife Refuges are administered "for the conservation, management, and where appropriate, restoration of the fish, wildlife, and plant resources and their habitats" (National Wildlife Refuge System Improvement Act of 1997). Wilderness land is "protected and managed so as to preserve its natural conditions" (Wilderness Act of 1964 [Public Law 88-577; 16 U.S.C. 1131–1136]). BLM lands are managed for "multiple use and sustained yield," meaning all uses, including extractive uses, must be considered equally and must supply a steady rate of output (Federal Land and Policy Management Act of 1976).

State of Alaska Land Management Law

The State of Alaska's natural resource management is primarily dictated by the state constitution, which encourages the settlement and development of resources and the sustained yield of all renewable resources. According to the Alaska State Constitution, Article 8 § 1, Statement of Policy, "It is the policy of the State to encourage settlement and the development of natural resources by making them available for maximum use consistent with the public interest." Article 8 § 4, Sustained Yield, states, "Fish, forests, wildlife, grasslands, and all other replenishable resources belonging to the State shall be utilized, developed, and maintained on the sustained yield principle, subject to preferences among beneficial uses." This extraction-oriented approach can lead to

BOX 5.4.1: UNITED STATES LAND MANAGEMENT LAWS (MAJOR EXAMPLES)

National Park Service Lands:	National Park Service Organic Act (16 U.S.C. 1) Redwoods National Park Expansion Act (Redwood Amendments) (Public Law 95-250; 92 Stat. 163)
National Wildlife Refuges:	National Wildlife Refuge System Administration Act (16 U.S.C. 668dd-668ee; 80 Stat. 927) National Wildlife Refuge System Improvement Act (Public Law 105-57)
Bureau of Land Management:	Federal Land Policy and Management Act (Public Law 94-579; 43 U.S.C. 1701 et seq.)
Wilderness Lands:	Wilderness Act of 1964 (Public Law 88-577)
All Federal Conservation Units:	Alaska National Interest Lands Conservation Act (Public Law 96-487; 94 Stat. 2371)

conflicts with the federal land units that place a greater emphasis on the protection of natural landscapes (such as parks, refuges, and wilderness).

Conflicts in Management

Disputes over the proper management of resources and the appropriate distribution of authority over those decisions continue to be a major challenge for resource managers and can extend into all areas of resource management. This is due, in part, to disagreements between the state and federal governments over wildlife management, mineral extraction, road construction, oil development, and other issues.

LAW AND POLICY DRIVERS IN NORTHWESTERN CANADA

The history of law and policy in the Canadian north generally is one of distant federal and Provincial government management, with little attention to local and Indigenous concerns. That began to change in the 1970s when the Yukon and Northwest Territories (NWT) gained some land and resource management powers (for example, over wildlife and water), and the long path toward settlement of Indigenous land claims began. In the Territories, land and resource regulation and permitting, environmental assessment, forests, minerals, among others remained federal responsibilities until 2003 in Yukon and 2014 in NWT when devolution of powers from federal to territorial governments occurred. In northern British Columbia, land and resources are managed as elsewhere in the province, although the northwest of the province has some special collaborative management through the Muskwa-Kechika Management Area Act ([SBC 1998] Chap. 38), and subsequent revisions (Muskwa-Kechika Advisory Board, 2016).

In the early 1990s, comprehensive land claims (claims by Indigenous peoples who had never signed a historical treaty for loss of land and resources and rights) in Yukon and NWT were largely settled, after a decades-long process. In Yukon, the territorywide

Umbrella Final Agreement (implemented by legislation) was followed by negotiation of local subagreements with each of the 14 First Nations (at the time of writing, three still have not been completed: two in the southeast Yukon, and one in the southwest). In Yukon and NWT, claims provided cash settlements to First Nations, as well as land selections, and an extensive system of comanagement boards at regional and territorial levels (Council of Yukon First Nations, 2016). These include the regional Renewable Resource Councils, Park Management Boards, and the territory-wide Yukon Fish and Wildlife Management Board (YFWMB), Yukon Land Use Planning Council, Yukon Environmental and Socioeconomic Assessment Board, and a Salmon Sub-Committee of the YFWMB. These boards include First Nations and non-First Nations representatives, and are advisory to territorial government decision makers, although it is generally expected that comanagement boards' advice will be accepted. In the NWT, there are about six traditional territories, several with comprehensive claims, several without, although the NWB region in the NWT is largely within settled claim regions. A similar structure of comanagement boards can be found in a slightly more integrated structure under the Mackenzie Valley Resource Management Act (S.C. 1998, c. 25; for example, Chamberlain and Haile, 2016).

However, with conservative governments in power territorially and federally, there has been increasing conflict between comanagement boards and territorial governments. This has resulted in multiple lawsuits, most expected to eventually reach the Supreme Court of Canada (for example, Slocombe and others, 2016). The Canadian federal government changed in late 2015, and the Yukon territorial government also changed in late 2016; changes in legal and policy positions and approaches quickly became apparent. The Supreme Court of Canada is increasingly playing a role in redefining the framework by which the lands and resources of northwestern Canada are managed, through its interpretations of claims by First Nations and other Indigenous people against provincial/territorial and federal governments. Those claims are rarely denied now, partly because aboriginal rights were enshrined in the Canadian Constitution of 1982. These Supreme Court findings, such as the 1997 Haida, 2004 Delgamuukw, and 2014 Tsilhqot'in decisions, increasingly have been recognized as the right of aboriginal people to be meaningfully consulted about development in their traditional territories, regardless of whether they have settled modern land claims. Several significant cases also are pending from Yukon and Northwest Territories.

These decisions are especially significant in northern British Columbia, where comprehensive claims have not been settled. Although there is a British Columbia Treaty Process and Commission, progress has been limited, and likely will remain so given the history of provincial government approaches and recent Supreme Court of Canada decisions.

In Yukon, recent government policy changes have made land more available and increased small-scale development on the landscape for subdivisions, tourism, mineral exploration, and agriculture. Large-scale mining has waxed and waned with the price of minerals and economic conditions (see Chapter 5.1). Demand for electricity is pushing attention to potential hydroelectric and other energy generation projects. Climate change is a growing issue for governments in Yukon, which have so far focused on the adaptation issues of communities and power supply, and attention to wildlife and parks yet to catch up with environmental change. Tourism, wildlife harvesting, and large-scale resource development are key contending activities that drive policy.

INFORMATION GAPS

- As legislatures begin to better understand the severity of the effects that are likely to be wrought by climate change, how this will affect existing land and wildlife management statutes or develop new statutory directives, the development of new international agreements for the protection of specific species or natural systems.
- How land management agencies will reinterpret their missions in light of new climate change realities. These reinterpretations may lead to management changes on the ground. Agencies also may amend their statutory interpretations for other reasons; for example, in the United States, the National Park Service's recent regulation change to implement state hunting rules on park lands.
- How ongoing court cases may alter the current interpretations of existing laws. For instance, the case of Sturgeon v. Frost (No. 14-1209, March 22, 2016) was remanded from the U.S. Supreme Court back to the lower courts. The decision in that case may affect the delineation of authority over public lands in Alaska.
- How changes in the political realm at the national and (or) state/provincial/territorial level may lead to legislative, regulatory, or enforcement changes that could affect any and all aspects of ecosystem and landscape management.
- Fuller understanding of comanagement processes, how to improve their effectiveness, and government responses to them, and implications of recent Supreme Court of Canada decisions about aboriginal land and resource management rights.
- Improving connectedness of land and resource comanagement processes to federal, and to local levels.
- Implications of climate change for species and ecosystems in the NWB region, and determining how to address them within comanagement and related processes.

5.5
VALUES AND ETHICS IN RESOURCE MANAGEMENT INSTITUTIONS

By Annette Watson[1] and Douglas Clark[2]

KEY FINDINGS

- Values and ethics not only inform managers about what species compositions to strive for within the boreal landscape but also inform individual choices that affect landscape change.
- Values not only define resource management issues, they actually drive what scientific information is generated in support of management actions.
- Managers need not be afraid of acknowledging the role of values in decision-making; it can help them facilitate ecosystem stewardship.
- Applied social science, particularly what is coming to be called "conservation social science," provides a capable toolbox for helping managers understand the values at play in the situations they must negotiate.
- Managers can be professionally effective by using all the tools at their disposal to help agencies, partners, and stakeholders to understand each other's values and to create situations where each has a substantive role in shaping and sharing the distribution of the values they all seek.
- Coping with environmental change is particularly difficult because values often are set based on what is, not what might be in the future.

INTRODUCTION

Values often are the unseen drivers of social-ecological systems. Human societies base their politics on, and self-organize through, ethical understandings of how the world should work. Values inform institutional missions and individuals' relationships with

1 College of Charleston, Charleston, South Carolina, USA.
2 University of Saskatchewan, Saskatoon, Saskatchewan, Canada.

ecosystems, thus natural resource managers must balance the demands of multiple stakeholders—often resulting in differing values about the use of these resources (Lee, 1994; Gunderson and others, 1995). At the same time, these values are not often made explicit during political struggles over land use and environmental management. Emerging research suggests that improved governance capability, especially capacity for healthy self-organizing responses, is essential for the adaptation of northern social-ecological systems (Arctic Council, 2013). Consequently, this chapter aims to show practitioners why and how values can be made more explicit, and how to harness such knowledge constructively.

Mattson and others (2012) observed that environmental management literature has a great deal to say about values, but often does not clearly define the term. They defined values as "fundamental non-linguistic ways that people orient to the world"; this idea can be operationalized by thinking about values as "physical and psychological indulgences that people desire or seek" (Mattson and others, 2012, p. 241). From this conceptual starting point, it follows that what individuals and societies do, and how it is done, is based on individual and collective values. In resource management decision-making contexts, managers aim to base decisions on "information" (Healy and Ascher, 1995), yet values often emerge as having greater influence on the debates and decisions of resource management. Values often have been shown to be an important but overlooked element of how the public more broadly receives and responds to scientific knowledge for policy-making (Ascher and others, 2010; Margerum, 2011). The work by cultural cognition scholars (see Box 5.5.1, "Cultural Cognition of Climate Change"), for example, shows that depending on the political orientation, a person may or may not "believe" in climate change—merely because they do not like the perceived solutions that would be possible. In this context, the role of values can be recognized as positive drivers that amplify change but also negative drivers that dampen change.

SCIENTISTS AND MANAGERS ARE ALSO "PUBLICS" WITH VALUES AND ETHICS

It is incorrect to suggest that only non-scientists are informed by their values. It is often argued that the practice of science strives to be objective, unlike the general public, and that any value-informed science is "junk science" that somehow smuggles in a subjective or advocacy perspective, but the relationship between values and science is far more complex (for example, Pielke, 2007; Foote and others, 2009). Indeed, the health sciences and conservation biology are overtly value-driven disciplines, yet still respected. Insights from the social studies of science show that the ideal of objectivity is merely an ideal, realized in the writing of peer review more than in the practices of conducting research (Latour and Woolgar, 1986; Latour, 1987; 1993; Harding, 1998). For example, recent work has shown that managing migratory geese in the boreal forest has, because of practices of wildlife biology that aim to mimic the surety of laboratory science, privileged aspatial calculations of abundance over assessments of

BOX 5.5.1: CULTURAL COGNITION OF CLIMATE CHANGE

According to Kahan and Braman (2006, p. 24), "The phenomenon of cultural cognition refers to a series of interlocking social and psychological mechanisms that induce individuals to conform their factual beliefs about contested policies to their cultural evaluations of the activities subject to regulation." Based on the work of Douglas (2004), scholars engaged in cultural cognition studies understand how individuals base their decisions along a continuum having two dimensions, between "Hierarchical" and "Egalitarian" preferences for organizing political life (Grid) and "Individualist" and "Solidarist/Communitarian" (Group) worldviews (Kahan and Braman, 2006, p. 151). This typology has more explanatory power than education level or scientific illiteracy when understanding perceptions of risk and the need for action (Kahan and others, 2011). From these cultural typologies, researchers have "mapped" out their relationship to various beliefs in risk management, including nuclear waste, nanotechnology, and climate change. For example, one study tested public response to a "climate change expert" and found a clear distinction between the "hierarchical/individualist" and the "egalitarian/communitarian" worldviews. Those who value political hierarchical structures and an individualist perspective are measurably less likely to believe in the risks of climate change than those who share values about equality and a community perspective (Kahan and others, 2011).

habitat undergoing climate change—thereby missing opportunities to engage with the Indigenous and local knowledges that first alerted managers to geese declines (Watson, 2013; Jos and Watson, 2016). As noted by one political scientist, "what we have learned from the constructivist contribution to science studies is that the defining and framing of the problem to be researched often implicitly rests on a particular understanding of a way of life, as opposed to other competing views" (Fischer, 2009, p. 147).

Put another way, all scientists individually pursue their work through the lens of the institutions within which they work, and the ethics they inherited from their academic disciplines. Framing a scientific perspective as "objective" can obscure the value-driven choices that underlie the selected framing (Cruikshank, 2005; Jos and Watson, 2016). It is thus incumbent on professionals to be clear about the values implicit in their standpoints with themselves and others (Clark, 2011). This is a difficult task because many resource management organizations base their mandate and credibility on an idealized image of scientific competence, but doing "good science" while still being transparent about one's values is possible to achieve (Clark and others, 2014). Pielke (2007) offers the characterization of "The Honest Broker of Policy Alternatives" as a useful heuristic for those seeking to situate the practice of science relative to the institutional values that surround it. In that formulation, Pielke (2007) stresses that the role of science should be to expand policy choices, not restrict them. However, he cautions that this does not absolve either scientists or policymakers from their shared responsibility to assess and communicate the significance of particular science for specific policy processes. Indeed, those who question orthodox doctrines of "objective science" often play critical roles in large organizations, especially by identifying pathways for adaptation during periods of rapid or unexpected change (Gunderson and others, 1995).

The actions of resource managers are also driven by the missions and ethics of the institutions within which they work. Perhaps some see their scientific training as instrumental in insulating their activities from the general public, but managers have been indoctrinated in particular ethics for which they are professionally responsible—and thus find themselves in conflict with other professionals and members of the public (Clark, 2011). That approach is a recipe for professional and personal frustration.

Scientists and managers, like all people, are only bounded rational: they apply science to understand only selected parts of a resource management issue/situation (Clark, 2011). This often backfires. For example, in the North Atlantic context, when managers privileged the theory of bioeconomics to manage New England fisheries, they inadvertently marginalized small-scale fishers and contributed to the overall public resistance to and mistrust of scientific management (St. Martin, 2001). Systematically applying certain scientific principles to all situations is merely an expansion of the scientific approach with which professionals are already familiar. The drawbacks of this approach are increasingly being recognized as agencies build capacity in "human dimensions" and with the value of interdisciplinary approaches to understanding and resolving environmental issues is understood (Bennett and others, 2016).

One of the most under-articulated values held by many environmental practitioners is the belief about policing the boundaries between humans and the natural world—an ethic born from recent human history, after witnessing some of the environmental effects of the Industrial Revolution. Responding to environmental degradation post-1900, for example, the U.S. Wilderness Act of 1964 (Public Law 88-577) says, "humans are a visitor who shall not remain." However, this codified ethic is based on an incorrect assumption of how modern North American ecologies developed; as shown by geographers and historians studying human-environment relations over much longer time scales, Indigenous peoples have radically altered many environments, through large-scale burning, hunting strategies, and other alterations of the "natural" world (Pyne, 1982; Denevan, 1992; Cronon, 1996). Tribes and First Nations have in fact "made" many of the landscapes that in the United States and Canada were thought of as "wilderness," even though—in both national contexts—these landscapes are managed to keep humans largely "separated" from the ecologies that persist within national park borders (Spence, 2000; Mortimer-Sanilands, 2009).

Assumptions about the ethical role of humans in ecosystems influence how individual managers are thinking about climate adaptation in such spaces. For example,

Magness and others (2012) surveyed managers across the U.S. Fish and Wildlife Service and found that ideas about "historic conditions" and "naturalness" not only varied among them but also implicated their own range of acceptable management options for landscapes undergoing change. Yet as noted by Bengtsson and others (2003), reserves are not interconnected enough to be functionally "wild"—and would not "naturally" remain in stasis were it not for restoration techniques. In fact, they argued that most of these landscapes are bound to change regardless of the strategies used to keep them the same, given projections of changes in climate.

Thus, how to adapt need not be centered on such ethics of what is "natural" as opposed to the "human," and some resource management agencies have missions that allow for these ethics. For example, in Alaska, lands managed by the state can use techniques of predator control (called "intensive" management) to cull wolf populations; most local tribes support this strategy, because most have lived in sedentary villages for only the last 50 years and are now enduring problems of food insecurity. Many of these Indigenous peoples suggest that intensive management can be used as an adaptive strategy, to compensate for these changes in their social system (Watson and others, 2014). However, there remains controversy over where such intensive human intervention should occur (Boertje and others, 2010), and the sometimes-random pattern of land ownership in Alaska has caused tensions between agencies that have different missions when it comes to the purposeful culling of wolves. A similar case involves bison relocation, showing the difference a national context can make when understanding values and ethics as drivers (see Box 5.5.1, "Wood Bison in the Boreal Forest: A Tale of Two Reintroductions").

UNDERSTANDING VALUES AND ETHICS AS A DRIVER

In the rapidly changing landscapes of the circumpolar north, managers risk failure unless they understand the ways in which values and ethics drive policy debates and selections about the use—or nonuse—of scientific tools to generate and apply knowledge (Ullsten and others, 2004; Arctic Council, 2013; Clark and others, 2014). Embracing or being transparent about the values and ethics that shape resource management will make the work of negotiating the politics of ecosystem management easier, because members of the public then already know that they are not the only ones with values (Cortner and Moote, 1999).

For scientists and scientifically oriented managers, the challenge is to not fall into the trap of trying to make values go away. Rather, it is a matter of deciding to become a more effective (and less frustrated) professional by building the skills to understand values and then work collaboratively with that knowledge to decide what social-ecological outcomes are in the common interest and how best to navigate toward them in a dynamic world. Managers need to become proficient at working across all the different boundaries shown in Figure 5.5.1. The emergence of what is

BOX 5.5.1: WOOD BISON IN THE BOREAL FOREST: A TALE OF TWO REINTRODUCTIONS

In the boreal region of Alaska, the role of values and ethics in different government agencies became apparent when a controversy arose in 1998. The Alaska Department of Fish and Game (ADF&G) developed a plan to locate wood bison (*Bison bison athabascae*) to areas where they had not lived for hundreds of years. The ADF&G proposed to relocate wood bison to Yukon Flats and Minto Flats to mitigate declining moose populations, a food source for many Alaskans, and to meet goals of broader species stability for wood bison across North America (Stephenson and Rogers, 2007). Significantly, the debate cleaved along ethical ideas about what is "natural" to the landscape: where managers with ADF&G managers called their efforts those of "reintroduction," managers from the U.S. Fish and Wildlife Service used the terminology of "introduction." Some argue that bison are not "natural" to the western boreal landscape because of the extreme lapse of constant occupation, and others cite Indigenous oral history and archaeological evidence to show the species is not "new" to the landscape. Some argue that bison can act as a positive driver on the boreal landscape, including effects from grazing intensity (Gates and others, 2010). However, what is "natural" within a changing boreal forest? It is an important question, especially as the boreal region evolves under a rapidly changing climate.

Moose might be understood as "invasive" as well, because moose populations were not present within the westernmost boreal forest until approximately the 1930s (Watson and others, 2014). As shown in Table 5.5.1, some of the effects of bison on the landscape cannot be wholly categorized as "bad" for these changing boreal landscapes, because at times they also may drive positive-feedback loops that affect ecosystem stability (for example, decreasing fire occurrence and spread by grazing fine fuels).

In contrast to Alaska, wood bison were reintroduced in the neighboring southwest Yukon from 1988 to 1992, and there was no significant debate about whether bison were "natural" (Clark and others, 2016). The only jurisdiction where this issue arose was Kluane National Park, which has not yet decided whether to accept immigrating bison or exclude them (Markel and Clark, 2012). Instead, the larger issue around this reintroduction was how to deal with the social and economic effects on First Nation livelihoods brought about by bison and bison hunting. Through participatory socioeconomic impact assessment, an empowered and innovative comanagement regime was able to systematically understand the range of values around bison. The comanagement regime then applied that knowledge to identify shared interests, defuse controversy, and ultimately develop a socially determined carrying capacity for this bison herd (Clark and others, 2016).

Table 5.5.1. Ecosystem influences of bison in the Northwest Boreal region.

Process	Description
Create patches	Grazing can produce a dynamic mosaic of vegetation patches that differ in seral stage and that differ due to variations in grazing intensity.
Enhance nutrient cycling rates	Grazing can enhance nutrient turnover and change dominant system mode from detritus-decomposition to consumption-defecation.
Enhance habitat quality	Grazing can increase habitat suitability for a variety of smaller birds and mammals, particularly those which prefer open and edge habitats.
Modify fire regimes	Bison consume fine fuels and create trails and trampled areas that reduce fire intensity and extent, and modify the effect of fire on vegetation heterogeneity.
Create disturbances	Trampling and wallows create seedbeds for some species; localized tree stands that are not tightly clumped are susceptible to major damage.
Stimulate primary production	Grazing removes senescent material from the sward and increases light penetration, nutrient availability, and growth.
Disperse plant seeds	Bison transport seeds in leg fur and gut, and may enhance establishment of native and exotic plants.
Maintain floral diversity	Grazing can result in greater grass and forb species diversity.
Support carnivores and scavengers	Bison are prey to some large carnivores, and bison.

being called "conservation social science" shows that not only is there demand for such analytical skills, but also that there is a solid collection of techniques from multiple disciplines that can be drawn upon to tackle complex socioecological problems more holistically (Bennett and Roth, 2015). For example, the Q-method (Brown, 1980) is an increasingly used tool for quantifying subjectivity in conservation science, helping diverse groups understand their own internal values and work better together (for example, Rutherford and others, 2009; Ray, 2011; Chamberlain and others, 2012).

The ability to treat values and ethics transparently is an essential skill set necessary to transition from being "top-down" managers of a steady-state system to being ecosystem stewards—enabling professionals to facilitate stakeholders to respond to and shape social-ecological resilience (Chapin, Carpenter, and others, 2010).

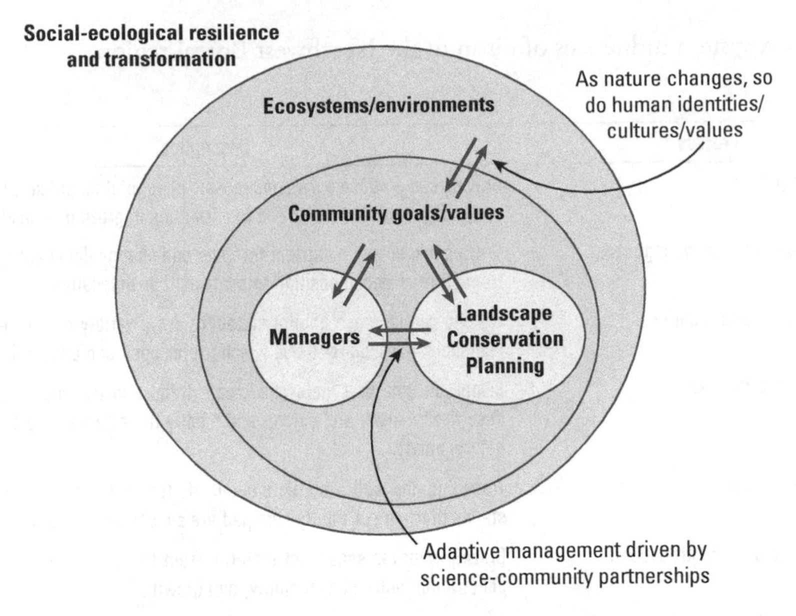

Figure 5.5.1. Schematic model of how human action and knowledge production is embedded in a world of both changing values and ethics and of changing ecosystems, while also affecting those ecosystems.

INFORMATION GAPS

- Adaptive management research designs are needed that incorporate a model of the assumed ethics of human-environment relationships.
- Assessments are needed to determine whether and how usefully the different analytical values schema developed by non-Indigenous researchers (Mattson and others, 2012) correspond with northern Indigenous peoples' values.
- Reconciling value conflicts across scales remains a significant challenge, for example, when a global public criticizes local communities' traditional resource management practices (Marker, 2006; Tyrrell and Clark, 2014).

6
INTERACTIONS AMONG DRIVERS

By Donald Reid[1]

INTRODUCTION

The mechanisms, characteristics, and rates of interaction among landscape-level drivers of ecological change are themselves in flux, because a changing climate is shifting a number of primary abiotic and biotic drivers, and because society continues to expand its use and extraction of natural resources. The novel regimes of interaction are inherently complex, reflecting the underlying interconnectedness of most ecosystem components. They are importantly understood in terms of positive (destabilizing), neutral, and negative (stabilizing) feedbacks. For example, increased temperatures and associated drying induce more frequent and severe forest fires and longer snow-free periods (reduced albedo) that collectively increase short-term greenhouse gas concentrations and absorption of solar radiation to the ground, thereby inducing more warming (a positive feedback). Increased melt of glaciers may counteract the downstream warming effect of higher air temperatures on streams and lakes (a negative feedback). To engage in ongoing stewardship and adaptation requires better quantification, characterization, and anticipation of these newly emerging regimes of interaction through monitoring, focused study, and projection modeling. As in the past, ongoing stewardship requires the assessment of the effects of novel human development activities (for example, mines, energy extraction, tourism traffic, and agricultural land conversion) on the socioeconomic system. Such assessments will vary depending on the relative emphasis placed on the various values (for example, subsistence use, water quality, government revenue) to be affected. These assessments and stewardship options are becoming more complicated, and potentially more constrained, because combinations of development activities have cumulative impacts in the same region. Cumulative impacts often are nonlinear, with thresholds in the response of an affected resource to a human activity (for example, stress levels in a wildlife species, economies of scale in resource extraction). These complex relationships reflect synergistic effects between types and intensities of activities and the affected resources. The Northwest Boreal (NWB)

1 Wildlife Conservation Society Canada, Whitehorse, Yukon, Canada.

region has a relatively low human footprint to date, and is well positioned to attempt comprehensive assessments of cumulative impacts with climate change projections included. This will require regional approaches to cooperative landscape planning, including regional strategic environmental assessments. The success of these endeavors will increase with better monitoring of the effects of existing developments, focused studies on values that are known to be vulnerable or crucial to human well-being, and improved projection models within which divergent scenarios of potential human activity and future climate can be assessed.

6.1
INTERACTIONS AMONG DRIVERS IN THE NORTHWEST BOREAL REGION

Nancy Fresco[1]

KEY FINDINGS

- Landscape changes do not occur in isolation. To understand and shape change, a holistic perspective on how climate, human uses, and ecological systems interact is helpful.
- These interactions are complex, involving multiple feedbacks that affect boreal ecosystems and those who rely on them, and can trigger threshold shifts and other unexpected changes.
- Although many of the mechanisms of these interactions and feedbacks have yet to be fully quantified, known conceptual linkages among drivers can help land and resource managers assess potential impacts to resources of concern.
- New data, tools, and models—such as the Integrated Ecosystem Model—are needed to help illustrate and quantify feedbacks and interactions between drivers, thereby improving our understanding of landscape processes, reducing uncertainty, and better addressing future threats to resilience.

LINKING DRIVERS OF CHANGE

The Northwest Boreal (NWB) region is undergoing multiple simultaneous types of landscape-level and ecological change driven by changes in climate and human uses of natural resources (Chapin, Lovecraft, and others, 2006). Each of these types of change (that is, human, climate, and ecological) includes numerous interrelated variables. Complex linkages among these variables can create feedback loops that are either positive (in the sense of each variable exacerbating the other, rather than in the sense of being good or desirable) or negative (that is, balancing or tending to cancel out

1 University of Alaska Fairbanks, Fairbanks, Alaska, USA.

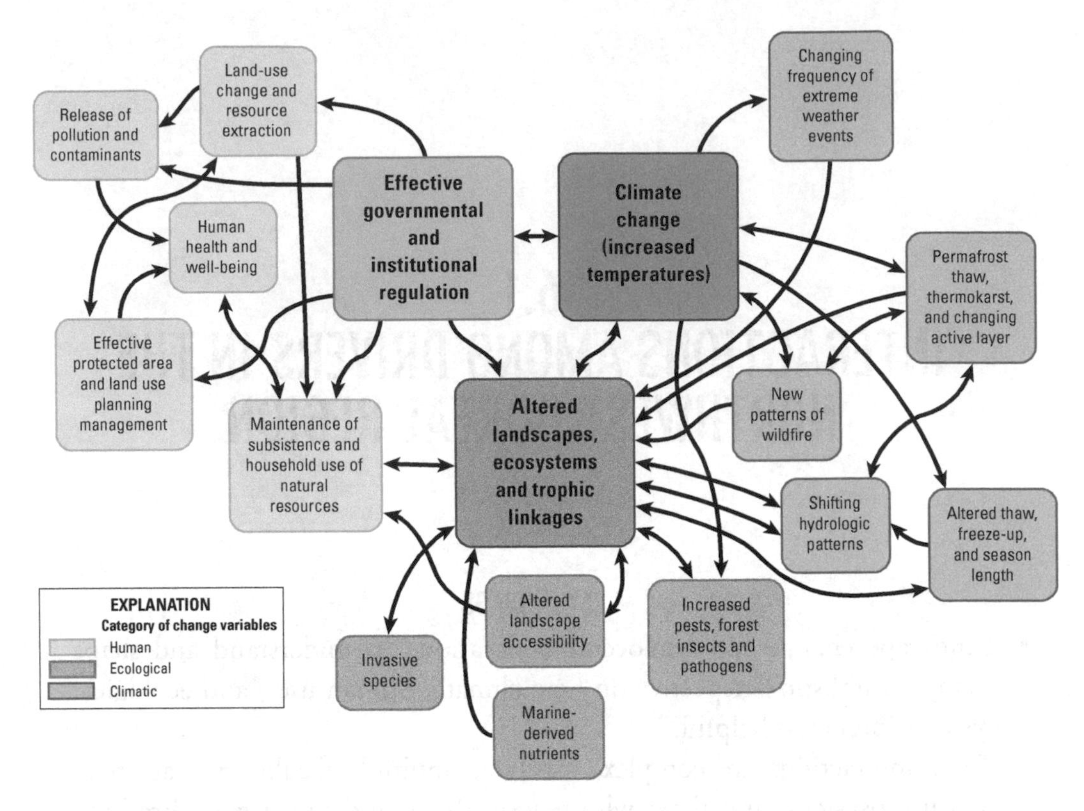

Figure 6.1.1. Interactions between primary and secondary drivers of change in three major categories, which include human, climatic, and ecological variables in the Northwest Boreal region.

one another). These linkages often are complex and only partially understood. Thus, in examining these drivers of change and the linkages among them, it is useful to first consider into which fundamental category or categories each driver falls, and then to consider the potential interactions between these categories, as well as the more specific connections between individual drivers.

Complex Interactions

The complex interactions among human, climate, and ecological variables, or drivers of change, are shown in Figure 6.1.1. Feedback occurs both within categories and between categories. Feedback also occurs between the three major, overarching drivers in the center of the diagram and the secondary or derivative drivers on the periphery.

Positive or Negative Feedback?

In some cases, the interactions depicted in Figure 6.1.1 are clearly negative, in other situations, they are clearly positive, and in some instances, the interactions could be either positive or negative depending on uncertainties, institutional decisions, or opposing effects among subcategories. For example, new patterns of wildfire already are affecting ecosystems and will continue to do so, particularly by shortening fire cycles and thereby increasing the relative proportion of early successional vegetation (Rupp and Springsteen, 2009; Beck and others, 2011). This change, in turn, is likely to affect relative abundances of mammals regionally (Simon and others, 1998; Maier and others, 2005; Joly and others, 2012), and thus subsistence opportunities. Subsistence species that rely on early successional habitat (for example, moose [*Alces alces*]) may increase in numbers, whereas those that rely on late-successional forest cover (for example, caribou [*Rangifer tarandus*]) may decline, making overall effects difficult to define (Nelson, Zavaleta, and Chapin, 2008). Fire behavior may be affected by pests and parasites that reduce the health of trees, making forests more flammable, as well as by management strategies aimed at reducing the risk of catastrophic fires after pest outbreaks (Lance and others, 2006). Other feedback loops associated with fires involve burning of both the forest canopy (which shades and cools the ground) and the organic soil layer (which insulates the ground) (Jafarov and others, 2013). More frequent and more severe fires decrease terrestrial carbon pools and increase the rate of carbon release to the atmosphere. They also exacerbate permafrost thaw in the short term, releasing yet more stored carbon and changing hydrological patterns.

Apart from fire, other climate-related factors drive complex interactions in the NWB region. Changing duration of snow cover affects phenologies (Hinzman and others, 2005; Brown and Robinson, 2011) with feedbacks on fire regimes. Changing precipitation and evaporation regimes affect hydrology and drought risk (Riordan and others, 2006; Rasouli and others, 2014) with feedbacks on tree survival and wildlife habitats. Increased permafrost thaw affects productivity and hydrology (Thormann and others, 1998; Allison and Treseder, 2011) with feedbacks on fish biology.

Positive, negative, neutral, mixed, and unknown interactions between drivers are shown in Table 6.1.1. Ecosystem changes and changes to subsistence are closely interrelated, and both include many mixed effects, when considered with regard to other drivers. Because ecosystem function is so complex, overall gains and losses are difficult to quantify.

Implications for Management

Management at the local level is not included as a variable in Table 6.1.1. However, management is an integral part of every feedback loop because the behavioral choices humans make are linked to both climate change and the alteration of landscapes, and these core variables are connected, directly or indirectly, to each additional variable. To understand what these interactions may imply for policy and management, it is crucial to lay out existing and potential policy and management choices for each of the connections between drivers of change.

Table 6.1.1. Interactions between drivers and variables with positive to negative reactions in the Northwest Boreal region. Interactions may be positive (+), negative (-), neutral (0), and/or unknown (?).

	Increased temperature, longer growing season, more extreme heat and less extreme cold	Wildfire frequency	Invasive species, parasites, and pathogens	Permafrost thaw and increased active layer depth	Surface water availabilty	Vegetation change and novel trophic assemblages	Marine derived nutrients	Availability of subsistence resources	Land use change, resource extraction, and pollution
Increased temperature, longer growing season, more extreme heat and less extreme cold		+	+	+	-	+	?	±	0
Wildfire frequency	+		+	+	?	+	0	±	0
Invasive species, parasites, and pathogens	±	0		0	0	+	0	-	0
Permafrost thaw and increased active layer depth	+	±	+		±	+	0	±	±
Surface water availabilty	±	-	±	+		±	?	±	±
Vegetation change and novel trophic assemblages	±	±	±	±	±		?	±	0
Marine derived nutrients	0	0	?	0	?	?		?	0
Availability of subsistence resources	0	0	0	0	0	±	?		+
Land use change, resource extraction, and pollution	+	?	+	?	?	?	-	-	

It also is important to recognize that the inherent uncertainty of projections, especially when tipping points come into play, underscores the importance of exploring the ramifications of multiple possible futures by scenario planning (Peterson and others, 2003) and by instituting adaptive governance to remain flexible to change (Kettle and Dow, 2014). Some ecological changes and feedbacks will result in ecological "surprises," which may best be identified and understood by maintaining long-term studies and monitoring, investigating the outcomes of disturbance events, and more detailed and realistic modeling (Lindenmayer and others, 2010).

Of particular note are interactions with strong positive feedback loops. At a national or international level, an example is the fact that a warming climate increases fire and permafrost thaw, which in turn releases carbon, thus exacerbating the warming climate. Drawing attention to this may influence policy. Locally, strong feedback loops between land management, subsistence use, shifting seasonality, and changing ecosystems may prove most crucial in shaping policy.

Likewise, policy changes could affect (and often should be designed to affect) the most detrimental of these social-ecological feedbacks. As in the beaver case study (see Box 6.1.1, "Case Study: Potential Interaction between Climate Change, Key Species Density, Hydrology, and Ecosystem Function"), finding thresholds in functional relationships can help managers assess alternative scenarios (Standish and others, 2014). For example, managers, when offered information on the projected dramatic increase in beaver populations, would be better equipped to plan for potential effects on trout, char, and salmon populations (Collen and Gibson, 2001).

INFORMATION GAPS

- Due to lack of research and historical experience, many interactions among drivers of change are poorly documented, not just in magnitude but also in direction.
- The functional interactions of drivers of change have rarely been quantified, including whether or not there are threshold effects or defined limits. Research is needed to quantify dominant interactions (for example, temperature and precipitation [drought] influencing vegetation change; novel vegetation patterns influencing opportunities for subsistence game harvest).
- Undocumented thresholds are likely to exist, may cause rapid and unprecedented shifts in system behavior, and need to be discovered where possible.
- Comprehensive models of the NWB social-ecological systems, including interactions of important drivers, need to be developed, refined, and tested with varying scenarios of future conditions.

- Although adaptive governance has been promoted and assessed in the context of specific resource management issues (for example, Kettle and Dow, 2014; Beier, 2011), further analysis of the efficacy of adaptive management in the face of high uncertainty is needed.

BOX 6.1.1: CASE STUDY: POTENTIAL INTERACTION BETWEEN CLIMATE CHANGE, KEY SPECIES DENSITY, HYDROLOGY, AND ECOSYSTEM FUNCTION

Researchers engaged in a Rapid Ecoregional Assessment in western Alaska (Bureau of Land Management, 2015) reviewed the available literature and data pertaining to changes in climate and in populations of key species in the far north. Research on the American beaver (*Castor canadensis*) (Jarema and others, 2009) showed that although beaver presence is possible above a relatively low threshold for mean summer temperature (7.9°C [46.2°F]), regional density is projected to increase non-linearly with increasing mean summer temperature (Fig. 6.1.2). Therefore, a summer temperature increase of only 4.5°C is projected to triple the number of beaver colonies (Bureau of Land Management, 2015).

Because beaver are themselves landscape engineers that create significant hydrogeomorphological impacts (Gurnell, 1998), a threshold shift in beaver populations may affect the hydrology and ecosystem balance of the region as a whole.

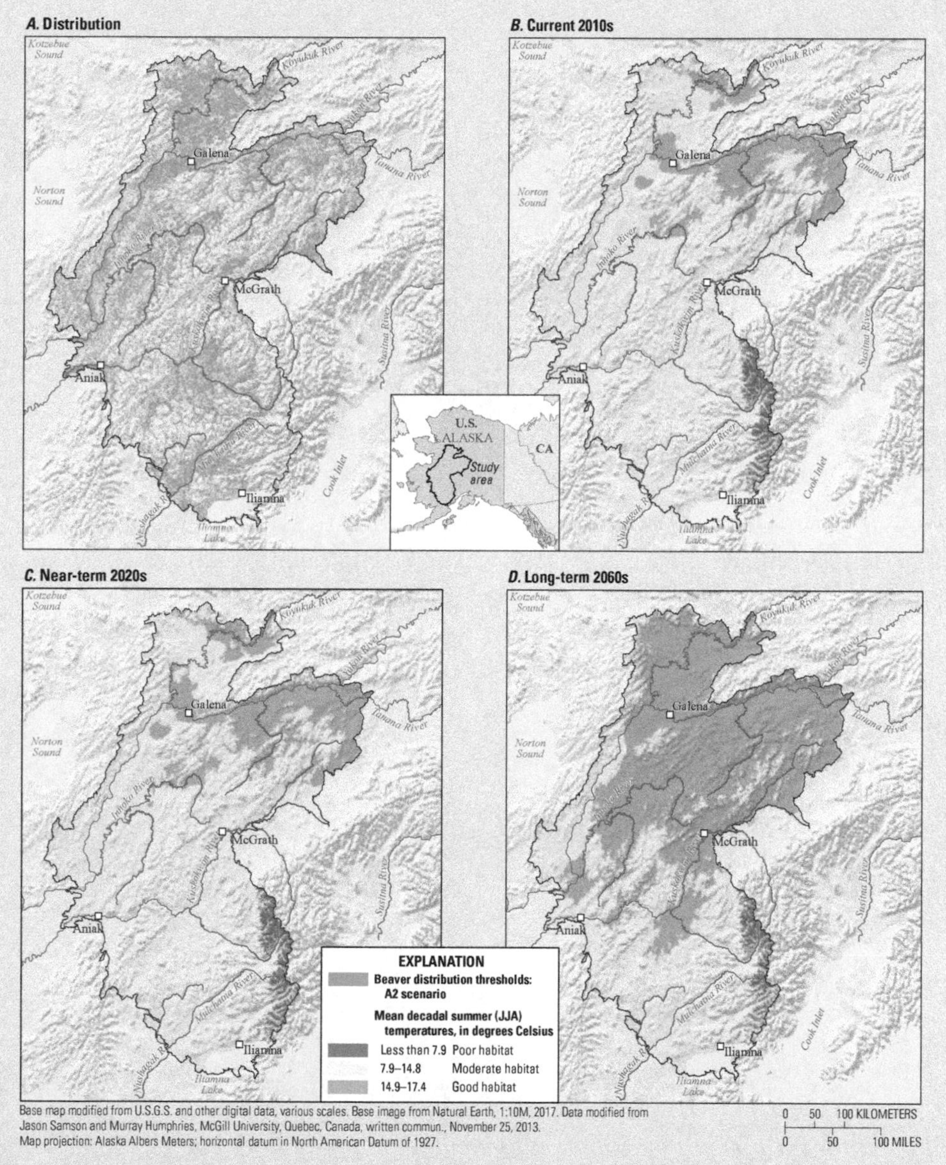

Figure 6.1.2. Temperature scenarios for (*A*) current beaver distribution; zones of differing beaver habitat suitability from (*B*) present (2010s); (*C*) through 2020s; and (*D*) 2060s based on projected mean summer (June, July, and August [JJA]) temperatures for the Yukon River lowlands–Kuskokwim Mountains rapid ecological assessment region of western Alaska (after Bureau of Land Management, 2015).

6.2
CUMULATIVE EFFECTS IN THE NORTHWEST BOREAL REGION

Chris J. Johnson[1] *and Wendy M. Loya*[2]

KEY FINDINGS

- The cumulative effects of human activities, acting as drivers of landscape change, primarily result in forest disturbances that modify forest composition and expand routes for human access in the NWB region.
- Identification and management of cumulative impacts, as results of landscape change, are premised on ecological, conservation, cultural, and utilitarian values, and include increased habitat loss and fragmentation, impacts to subsistence uses, and diminished ecosystem services.
- Project-specific regulatory approvals (for example, environmental assessment) are inadequate to quantify and manage cumulative impacts due to the complexity of the systems, interactions, and synergies among individual effects, and the limited ecological data to guide decision-making.
- Given the relatively low levels of cumulative impacts across the northwest boreal region, there is opportunity to proactively address change at broad spatial and temporal scales through regional (for example, regional strategic environmental assessments), cooperative approaches to landscape planning.

Compared with much of North America, the extent and magnitude of development and landscape change in the Northwest Boreal (NWB) region of Alaska and Canada, remains modest. Much of the land is held in conservation areas or has been too remote for industrial development. Large areas of the Alaskan component of this ecoregion are managed for wildlife habitat or wilderness values within the National Wildlife Refuge and National Park Service systems, in addition to federal mining, recreation,

1 University of Northern British Columbia, Prince George, British Columbia, Canada.
2 Formally with The Wilderness Society, Anchorage, Alaska, USA.

and conservation lands under the Bureau of Land Management. The surrounding state, Native-owned, and private lands host communities and industrial development, but overall the current footprints are small. In the Canadian part of the northwest boreal region, less area is focused on parks and conservation values, but, as in Alaska, industrial development is relatively limited when compared to southern boreal forests. Therefore, there are still opportunities to define and achieve cumulative impacts for planning and management in the region.

Increasing land use and climate change will result in considerable challenges to the effective management and conservation of the NWB region (Kellomäki, 2016; Hebblewhite, 2017). The current anthropogenic drivers of land use change include resource extraction (primarily oil and gas pipeline development, mining, and forestry, see Chapter 5.1), agriculture, military training, tourism, recreation, transport infrastructure, and community expansion. These drivers are strongly coupled to, and therefore influence, natural disturbances in the northwest boreal region, including extreme weather, fire, permafrost thaw, and soil erosion.

DEFINING CUMULATIVE EFFECTS AND IMPACTS

Cumulative *effects* are the sum of changes to the environment resulting from past, present, and future actions. In turn, cumulative *impacts* represent the consequences of those changes, which often are described in terms of environmental or socioeconomic values. Cumulative impacts can result from the additive and synergistic effects associated with anthropogenic activities while taking into account the natural forces that shape the landscape (Duinker and others, 2013). At the landscape scale, cumulative effects can be more readily estimated as the footprint of human and natural drivers of change. Impacts, however, are a values-based construct and they can vary across ecological scales. For instance, a population or ecosystem process may be unaffected by actions that occur infrequently or over a small area, but when those same actions occur across entire landscapes, the impacts may be significant (Johnson and St-Laurent, 2011). Furthermore, a time lag or nonlinearity in system response can have a significant role in determining the impact of many small aggregate effects.

Across the NWB region, differences in how ecosystem services or biodiversity are valued by human communities should be expected. For example, some communities may be seeking growth in their cash economy, whereas others may be seeking to maintain direct harvest from the ecosystem, such as subsistence hunting. The levels of risk and resilience of the various components of the ecosystem also will vary depending on sensitivity to the drivers of change as well as the magnitude of current and future change. Without a framework to identify and formally recognize these values, it is not possible to ensure that the various elements and ecosystems of the northwest boreal region will remain unthreatened by the cumulative effects of change. Such a framework is important for projecting, discussing, and defining a vision for levels of change, and

Table 6.2.1. Measures and associated thresholds for assessing cumulative impacts for terrestrial and avian wildlife species in the Northwest Boreal (NWB) region.

[Partly after Axys (2001). Although measures were phrased as potential thresholds for monitoring the distribution and abundance of key species or limiting the negative effects of human development and activity, few empirical stop-points were reported. Further research is required to identify an empirical threshold for each of these measures (see Johnson, 2013). This work (Axys, 2001) illustrates the relative ease of identifying measures for monitoring cumulative effects (for example, road density) and impacts (for example, habitat change), but the difficulty in deriving empirical thresholds or stop points for the range of species found across the NWB.]

	Habitat measures	Population measures	Land and resource practices
Grizzly Bear	**Habitat effectiveness** Measure: Predict realized habitat quality as a function of seasonal forage and human disturbance Threshold: Minimum habitat effectiveness by planning area	**Mortality** Measure: Monitor human-caused mortality Threshold: Maximum human-caused mortality before high risk of population decline	**Road density** Measure: Regional road density Threshold: Maximum road density across planning units before high risk of population decline **Human visitation** Measure: Seasonal activity/visitation rates of humans Threshold: Maximum visitor activity in planning units
Woodland Caribou	**Habitat availability** Measure: Classify availability of winter habitat according to lichen abundance and human disturbance (zone of influence and disturbance magnitude) Threshold: Minimum habitat availability by planning area	**Population change** Measure: Indices (cow/calf ratio) or direct measures of population change Threshold: Minimum recruitment of caribou to maintain population Measure: Model energetic and reproductive cost of altered movement and foraging patterns related to human disturbance Threshold: Maximum level of disturbance before reproductive potential of female is reduced	**Controls on access and exploration** Measure: Linear road development and area of clearing associated with oil and gas activities Threshold: Maximum linear development and clearing across planning units before high risk of population decline
Land birds	**Habitat availability and configuration** Measure: Landscape fragmentation, size and distribution of forest gaps; linear extent of forest edge Threshold: Minimum patch size or configuration of suitable habitat; size frequency and distribution of forest disturbances; distance to forest edge before key species are known to decline	**Population change** Measure: Assess annual trends (for example, breeding bird survey) in observed numbers of key species Threshold: Maintain range of natural variability of population	**Retention of habitat features for nesting birds** Measure: Protect and buffer cavity nest trees and the repeatedly used nest sites of raptors and owls; riparian and wetland retention zones/buffers; spatial scale and regional distribution of different stages of forest succession within range of natural variability; minimize disturbance. Threshold: No-harvest and no-disturbance habitat retention in minimum-diameter or minimum-width buffer zones (nests and riparian habitats); minimum/maximum limits to cutblock size; maximum limits to cumulative early successional disturbed habitats; minimum density of snags and cavity-recruitment trees in cutblocks

managing future impacts (Gunn and Noble, 2009a; Chapin Carpenter, and others, 2010).

UNDERSTANDING CUMULATIVE IMPACTS

Quantifying the impacts of land use and climate change is a difficult task in that systems are complex and have natural ranges of variability, often at long temporal scales. Ecological monitoring is recognized as one strategy for assessing many small, but cumulative, impacts or subtle trends in environmental change (Squires and others, 2010; Burton and others, 2014).

Advances in the collection and application of large-scale spatial data have allowed for more sophisticated analyses of the long-term supply of resources as well as the corresponding impacts of such developments (Schneider and others, 2003). For many wildlife species, it is possible to quantify the zones of impact associated with multiple disturbance types, relate those zones to habitat degradation (Table 6.2.1), and then build that understanding into decision-making or planning processes (Martin and others, 2009).

In the face of climate change, the analysis becomes an iterative process to account for ecosystem transitions that may differ in scale or speed compared to historical rates of succession and disturbance for boreal ecosystems and which require complex projection models such as the Integrated Ecosystem Model (https://csc.alaska.edu/resource/integrated-ecosystem-model) developed for this region. Such projections facilitate evaluation of the effects of planning decisions beyond short-term interests or conditions (Veldkamp and Lambin, 2001). Further, because of the interdependent effects of some anthropogenic changes (that is, one development often facilitates further development—growth inducing effects), it is necessary to consider multiple scenarios of development to account for uncertainty in what, when, and where changes will occur (Wilson and others, 2013).

MANAGING FUTURE CUMULATIVE IMPACTS

Despite a long history of grappling with the process of evaluating and minimizing cumulative impacts, regulatory approaches often have been criticized (Ross, 1998; Duinker and Greig, 2006; Gunn and Noble, 2011). Cumulative effects must be considered during the assessment of individual large developments with the potential for significant impacts. This still excludes many small projects where impacts can act cumulatively to have large impacts, the metaphorical "death by a thousand cuts" (Johnson, 2011). Further, regulatory environmental assessment generally lacks a *quantitative* framework to consider future developments that result in multiple disturbance drivers across landscapes. Even when plausible development actions are considered together, there often are no well-established or enforceable targets to guide a sustainable level of development (Johnson, 2013).

BOX 6.2.1: CUMULATIVE IMPACTS AND THE DECLINE OF CARIBOU IN BOREAL REGIONS

Caribou (*Rangifer* spp.) is an important keystone species from both a cultural and an ecological perspective (Gau and others, 2002; Garibaldi and Turner, 2004). The northwest boreal region is fortunate to retain populations of both barren-ground (*Rangifer tarandus groenlandicus*) and woodland caribou (*Rangifer tarandus caribou*) that appear to be within their natural range of variation in both distribution and abundance. This is not the case throughout the southern and central mountains of British Columbia and much of the boreal across the remainder of Canada (Festa-Bianchet and others, 2011).

The drivers of landscape change for caribou are both obvious and subtle, but certainly cumulative. Depending on location, a combination of forest harvesting, oil and gas exploration and development, mining, and recreation has resulted in less habitat and altered predator-prey communities (Fig. 6.2.1) (Festa-Bianchet and others, 2011; Latham and others, 2011). A changing climate will potentially disrupt migratory patterns related to plant phenology, limit access to winter

Figure 6.2.1. Aerial view of a network of oil and gas exploration and development infrastructure, including seismic lines, well pads, pipelines, and compressor stations, in the range of the boreal caribou. Linear features provide enhanced networks for predators such as the grey wolf (*Canis lupus*) to search for and detect prey such as caribou. Woodland caribou in this landscape are in rapid decline (Johnson and others, 2015). Photograph by Dale Seip, British Columbia Ministry of Environment and Climate Change Strategy.

forage, and result in the introduction or greater prevalence of parasites and disease (Brotton and Wall, 1997; Post and Forchhammer, 2008; Yannic and others, 2014). The negative relationship between caribou presence and human activities is long-standing, but it is now becoming more obvious as herds continue to disappear (Johnson and others, 2015). Empirical evidence demonstrates a strong south to north cline in conservation risk and population decline (Schaefer, 2003; van Oort and others, 2010).

For many locations in southern Canada, conservation is dependent on expensive or high-risk activities such as maternal penning, population augmentation, and the cessation of Aboriginal harvest. The cumulative impacts of human activities are now too deeply engrained in those landscapes to simply rely on habitat restoration or caribou-friendly resource management practices (Table 6.2.1). The socioeconomic and natural environment across the NWB is different—there are still opportunities to maintain large and resilient populations of caribou.

However, as the decline of the species across the south has demonstrated, the time to act is now. Landscape planning for the persistence of caribou and other wide-ranging species must occur while there are opportunities to make low-cost decisions that support cultural, subsistence, and conservation values (Timoney and Lee 2001).

Regional strategic environmental assessment (R-SEA) is one method for managing cumulative effects and impacts at regional scales. Inherently flexible, R-SEA is "designed to systematically assess the potential environmental impacts including cumulative effects, of alternative strategic initiatives, policies, plans, or programs for a particular region" (Canadian Council of Ministers of the Environment, 2009, p. 6). A regional strategic environmental assessment is designed to improve the overall management of cumulative effects in the context of sustainability, increase the effectiveness of project-specific assessments, and identify strategic directions and objectives for regional areas relative to potential future cumulative impacts (Gunn and Noble, 2009a).

In Canada, R-SEAs are not a regulatory requirement; most jurisdictions have project-specific environmental impact assessment regulatory regimes. The defining elements of regional strategic environmental assessment have been put into practice with limited success in Canada (Gunn and Noble, 2009b), but government-mandated regional land use planning can cover some of the same goals. For example, the North Yukon Regional Land Use Plan has defined cautionary and critical levels of both linear density and surface disturbance from human infrastructure to direct and control intensity of hydrocarbon exploration and development, and subsequent activities, in a taiga/tundra region (Francis and Hamm, 2011). There also are proposals for the application of regional strategic environmental assessment to northern regions with similar challenges (Chetkiewicz and Lintner, 2014).

Lessons learned from observations of cumulative impacts in other landscapes suggest that management challenges will only increase as development continues across the NWB region and climate change moves systems outside their historical range of variation (see Box 6.2.1, "Cumulative Impacts and the Decline of Caribou in Boreal Regions"; Aumann and others, 2007; Parlee and others, 2012). Given the political realities of this area—including multiple levels of governments, a wide range of conservation values, and many drivers of change—effective governance will be a key element for success. For such an approach to succeed, significant leadership will be needed, along with a strong demonstration of effective conservation and management actions that address many small drivers of change that may limit the resulting effects across landscapes (Gunn and Noble, 2009b).

INFORMATION GAPS

- Assessment and management of cumulative impacts is needed, ideally before actions are proposed, involving broad-scale and collaborative involvement of governments, community groups, and industry to identify ecologically or culturally important values, ecosystem services, and species.
- Precise and repeatable measurements are needed that can identify and monitor important species or ecosystem services that are vulnerable to cumulative impacts
- Thresholds are needed to accompany measures of effect and impact, although such thresholds often are imprecise and adapted to socioeconomic realities suggesting that they are best labeled as regulatory or management limits.

7
PRACTICES OF COPRODUCTION

By Dawn R. Magness[1]

INTRODUCTION

This is a time of rapid change across the Northwest Boreal (NWB) region. The problems that need to be addressed span multiple scales, including global issues such as climate change. Over the past 50 years, humans have changed ecosystems faster and across larger scales than any other time in history, leading to gains in net human well-being and economic development often at the cost of ecosystem degradation and further impoverishment of certain populations (Millennium Ecosystem Assessment, 2005). New approaches are needed to address these problems that can be complex, interdisciplinary, and multijurisdictional. To help solve and implement solutions for complex environmental problems, an integrated understanding of science, policies and political process, institutions, and the values and knowledge of diverse stakeholders may be needed (Millennium Ecosystem Assessment, 2005; Chapin, Carpenter, and others, 2010; Clark, 2011).

Coproduction is the interplay between knowledge and decisions in adaptive governance (Wyborn, 2015). Action in a multi-jurisdictional landscape with diverse stakeholders will require coproduction to align a collective vision of what should and can be done. Adaptive governance is a term applied to flexible, collaborative decision-making that links social, political, economic, and ecological systems and allows for an iterative learning process. Adaptive governance structures are more likely to adjust to new situations and react more nimbly to surprises.

In this chapter, the focus is on practices that may assist in the likelihood of coproduction of knowledge and decision-making across the Northwest Boreal region. Communities will need to be engaged in different ways (Chapter 7.1), and scientists and managers will need to take on different roles (Chapter 7.2) in the production and use of knowledge. Directional change will require new assessment tools to facilitate adaptive learning (Chapter 7.3). How these data (Chapter 7.4) and other outreach

1 U.S. Fish and Wildlife Service, Kenai National Wildlife Refuge, Soldotna, Alaska, USA.

efforts are packaged and disseminated from the NWB Landscape Conservation Cooperative (LCC) (Chapter 7.5) also can facilitate coproduction.

7.1
MEANINGFULLY ENGAGING COMMUNITIES

By Annette Watson[1] and David Natcher[2]

KEY FINDINGS

- Trust is the fundamental condition of meaningful collaboration.
- Research relationships between scientists and communities historically have been premised on colonial power relationships that often have caused Indigenous communities to mistrust researchers and research products.
- Traditional Ecological Knowledge (TEK) or Indigenous Knowledge (IK) systems operate through different understandings of the relationship between humans and non-humans than assumed in many Western science traditions, with effects on research questions and understanding causal relationships.
- Allowing space in the research procedure for community input to shape the research questions and the agenda is an important element of a meaningful research relationship.
- "Iterative" research designs are more rigorous because they allow for refining research questions and methods to best serve the communities being engaged.
- The differences between research-community relationships in Alaska and Canada stem from differences in the land claims processes and governing bodies created to administer to the needs of Alaska Natives and First Nation, Metis, and Inuit communities.
- Research budgets should ensure that communities fiscally benefit from the process of research, at individual and tribal and Indigenous scales.

1 Department of Political Science, College of Charleston, Charleston, South Carolina, USA.
2 Department of Agriculture and Resource Economics, University of Saskatchewan, Saskatoon, Saskatchewan, Canada.

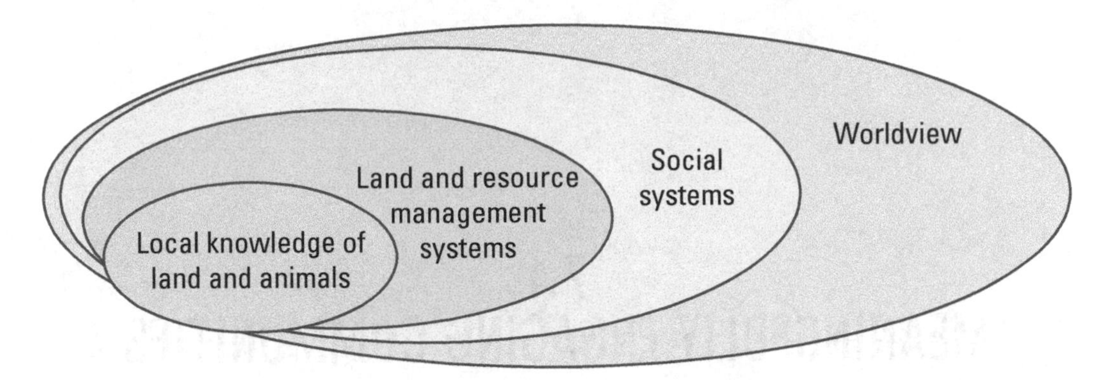

Figure 7.1.1. Knowledge and management strategies embedded within worldviews. Although this figure was created to show the levels of analysis in traditional knowledge systems, not only is Traditional Ecological Knowledge founded in its assumptions about worldview, all knowledge systems connect their cultural worldview to their systems of land and resource management. From Berkes (1999, p. 13).

INTRODUCTION

Recent research from social and natural sciences has shown that to engage Native or Aboriginal communities in meaningful ways, researchers can no longer merely "deliver" information and products to local peoples. Although some research "delivered to" tribes, such as the results stemming from environmental contaminant studies, have provided necessary data to support various claims by tribes, the history of science is replete with examples of how academic research has been used to marginalize and disempower Indigenous peoples. Social scientists have empirically documented how Western scientific disciplines are a product of culture and politics, rather than objective witnesses of the physical and social world (Haraway, 1988; Nader, 1996; Harding, 1998; Bravo and Sörlin, 2002). Many have argued that resource management sciences assume a false division between facts and values (Jasanoff, 1995), and have been used as a tool by Western colonial powers to control the resources and lands historically occupied, used, and cared for by Indigenous peoples (Howitt, 2001).

For many Indigenous communities, research is viewed as a colonial intrusion into their daily lives, and communities in Alaska and Canada have grown resentful of being the "subjects" of research rather than partners (Tuhiwai Smith, 2012). Numerous examples exist of the conflicts that occur when researchers make claims of scientific objectivity, and then deny the value of Indigenous management practices and values based on TEK (Nadasdy, 2003; Andersen and Nuttall, 2004; Watson, 2013) (Fig. 7.1.1 and Table 7.1.1). Because Indigenous peoples were rarely involved in the formulation of research questions or the interpretation of research results, scientific work has generated considerable mistrust of what is perceived to be unethical and unjust research practices.

Table 7.1.1. Differences between the underlying values of Western Studies and Traditional Ecological Knowledge.

Western studies	Traditional ecological knowledges
Spirituality centered in a single Supreme Being	Spirituality embedded in all elements of the cosmos
Humans exercise dominion over nature to use it for personal and economic gain	Humans have responsibility for maintaining harmonious relationships with the natural world
Spiritual practices are set apart from daily life	Nature is honored routinely through daily spiritual practice
Human reason transcends the natural world and can produce insights independently	Wisdom and ethics are derived from direct experience with the natural world
Universe is made up of an array of static physical objects	Universe is made up of dynamic, ever-changing natural forces
Universe is compartmentalized in dualistic forms and reduced progressively smaller conceptual parts	Universe is viewed as a holistic, integrative system with a unifying life force
Time is a linear chronology of "human progress"	Time is circular with natural cycles that sustain all life
Nature is completely decipherable to the rational human mind	Nature will always possess unfathomable mysteries
Human thought, feeling and words are formed apart from the surrounding world	Human thought, feelings, and words are inextricably bound to all other aspects of the universe
Human role is to dissect, analyze, and manipulate nature for own ends	Human role is to participate in the orderly designs of nature
Sense of separateness from and superiority over other forms of life	Sense of empathy and kinship with other forms of life
View relationship of humans to nature as a one-way, hierarchical imperative	View proper human relationship with nature as a continuous two way, transactional dialogue

DEFINING THE RELATIONSHIP BETWEEN COMMUNITIES AND RESEARCHERS

With the political empowerment of Indigenous peoples, the relationship between communities and academic researchers has undergone considerable change. This has been particularly apparent in the Northwest Boreal (NWB) region of Alaska and Canada where comprehensive land claims have been settled with Alaska Native and First Nation governments. In 1971, the first comprehensive land claim in North America was signed between Alaska Natives and the United States government. In 1971, the Alaska Native Claims Settlement Act (Public Law 92-203 [85 Stat. 688]) also set into motion a political transformation across Canada that has resulted in the settlement of 26 comprehensive land claims, including 11 in the Yukon. Although the political empowerment of Indigenous peoples in Alaska and Canada has allowed for the recasting of government-to-government relationships, this political resurgence has led to a reorientation of how research is done in northern Indigenous communities. No longer excluded from much of research being conducted, Indigenous peoples and their representative governments have assumed their rightful place as research collaborators, facilitators, and administrators. Indigenous governments in Alaska and Canada now have the authority to approve, reject, and set the conditions for how research will be done within their communities and settlement lands.

For example, the Council of Yukon First Nations is explicit about the relationship between Aboriginal self-government and research by noting that any researcher hoping to work on First Nation Settlement Land must first obtain the permission of the respective First Nation. Similarly, all research proposals concerning the territory of the Tetlit Gwich'in (northeast Yukon) must first be approved by the Gwich'in Tribal Council, as specified in their land claims agreement. In addition to describing the objectives and research methodology, researchers can benefit from the local knowledge of the Tetlit Gwich'in people that they can provide.

Various academic institutions have formally recognized the role of Indigenous peoples in these reformulated research protocols (Indigenous Peoples Specialty Group, 2010; National Science Foundation, 2013). Both governmental research initiatives (for example, Miraglia, 1998; Berkes and others, 2006) and Native institutions have emerged that provide key guidance on how to develop meaningful research engagements with communities. A basic protocol is articulated by the Alaska Federation of Natives (see Box 7.1.1, "Alaska Federation of Natives Guidelines for Research"). Scholars across the social and natural sciences have responded to these changes by engaging Indigenous and other local communities as equal partners in the research. This includes opportunities for communities to help determine the purpose, methodologies, and benefits of proposed research (Louis, 2007; Chapin and others, 2013; Cochran and others, 2013; Watson and Huntington, 2014).

DIFFERENT ASSUMPTIONS

Meaningful engagement has not been easy or unproblematic. Because Western sciences and "Native Science" (Cajete, 2000) assume different metaphysical universes, understanding contrasting notions of the physical environment, and humans' proper relationships within that environment, has been challenging. Some scholars have defined TEK as a "sacred ecology" and emphasize its holistic perspective compared to the more reductionist scientific method (Kawagley and Barnhart, 1998; Berkes, 1999; Kawagley, 2006). Others have articulated how Western sciences are based on the assumptions of "Enlightenment humanism," whereas TEK predates the Enlightenment—and therefore, the philosophies that ground these ways of knowing are fundamentally different (Watson and Huntington, 2008). Table 7.1.1 summarizes different assumptions and values. See Box 7.1.2, "Similarities and Differences between Western Sciences and Traditional Ecological Knowledges" for similarities and differences between Western sciences and TEK.

One often-cited difference is the understood relationship between humans and nonhumans, which affects management procedures and actions. Enlightenment philosophy forms the basis of the scientific understanding of the relationship between

BOX 7.1.1: ALASKA FEDERATION OF NATIVES GUIDELINES FOR RESEARCH

The Alaska Federation of Natives conveys to all scientists and researchers who plan to do studies among Alaska Natives that they must comply with the following research principles:

- Advise Native people who are to be affected by the study of the purpose, goals, and timeframe of the research, the data gathering techniques, the positive and negative implications and effects of the research.
- Obtain informed consent of the appropriate governing body.
- Fund the support of a Native Research Committee appointed by the local community to assess and monitor the research project and ensure compliance with the expressed wishes of Native people.
- Protect the sacred knowledge and cultural/intellectual property of Native people.
- Hire and train Native people to assist in the study.
- Use Native languages whenever English is the second language.
- Guarantee confidentiality of surveys and sensitive material.
- Include Native viewpoints in the final study.
- Acknowledge the contributions of Native resource people.
- Inform the Native Research Committee in a summary and in nontechnical language of the major findings of the study.
- Provide copies of the study to the local people.

humans and "animals"; humans act because they are able to think and learn; animals act due to behaviors and responses to stimuli. Aboriginal peoples, by contrast, maintain notions of "relational sustainability" (Langdon, 2006) where a "moral charter" (McIntosh and others, 2000) is observed with animals that gave rise to specific cultural behaviors exhibited through ceremony and practice (Dale and Natcher, 2014). Many Native cultures believe that animals "give themselves" to respectful hunters, and these animals can think, learn, and react in response to human behavior and to environmental changes (Salmón, 2000; Anderson and Nutall, 2004; Wenzel, 2004; Natcher and others, 2007; Watson and Huntington, 2008; Watson, 2013). These different assumptions about human and nonhuman action have implications to the causal systems that a research project may assume, in addition to the selected research methods, analysis of results, and subsequent management actions (see Box 7.1.3, "Ways of Knowing and Managing Land and Animals," and Figs. 7.1.1 and 7.1.2).

Therefore, although Table 7.1.1 outlines some key philosophical differences between Enlightenment-based and IK systems, neither TEK nor Western sciences are static, and newer interpretations are continually emerging that may more fruitfully

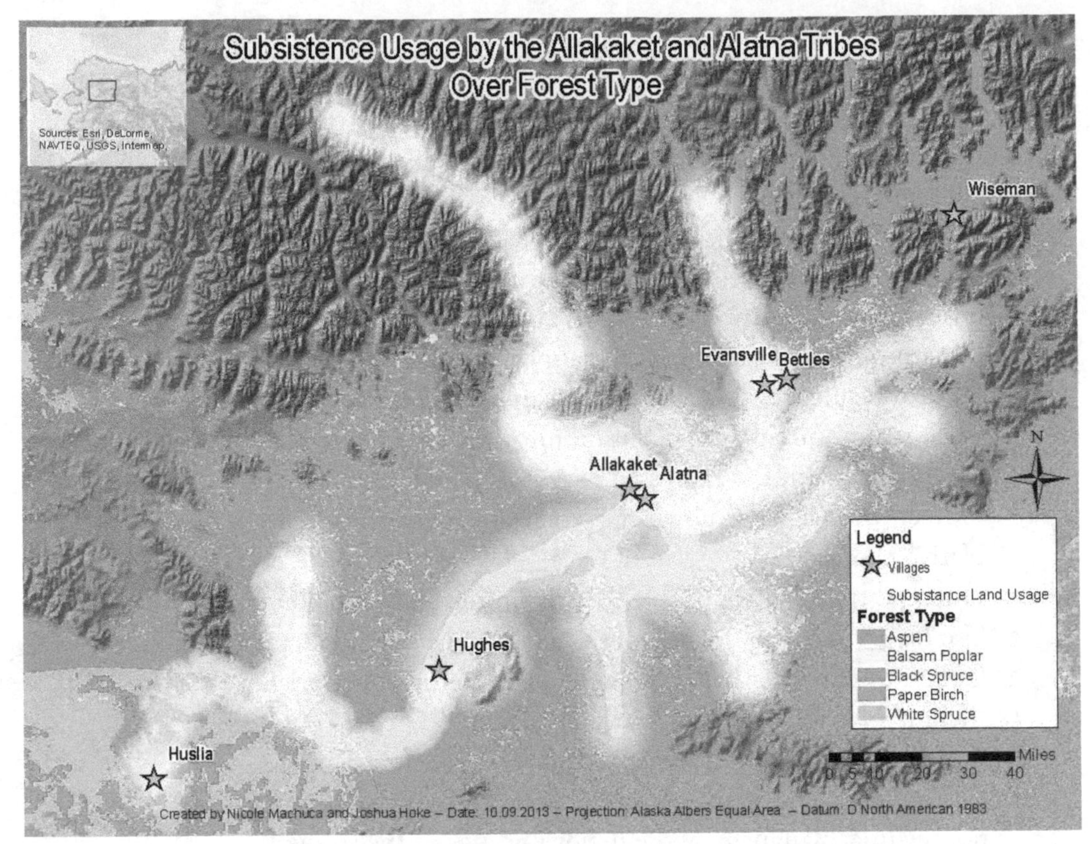

Figure 7.1.2. Subsistence use of the Allakaket and Alatna Tribes and forest diversity. Map shows two layers, one showing forest type, and the other showing the primary areas of subsistence use by these two tribes. During the "community review," researchers noted that people tended to practice subsistence where the forest was most diverse—but causally, according to elders, the forest also may be more diverse where people practice subsistence, because the act of subsistence is the "taking care" of the land, including actions that would increase forest diversity (from Watson and others, 2014).

bridge Western sciences and TEK. What is interesting to note when examining Table 7.1.1 is that recent theories of "social-ecological systems" do not necessarily fit the characteristics that outline "Western Knowledges." Not only are these theories attempting to conceptualize dynamic systems that do not separate the social from the natural, they also recognize nonlinear change (Scheffer and others, 2001; Gunderson and Holling 2002; Walker and others, 2004). Further, scholars are beginning to explore other approaches in the environmental sciences, including "fuzzy" logic and complexity theory, which seem to hold some similar assumptions to TEK (for example, Berkes and Berkes, 2009). In these ways, from the social and natural sciences, there may be many avenues to align the interests, methods, and knowledges of the science community and local Indigenous communities.

BOX 7.1.2: SIMILARITIES AND DIFFERENCES BETWEEN WESTERN SCIENCES AND TRADITIONAL ECOLOGICAL KNOWLEDGES

Common to Western sciences and TEK is that not all people are considered experts in their respective knowledge systems. Although expertise in Western sciences is affirmed through formal education and advanced degrees, expertise in Indigenous societies is limited to elders, subsistence practitioners, and recognized culture-bearers. This guides the methods used to elicit TEK; for example, a random sample of Indigenous peoples will not procure the traditional knowledge of the culture, just as a random sample of non-Natives will not produce clear understanding of scientific knowledge. This necessitates including various qualitative social science techniques in addition to "hard" sciences and research methods used to understand the TEK of ecosystem processes. The field of Indigenous studies also recently emerged to provide a theoretical base to understand TEK, and innovations within qualitative methods have evolved to engage with TEK (Kovach, 2010; Tuhiwai Smith, 2012; Watson and Huntington, 2014).

BRIDGING TRADITIONAL KNOWLEDGE AND WESTERN SCIENCE

Notwithstanding the challenges, many Indigenous governments and communities recognize the importance of bringing traditional knowledge and Western science together to inform environmental decision-making. The Council of Athabaskan Tribal Government's Statement on Research Protocols is particularly illustrative of this approach: "Through an Indigenous Research Paradigm, and a community participatory approach, the Council of Athabaskan Tribal Governments demonstrates its self-governance through the use of science and local knowledge that together guides the management of natural and cultural resources in the Traditional Territory of the Gwich'in Nation."

The model science and information exchange within the NWB region is ongoing and aims to bridge TEK with western sciences and resource management with the goal that communities and end users of information be part of a collaborative knowledge generating process (knowledge coproduction). Social and natural scientists in the boreal region have been actively building bridges of understanding between Natives and non-Natives, managers, and the public through new models of research.

These newer conditions for engaging in research with tribal and Indigenous communities should be viewed not as obstacles but rather as opportunities for more meaningful collaboration. This has been the case for researchers working with the Vuntut Gwitchin in Old Crow, Yukon (see Box 7.1.4, "Yeendoo Nanh Nakhweenjit K'atr'ahanahtyaa (YNNK) or Environmental Change and Traditional Use of the Old Crow Flats in Northern Canada") as well as for researchers and institutions collaborating

BOX 7.1.3: WAYS OF KNOWING AND MANAGING LAND AND ANIMALS

To non-First Nation representatives of the Carmacks Renewable Resource Council (CRRC), their primary role is to protect and allocate wildlife resources between competing interests. Adhering to a utilitarian worldview, non-First Nation representatives feel that uncertainty, either ecological or social, can be overcome through effective planning. Mediating their relationship to the environment through resource inventories, wildlife assessments, management programs, policy initiatives, and administrative procedures, non-First Nation representatives may feel distinctively removed from some environmental decisions. The result is a centralized approach to management that compartmentalizes individual categories of the environment (that is, fisheries, wildlife, and forests). This ideological position is further reflected in Chapter 16 of the Little Salmon Carmacks First Nation Final Agreement (Minister of Indian Affairs and Northern Development, 1998), where the CRRC has been given the mandate to make recommendations concerning the anticipated harvest of moose and woodland caribou (provision 16.6.10.12), the management of fur-bearing species (provision 16.6.10.7), the content of wildlife management plans (including harvesting plans and total allowable harvests [provision 16.6.10.1]), and the allocation and harvest of commercial salmon (provision 16.6.10.3) (Minister of Indian Affairs and Northern Development, 1998). Supporting each of these provisions is a fundamental ideology that remains firmly grounded in the western intellectual tradition.

In contrast, the relationship that First Nation representatives share with the environment is one manifestly bound in shared norms and customs. Mediated through a form of traditional law referred to locally as Doo'Li, First Nation members adhere to a moral system that is used to govern an individual's relationship with the environment. Based on reciprocity and exchange, this moral conception of the environment is articulated through rules and codes of conduct that remain grounded in Northern Tutchone culture. Rather than an overt form of environmental management, Doo'Li is a means by which social relationships, both human and non-human, are maintained and respected. Although challenged by a century of colonial institutions—governance, health, education—First Nation members continue to observe many of the traditional laws that govern their relationship to the land" (Natcher and others, 2005, p. 245).

with the tribes of Allakaket and Alatna, Alaska (see Box 7.1.5, "Traditional Ecological Knowledge of Moose and Climate Change in Allakaket/Alatna, Alaska"). When communities are engaged in the generation of information, it not only increases the

BOX 7.1.4: YEENDOO NANH NAKHWEENJIT K'ATR'AHANAHTYAA (YNNK) OR ENVIRONMENTAL CHANGE AND TRADITIONAL USE OF THE OLD CROW FLATS IN NORTHERN CANADA

Through this project, a team of researchers, representing the Vuntut Gwitchin, Yukon Department of Environment, Parks Canada, and a multidisciplinary team of southern-based researchers, are exploring how the complexities of climate change are affecting the Old Crow Flats. Included among the many outcomes of this project is the codevelopment of a long-term environmental monitoring program for the Old Crow Flats that is based on Western science and Gwitchin traditional knowledge. As noted by the project participants, this form of collaboration represents a "new research paradigm in northern Canada—one that is collaborative, interdisciplinary, reflective of northern priorities, and policy-oriented" (Wolfe and others, 2011).

chances that the information will be applied, it incorporates diverse knowledge that will contribute to understanding ecosystem processes and change over time.

Key Factors to the Success of the YNNK Experience: (1) community research experience and capacity, (2) community consultation prior to proposal writing, (3) legitimate convergence of community priorities and researcher interests, (4) funding agency guidelines that reward innovative but costly approaches to community-based research, (5) ongoing communication between researchers and the community at all project stages, (6) planning for contingencies, including the timing of funding deadlines, (7) informal social networks that populate the northern research landscape, and (8) personal relationships and trust that emerge from working closely together on shared interests (Wolfe and others, 2011).

Meaningful collaboration can be achieved through both administrative and intellectual responsibilities by researchers and communities; the practice of research to achieve this is often referred to as "iterative" in the social sciences (see Boxes 7.1.4, "Yeendoo Nanh Nakhweenjit K'atr'ahanahtyaa (YNNK) or Environmental Change and Traditional Use of the Old Crow Flats in Northern Canada," and 7.1.5, "Traditional Ecological Knowledge of Moose and Climate Change in Allakaket/Alatna, Alaska"). Although in "hard" sciences the scientific method dictates that the researcher design research to test a hypothesis, with a carefully orchestrated beginning, middle, and end of an experiment or protocol, community-based work requires constant revisiting of assumptions and nuanced understanding of the contexts within which the discussions take place (for example, Huntington and others, 2006; Watson and Huntington, 2014). Iterative procedures to check the assumptions of the researchers, institutions, and the community participants will only bring more rigor to a process—it is better

BOX 7.1.5: TRADITIONAL ECOLOGICAL KNOWLEDGE OF MOOSE AND CLIMATE CHANGE IN ALLAKAKET/ALATNA, ALASKA

This project (Watson and others, 2014) was conducted through collaboration between the National Park Service (NPS) (the funding organization) and the Alaska Native regional non-profit Tanana Chiefs Conference (TCC), who facilitated the working relationship between the tribes and researchers. The purpose was to understand subsistence use in the area, as well as to collect local and traditional knowledge of climate change and moose ecology as it relates to wildlife management for the area. Important to the success of this collaboration was the involvement of the regional non-profit organization. Staff from TCC as well as the researchers visited these tribes in multiple trips, first, to make the formal introduction of the research project, and second to collect feedback about the project to be incorporated in the research design and project budget. Staff also participated in fieldwork on subsequent data-collection trips and the presentation of results to the communities. During each of these trips, tribal leaders and community members could contribute ideas for the use of this project to meet their needs. In this case, the communities desired to participate in documenting their subsistence patterns because of a planned development project that could affect their subsistence, and thus charged the researchers with collecting additional data that could contribute to a future cultural and economic impact statement, as well as environmental impact statements. Significantly, the "presentation of results" was not the end of the project, but a final community review—an opportunity for the tribes to "check" the analysis by the researchers (Fig. 7.1.2). Each visit was an opportunity for the researchers to be transparent about the ways that the data will be used in land management—*before* written results were finalized. Such procedures should not be seen as a practice of "censorship" by tribes, but as a way to build mutual understanding and trust between collaborators.

to work through potential conflicts at an early stage of research, not after researchers publish their results.

Common to these examples is a reformulation of the political economy of research monies; all research partners need the fiscal capacity to work together. However, the capacity to administer research protocols on the part of tribes and other Native-run institutions has been inconsistent; in the United States as a whole, compared to Canada, there is less funding of offices or roles within tribal administration for working with researchers. In these ways, it is apparent that the land claims processes in Canada and the United States have produced different capacities for tribes to work with researchers.

The transformation in research relationships is not universal or complete. Rather, depending on a range of conditions, including the capacity of small communities, institutional inertia, and intellectual biases that continue to prioritize western scientific knowledge over traditional knowledge systems, there remains considerable work to be done. Nonetheless, positive changes are being made, and collaboration between researchers and communities is becoming the norm rather than the exception.

INFORMATION GAPS

- Researchers need to assess whether tribal and other Indigenous organizations have the fiscal capacity to administer research collaborative relationships.
- Traditional management practices and TEK in the different areas of the NWB region need to be understood prior to developing research questions.
- Funding cycles and grant-writing procedures need to be reformulated to encourage iterative research designs, knowledge of coproduction, and to reflect the time it takes to do rigorous community-based work.
- Researchers should carefully align their theories with the philosophical assumptions of the local culture; for example, research on non-linear social-ecological systems often is more compatible with TEK than research assuming linear change. Yet more work is needed to bridge these approaches to knowledge of ecosystem management.

7.2
EVOLVING ROLES OF SCIENTISTS AND MANAGERS

By Dawn R. Magness[1] and Douglas A. Clark[2]

KEY FINDINGS

- Many professional ideals predominant in natural resource management, such as historical condition, naturalness, the "scientific management" approach, and ecological integrity, cannot be assumed to be viable management goals under directionally changing social and environmental conditions.
- New values that are compatible with managing future conditions link ecological and social systems and aim to include multiple perspectives. These values include human dignity, human well-being, adaptive capacity, resilience, and transformability.
- Scientists will need to engage in iterative collaborations with managers, communities, and other stakeholders to coproduce scientific knowledge that is relevant to the problem context.
- Managers will need to engage in participatory processes that help to balance and integrate multiple interests in order to effectively respond to and shape ecological change.

ECOSYSTEM STEWARDSHIP REQUIRES DIFFERENT ASSUMPTIONS

Until recently, natural resource management theories and practices were developed under the assumption of a stable environment (Gunderson and Holling, 2002; Chapin, Carpenter, and others, 2010; Cole and Yung, 2010). Managers were trained to use past conditions (historical baselines) to guide their vision of how the ecosystems should be (Millar and Woolfenden, 1999; Cole and Yung, 2010; Magness and others, 2012). In a changing world, ecosystem stewardship has emerged as a new paradigm for ecosystem and natural resource management. Ecosystems are now thought to be path dependent, directional, and always changing (Gunderson and Holling, 2002; Walker and others, 2004). Some ecosystem states (such as those that are robust or resilient) can

1 U.S. Fish and Wildlife Refuge, Kenai National Wildlife Refuge, Soldotna, Alaska, USA.
2 School of Environment and Sustainability, University of Saskatchewan, Saskatchewan, Canada.

occur—and more than one may exist for a given system—but are subject to ecological reorganization after critical thresholds have been surpassed (transformation; Scheffer and others, 2001; Walker and others, 2004; Chapin, Carpenter, and others, 2010; Walker and Salt, 2012). Therefore, management goals such as historical condition, which presume environmental stability, are no longer viable (Millar and Woolfenden, 1999; Ruhl, 2008; Camacho, 2009; Garmestani and Allen, 2014). Pursuing such stasis-oriented goals in a changing environment can foreclose future options for adaptation and cause actual damage (Adger and others, 2011).

To understand path-dependent ecological change, the emerging field of sustainability science has refined the interpretation of human and natural systems differently (Gunderson and Holling, 2002; Millennium Ecosystem Assessment, 2003). In the past, many management practices considered social systems as separate from natural systems (see Box 7.2.1, "Values that Inform Management Approaches"). For example, the concepts of naturalness and ecological integrity use the absence of human influence as the measure of management success for parks and protected areas (Anderson, 1991; Clark and others, 2008; Cole and Yung, 2010; Magness and others, 2012). As humans alter climate and other global systems, most ecosystems can no longer be considered outside the sphere of human influence (Vitousek and others, 1997). Many scientists assert that we have entered the Anthropocene, a new geological epoch where humans are the most significant driver of Earth's systems (Steffen and others, 2007). Additionally, to manage and guide future conditions, human choices must be explicitly considered because multiple ecological trajectories and outcomes may be possible (as opposed to one historical condition), including novel system states (Chapter 4.2). Management actions may tip the balance toward a favored outcome, and these choices will affect the resources available to local communities (Chapin and others, 2009; Hobbs and others, 2013). Recently advanced ideas for managing ecological trajectories embrace change and uncertainty, explicitly link human and ecological systems, and include stakeholders in collaborative learning and decision processes (see Box 7.2.1, "New Values Aligned with Ecosystem Stewardship").

HOW DOES ECOSYSTEM STEWARDSHIP CHANGE SCIENTIFIC AND MANAGEMENT PRACTICES?

Managing for future ecological conditions and novel ecosystem conditions is a new challenge for natural resource managers and scientists. The traditional role of the researcher has been to transfer information about past or current ecological conditions to managers who take action to maintain or restore those conditions (Chapin, Carpenter, and others, 2010). Making management choices about future conditions requires some understanding of the range of possible futures, and these futures often are dictated as much by social processes as ecological processes. Furthermore, it is not clear who should decide which future is preferred. The need to understand social drivers and ensure community autonomy when choosing between alternative visions

BOX 7.2.1: NEW VALUES THAT INFORM MANAGEMENT APPROACHES

Human Dignity: As many people as possible participate in and benefit from the shaping and sharing of essential values, signified by widespread experiences of well-being. This requires a healthy biophysical environment (Mattson and Clark, 2011).

Human Well-Being: "The quality of life in terms of material needs, freedom and choice, good social relations and personal security" (Chapin, Carpenter, and others, 2010, p. 241).

Adaptive Capacity: The capacity of human and ecological systems to respond to, create, and shape variability and change in the state of the system and maintain ecosystem services is important. Adaptive capacity generally is enhanced by diversity, continued learning, flexibility, and creating or maintaining opportunity (Chapin, Carpenter, and others, 2010).

Resilience: The ability of a social-ecological system to cope with change while retaining its structure and functions (Walker and Salt, 2012).

Transformability: "The capacity to create a fundamentally new system when ecological, economic, and/or social conditions make the existing system untenable" (Walker and Salt, 2012, p. 216).

Values Aligned for Managing with Stable Ecological Conditions

Historical Condition: Composition, structure, and function, including natural variability, of an ecosystem, constructed for a past reference period that is considered to have had less human influence, such as the Medieval Warm Period or Pre-European settlement (Millar and Woolfenden, 1999).

Naturalness: Ecosystem composition, structure, and function in the absence of human influence (Anderson, 1991).

Ecosystem Integrity: Although numerous formalized definitions exist, most simply reflect maintaining the ecological structures and functions that would occur in the absence of human influence, presuming that ecosystems are static and that all human activities, except management interventions, are stressors (Clark and others, 2008).

of the future will require participatory processes. In the changing world, scientists need to move from single-way communication and mechanistic policy implementation to full collaboration through adaptive management and colearning (Chapin, Carpenter, and others, 2010). Likewise, the role of the resource manager is shifting from that of the decision maker who uses management actions to restore past conditions or ensures that current conditions are sustainable to a facilitator who engages stakeholder groups and works across professional and cultural boundaries to make choices about how to respond to, and shape, social-ecological change (Chapin, Carpenter, and others, 2010).

Another challenge is the complex social-political context in which natural resource management unfolds. "Scientific management" is a term applied to the central organizing concept in natural resource management for more than 100 years. This concept essentially presumes that science is an objective, neutral source of reliable insights that has been implemented by top-down, centralized bureaucracies (Brunner and others, 2005).

Unsurprisingly, scientific management is ineffective at providing solutions to complex problems that have social and ecological dimensions, typically producing conflict and gridlock. Today, most natural resource problems must be addressed in multi-jurisdictional landscapes that include complex social dynamics and multiple, competing value demands that must be reconciled. In this arena, scientific information is never comprehensive and can be "cherry-picked" to justify divergent agendas rooted in different interests and worldviews (Sarewitz, 2004). When the definition of a problem by experts does not match the stakeholder or communities' values, outcomes can include eroded public trust in scientific information or the management regime, and even failure of management initiatives (Brunner and others, 2005; St. Martin, 2006; Watson, 2013; Clark and others, 2014). "Adaptive governance" is the term applied to a new approach emerging in the wake of scientific management's shortcomings that integrates the social and political context with scientific and other expertise (Brunner and others, 2005; Brunner and Lynch, 2010). Adaptive governance is the "evolving and locally context-specific balancing and integration of alternative interests through participatory engagement between governments and communities facilitated by the integration of local and scientific knowledge" (Nelson, Howden, and Smith, 2008, p. 4). For adaptive governance, scientists should iteratively work with managers and communities to co-produce knowledge that addresses meaningful problems, definitions, and solutions. Some scientists and managers believe in an idealized linear relationship between science and policy: that good information or more education will lead to good outcomes. This is a false hope (Sarewitz, 2004; Pielke, 2007). Without iterative collaboration, scientists' assumptions about what information is useful may not match the needs of decision makers (Dilling and Lemos, 2011). Managers need to facilitate discussion between stakeholders with different values and clearly communicate the values of the institutions they represent. Failure to do so can lead to mistrust and ineffectiveness, with negative consequences (see Boxes 7.2.2, "Greater White-Fronted Goose Management: Who Decides What Information Is Important to the Management Regime?," and Box 7.2.3, "The Peel Watershed Land Use Plan: How Not to Manage a Participatory Process").

Taking on these more collaborative and adaptive roles will be out of the comfort zone of many scientists and managers who have been trained to see themselves as value-free, objective, and outside of political and decision-making processes (Mattson and others, 2012). Considerable empirical evidence from the social sciences refutes the notion that any human being—including scientists and managers—can be fully objective or even work outside of a problem context (Sabatier and Jenkins-Smith, 1999; Howitt, 2001; Clark, 2011). All humans have their own personal, cultural, and institutional backgrounds and value systems. In the natural sciences, key worldviews or values are reinforced by training and employer culture (see Box 7.2.1, "New Values Aligned with Ecosystem Stewardship"). Often these underlying values are invisible within science agencies, but can cause mistrust and contested decisions when they are not clearly articulated (Watson, 2013; Clark and Slocombe, 2009; Clark and others, 2014). The values of individuals and institutions, even when they are unstated, influence how problems are defined, and problem definition is central to both research design and the proposed management solutions (Watson, 2013; Chapter 7.1). Therefore, science may be rigorous (for example, transparent and repeatable) but not necessarily objective, because choices about the problem, hypotheses tested, and metrics all flow from what is valued by individuals or institutions. Assuming otherwise is a recipe for frustration, conflict, and professional ineffectiveness.

The challenge for managers and the scientists who support them is not so much to produce "good science," which compels decisions, as it is to discover rational, feasible, and justifiable ways to address the full dimensions of the problems. Therefore, shifting to colearning will require more meaningful and iterative interaction between scientists, managers, and involved communities (Wilkinson and others, 2007; Margerum, 2011): a view supported by reflective managers from across the Northwest Boreal (NWB) region (for example, Smith, 2004; Wolfe, 2006; Urquhart, 2012).

OPPORTUNITIES FOR THE NORTHWEST BOREAL LANDSCAPE CONSERVATION COOPERATIVE

The transition toward ecosystem stewardship is challenging for managers trained in scientific management, both as individuals and within institutions that were organized around previous approaches (Clark and others, 2014). The NWB LCC could serve these agencies and affected communities by building capacity and opportunities to engage in adaptive governance practices such as fostering iterative learning environments and participatory group processes. These processes can be used to align values, which is a necessary step toward facilitating strategic planning at broad spatial and institutional scales.

BOX 7.2.2: GREATER WHITE-FRONTED GOOSE MANAGEMENT: WHO DECIDES WHAT INFORMATION IS IMPORTANT TO THE MANAGEMENT REGIME?

Greater-white fronted geese (*Anser albifrons*) are a transnational species that breed in Alaska and Canada and winter along the Pacific coast of the United States, Texas, and Mexico. In Alaska, greater-white fronted geese are an important food source for subsistence hunters in spring because they are one of the first migrants to come back at a time of year when other subsistence foods are depleted. In 1998, Alaska Natives legalized their subsistence practices through treaty negotiation, which resulted in the creation of the Alaska Migratory Bird Co-Management Council (AMBCC). The AMBCC consists of one U.S. Fish and Wildlife Service (FWS), 1 Alaska Department of Fish and Game, and ten Alaska Native representatives who negotiate recommendations for subsistence harvest regulations. These recommendations are given to the national FWS committee that decides if the recommendations should be passed. The AMBCC is touted as comanagement, but Alaska Natives often express that they feel disempowered and not represented by the process.

Although Alaska Native representatives have democratic access to decision-making, the boundaries of the management problem are not open for discussion. The framing of the comanagement regime is based on the legal obligations of the FWS to protect wildlife populations from overexploitation. With this focus, the global population size and mortality rates attributed to regions with variable hunting regulations are the most important considerations. This focus flows from assumptions that are deeply tied to Western history and cultural identity (Chapter 8, Table 8.1).

Although the considerations of the FWS are valid, other considerations that come from other cultural perspectives are missed with this framing. Alaska Natives believe animals are ontologically the same as humans. Geese choose to give themselves to respectful hunters who kill quickly and take care of the landscape that supports them. Traditional Ecological Knowledge is not just empirical observation to be used as data; it also is an ethical knowledge about how to treat animals. The focus of information for management is more toward "thinking like a goose" to understand where and why they choose to be on the landscape.

At best, the current comanagement system misses opportunities for communities to be more actively engaged with goose management at relevant scales below the global population level. At worst, it gives the illusion of participation while not fully embracing the cultural identity of Alaska Natives.

BOX 7.2.3: THE PEEL WATERSHED LAND USE PLAN: HOW NOT TO MANAGE A PARTICIPATORY PROCESS

From 2004 to 2011, a watershed planning commission made up of the Yukon Government and four First Nation governments worked together to create a land use plan for the Peel Watershed of Yukon in a process mandated by the 1993 Umbrella Final Agreement Between the Government of Canada, the Council for Yukon Indians and the Government of Yukon. The Government of Yukon, however, felt the recommendations for environmental protection were too restrictive for industry, and modified their final recommended plan unilaterally. The recommendations in the original plan seemed to have had significant public support, and this unilateral decision drew fierce criticism. A challenge by two First Nations and two environmental organizations (with a third First Nation as intervenor) to the government's actions was heard in Yukon Supreme Court in July 2014. As Thomas Berger, lawyer for the plaintiffs, characterized the Yukon government's actions, "they took the ball and went home with it."

Staples and others (2013) analyzed publicly available data on the decision-making process led by the Yukon Government following the commission's submission of their final recommended plan. Those authors determined that the Yukon Government failed to effectively reconcile different perspectives and values through the ostensibly participatory decision-making process it ran. They further contended that it was not the polarizing nature of these perspectives that caused land-use planning for the Peel region to break down; rather, it was a "broken" decision-making process that had failed to secure the common interest. This failure left many of those involved in the planning process with the perception that their voices were no longer being heard in this process.

This cautionary case study offers a number of lessons for participatory decision-making processes in the north. Clarifying ground rules early and giving them authority would minimize opportunities for parties seeking to advance special interests at the expense of the common interest. Establishing a means for resolving conflicts about adoption of a final plan would offer the greatest chances that a plan's policy prescriptions are acted upon. Common ground between parties needs to be established early on through systematic clarification of participant and stakeholder values, moving away from the parties' positions and toward their actual interests.

Mechanisms are needed for implementing and evaluating plans and should preserve the parties' authority and control, yet create a sustained commitment to justifiable, rational, and feasible implementation in the common interest.

Managers and field operations may be increasingly caught in dilemmas between undertaking contextual adaptive actions that "make sense" on the ground, and adhering to fixed but maladaptive institutional policies, regulations, and imperatives. Many resource regimes and institutions currently focus on stable or past conditions, and managers lack flexibility because these assumptions are formalized in law and policy (McNeeley, 2011; Magness and others, 2012). Most natural resource institutions currently lack sufficient mechanisms for iterative learning (double- and triple-loop) and adaption at the levels that are required (Garmestani and Allen, 2014). These problems will cause increased need and opportunity for interjurisdictional learning, enabled in situations of decentralized and overlapping authority, such as the NWB LCC. Additionally, managers generally are trained in the conterminous United States and southern Canada, so they come to this region without sufficient understanding of its social, biological, and physical context to be fully effective. The NWB LCC could enhance the capacity of managers and scientists to understand regional context, including identifying conflicting problem definitions and solutions across multiple scales and stakeholders.

INFORMATION GAPS

- No published appraisals of management performance are available across or within jurisdictions, over time, for the NWB LCC.
- Nothing has been published about how management regimes or agencies cope with and adapt to biophysical changes in the NWB region.
- Critiques of comanagement should be updated, as they do not reflect the nearly two decades of evolution since (for example, Nadasdy, 2003).
- Comparative studies of management regimes and collaboration are few and limited in geographic scope (Danby and Slocombe, 2005).
- Literature on resilience and legal/policy regimes under changing environmental conditions documents many challenges but is new, limited in detailed insights, and, so far, focuses mainly on the United States (that is, not yet in Canada [Garmestani and others, 2013; Garmestani and Allen, 2014]).

7.3 ASSESSING AND PROJECTING CHANGE

Erika L. Rowland[1] and Matt Carlson[2]

KEY FINDINGS

- The multifaceted nature of ecosystem response to diverse drivers of change requires tools for holistic change assessment and requires assessments closely linked to decision-making needs.
- Projecting change and assessing its consequences entail uncertainties. Scenario analysis, ecological monitoring, and management experiments are valuable approaches for assessing change in the face of uncertainty
- The uncertainties resulting from the interactions between biophysical and social drivers justify the expanded application of iterative and adaptive management practices.

Change in boreal ecosystems results from diverse biophysical and anthropogenic drivers or processes unfolding over large regions and long time periods. Effective conservation action must be grounded in an understanding of past and potential change caused by these drivers. Some drivers of change have an element of proactive human control as they represent the outcomes of resource management decisions (for example, Chapter 5). To other drivers of change, humans can react only through adaptation and management aimed at seizing opportunities and minimizing risks posed by the shifting conditions under which management decisions are made (Chapter 8). The multifaceted nature of change, resulting from diverse drivers and responses, demands assessment tools that deal with full ecosystem complexity rather than just single elements.

In the face of ongoing flux, decision-making must be adept at assessing and responding to change. Well-suited in this regard is the *adaptive management* framework, which focuses on iterative learning enhanced by change assessment to improve understanding of systems and management effectiveness (Fig. 7.3.1). A focus of this

1 Formally with the Wildlife Conservation Society, Bozeman, Montana, USA.
2 ALCES Group, Calgary, Alberta, Canada.

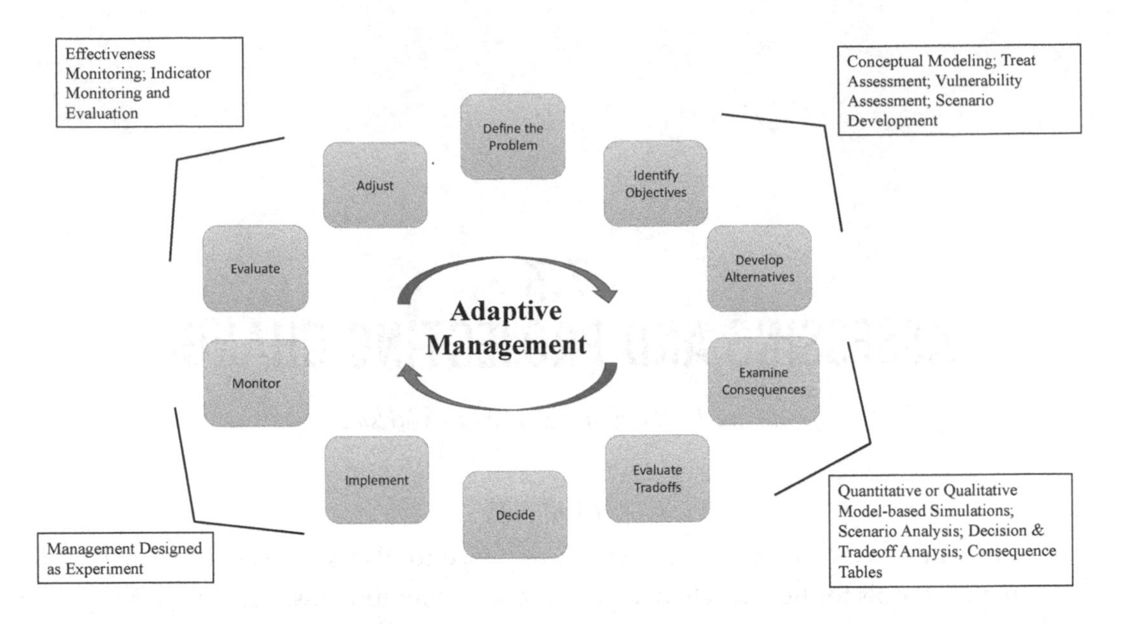

Figure 7.3.1. The adaptive management process. Boxes indicate the processes of various assessment and projection tools and which processes can be used. Modified from Allen and others (2011).

structured decision-making and implementation process is the identification and reduction of uncertainty, ideally through management experiments, which accelerate learning (Allen and others, 2011). With a long history of application in resource management, adaptive management is receiving additional attention in the context of climate change (Lawler and others, 2010; Caves and others, 2013). Climate change adaptation planning frequently uses adaptive management, along with some other planning and decision frameworks (for example, Gleeson and others, 2011; Weeks and others, 2011; Cross and others, 2012; Swanston and Janowiak, 2012). Until recently, climate was considered part of the management context, but the current rate and magnitude of change in climate and related variables have shifted this perspective (Millar and others, 2007; Chapin, McGuire, and others, 2010). Adaptation planning now examines how climate change might influence conservation and management goals, objectives, strategies, and actions, and their consequent monitoring and evaluation (Stein and others, 2014).

ASSESSMENT AND PROJECTION TOOLS

Change-assessment tools are summarized within the context of the adaptive management process (Fig. 7.3.1), focusing on climate change due to its pervasive influence on Northwest Boreal (NWB) ecosystems. Most interpretations of adaptive management involve several steps that address (1) definition of the problem, including setting objectives and identifying management options; (2) examination of the consequences

Table 7.3.1. Change assessment approaches and their place in the adaptive management process.

Approach	Adaptive management step
Conceptual models	Define problem, Identify objectives, Develop alternatives.
Scenario development	
Threat assessment	
Vulnerability assessment	
Model-based simulation of management and climate scenarios	**Examine consequences, evaluate tradeoffs**
Multidisciplinary management experiments—effectiveness monitoring	Monitor and evaluate implemented strategies and actions
Ecological monitoring	
Climate-related monitoring	

of management options and evaluating tradeoffs; and (3) implementation and evaluation of management. Although change assessment is fundamental throughout the adaptive management process, specific tools are applied at different steps (Table 7.3.1). Integration of existing knowledge through conceptual modeling, and threat or vulnerability assessment, commonly is undertaken during the problem definition phase to identify important issues, explore drivers of change, and consider management alternatives in response. Quantitative scenario analysis commonly is applied to examine the consequences of management options. Experiments and monitoring are used to evaluate the effectiveness of management actions and reduce uncertainty. Each of these types of change assessment should engage decision makers so that adaptive management fulfills its role of linking management and research.

TOOLS TO DEFINE PROBLEMS, IDENTIFY OBJECTIVES, AND DEVELOP ALTERNATIVES

A strategic formulation of the management problem requires a holistic account of ecosystem response to perturbation. Conceptual models use diagrams and narrative to articulate causal relationships between drivers and ecosystem attributes. By focusing attention on important processes, a conceptual model helps to build a common understanding of system dynamics and to identify focal issues and indicators (Parker, 2003; Lindenmayer and Likens, 2010; Woodward and Beever, 2011). When combined with indicator status and trend information, conceptual models provide a framework for threat assessment to prioritize management issues (for example, Henry and others, 2008). Assessing the effects of climate change on resources of interest is important at early stages in the planning process, and can draw from many sources of information (Rowland and others, 2011). Examining ecosystem response to changes in climate in recent decades (for example, Danby and Hik, 2007; Beck and others, 2011; Wolken and others, 2011) and the distant past (that is, hundreds to thousands of years; for example, Lynch and others, 2002) can assist with projections of future ecosystem responses (Guisan and others, 2013; Hamann and Aitken, 2013;, Littell and others, 2011) to develop an understanding of climate change vulnerability (Glick and others, 2011). Although climate change exposure and sensitivity indicate potential effects to a natural resource, adaptive capacity—the ability of a resource to cope with or adjust by ameliorating exposure and (or) sensitivity—also is a necessary factor to consider when fully evaluating vulnerability (Glick and others, 2011; Beever and others, 2015).

TOOLS TO EXAMINE CONSEQUENCES AND EVALUATE TRADEOFFS

Planning is aided by quantitative projection of management outcomes, the scope of which should match that of the policy question. In the context of land-use planning, this generally requires simulating the cumulative effects of multiple land uses and ecological processes, using a diversity of indicators to inform tradeoff analysis (Carlson and others, 2010; Francis and Hamm, 2011; Thompson and others, 2014). To influence decision-making, projections should be accessible, and web delivery of simulations can help (Carlson and others, 2014). The intent of land-use simulation models, and other scenario assessments, is not prediction, which is unattainable due to uncertainty and contingency (Peterson and others, 2003). Rather, a range of scenarios should be assessed to understand the benefits and liabilities of options and the implications of uncertain parameters (Duinker and Greig, 2007; Thompson and others, 2012).

Scenario analysis is increasingly common in climate change adaptation planning, and can integrate climate and other drivers of change (Peterson and others, 2003; Mahmoud and others, 2009; Caves and others, 2013). In addition to assessing the consequences of management interventions, exploratory scenarios can offer insights toward shaping decision objectives, understanding vulnerabilities, and identifying management alternatives. Guidance is available for selecting appropriate climate model(s) (that is, climate scenarios) (Mote and others, 2011; Snover and others, 2013) and for developing integrated scenarios (Wiseman and others, 2011; National Park Service, 2013; Rowland and others, 2014).

EXPERIMENTS AND MONITORING

Projection models often are limited by uncertain knowledge of parameter interactions, so new knowledge on ecosystem responses to management, climate, and other drivers of change needs to be derived frequently. Novel analysis of historical data should be explored, especially if multiple initiatives can be integrated to expand sample size and spatial and (or) temporal scope (for example, The Boreal Avian Modelling Project, http://www.borealbirds.ca). Often, however, new data must be collected through monitoring (repeated measurement to detect change) and experiments (assessing response to interventions). Although monitoring is vital for addressing uncertainties in projects implemented at varying spatial scales, ecosystem response to cumulative effects may be gradual and is of greatest concern at regional scales. This highlights the importance of the broader context when taking action (Groves and others, 2012), and the need to sample across large spatial and temporal scales with sufficient intensity for statistical power (Carlson and Schmiegelow, 2002). Examples of large-scale, comprehensive ecological monitoring initiatives include the Alberta Biodiversity Monitoring Institute (http://www.abmi.ca), the upcoming National Ecological Observatory Network (http://www.neoninc.org), and the USA National Phenology Network (http://www.usanpn.org). A beneficial attribute of these programs is their commitment to public dissemination of data.

Although monitoring is invaluable for detecting change, attributing causation often requires experiments that apply management interventions (Walters and Holling, 1990). These require collaboration with agencies capable of applying land-use treatments and protection of landscapes sufficiently intact to act as experimental controls (Schmiegelow, 2007). Large research programs such as the Kluane Boreal Forest Ecosystem Project (Krebs and others, 2001) and the Experimental Lakes Area (Stokstad, 2008) are especially useful for elucidating ecosystem behavior due to the ability of multidisciplinary investigation to assess interactions among components and due to insights gained through long-term study (Lindenmayer and others, 2010).

INFORMATION GAPS

- There is a general need for more land use and natural resource management decision-making processes to better incorporate climate change assessments into the processes (Guisan and others, 2013; Snover and others, 2013).
- The implementation of change assessments is experiencing the following gaps:
    - Integration of climate change and land-use change as combined drivers of ecosystem change;
    - Ability to link derivative ecosystem attributes, especially ecosystem services, to land use intensity and landscape composition;
    - Robust ability to quantify and project the social and economic consequences (for example, health, crime, employment) of climate and land use change;
    - Quantitative scenarios that describe expected long-term (decadal) land-use patterns; and
    - Improved inventories or models of current and future landscape attributes, such as forest age and hydrological regimes, in order to parameterize landscape models.

BOX 7.3.1: CASE STUDY: ASSESSING THE CUMULATIVE EFFECTS OF LAND USE IN WESTERN BOREAL CANADA

The western boreal region of Canada is being transformed by development of its natural resources, including oil, bitumen, gas, and timber. To provide a synoptic understanding of the consequences of future development, the Canadian Wildlife Federation applied the ALCES model (Carlson and others, 2014) to simulate the cumulative effects of multiple land uses (for example, energy, forestry, agriculture, settlements) to a 693,000 km^2 western boreal region over the next five decades (Carlson and Browne, 2015). Simulations contrasted the consequences of various development rates, management practices, and levels of protection to wildlife (caribou, moose, fisher, songbirds, fish), ecosystem services (carbon storage, water quality, air quality), and the economy (gross domestic product, employment). The expected rate of natural resource development generated rapid economic growth, but the associated northward spread of the industrial footprint had negative consequences for many environmental indicators (Fig. 7.3.2). Expansion of the protected areas network was the most successful approach for improving environmental performance, and simulations explored environmental and economic trade-offs across a range of protection levels. Such information is useful to managers and stakeholders when identifying land-use strategies that balance economic and environmental objectives.

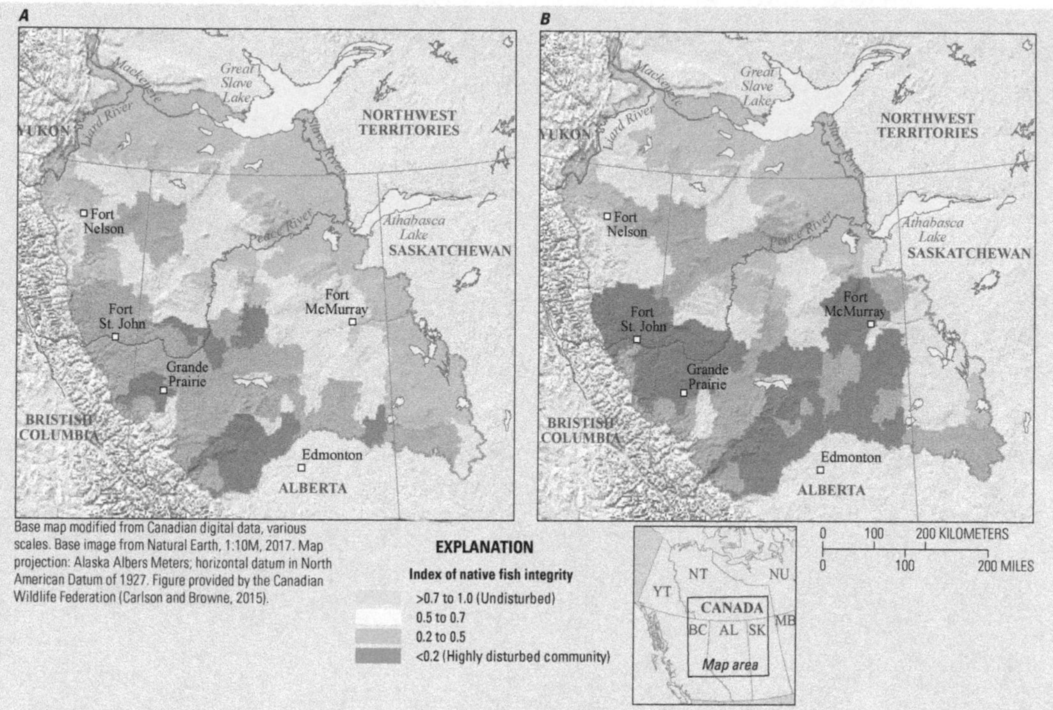

Figure 7.3.2. The cumulative effects of land use to the native fish community in a Canadian western boreal landscape over the next 50 years, as simulated using ALCES software (Carlson and others, 2014). The 693,345 km^2 region extends across parts of boreal Alberta, Saskatchewan, British Columbia, and the Northwest Territories. Maps present the (*A*) status of a current index of native fish integrity and (*B*) projected native fish integrity 5 decades in the future. The index of native fish integrity conveys changes in abundance and composition of fish species with a value ranging from 1 (undisturbed community) to 0 (highly disturbed community). Figure provided by the Canadian Wildlife Federation (Carlson and Browne, 2015).

7.4
DATA MANAGEMENT AND DISSEMINATION

Donald G. Morgan[1], Tosha Comendant[2], and Katie O'Connor[3]

KEY FINDINGS

- Decision makers at various levels, including provincial, state, municipal, Indigenous governments, corporate, university, and nongovernment entities, typically only consider data and analysis they trust.
- Trust in data cataloging and holdings is built through co-ownership and shared governance processes.
- Data quality and usability for conservation assessments and planning decisions are improved with standard documentation and adherence to data life cycles.
- Challenges to sustaining unified data include time lags between updates, scale mismatches in data collection between and across jurisdictions, and the loss of data that does not fit in regional frameworks.
- Technological advances and broad availability of online resources allow large spatial datasets and information to be better documented, visualized, integrated, and accessed.
- Platforms, viewers, and tools are supporting broad partnerships, facilitating new cross-disciplinary collaborations, and encouraging bigger and more innovative solutions.

Meeting conservation goals and regulations at landscape scales increasingly require collaboration among diverse partners, multiple agencies or organizations, and stakeholders that have complex interests (Groves, and others, 2002; National Academy of Sciences, Engineering, and Medicine, 2015). In practice, many conservation initiatives face sizeable challenges to efficiently collaborate due to organizational challenges, lack

1 Ecosystems Branch, Conservation Sciences, Ministry of Environment and Climate Change Strategy, Smithers, British Columbia, Canada.
2 Conservation Biology Institute, Corvallis, Oregon, USA.
3 Conservation Biology Institute, Corvallis, Oregon, USA.

of capacity to develop and use relevant tools and technologies, insufficient resources or policies for integration, and difficulty managing and scaling data and information from local to regional to national to global scales or across boundaries (Salafsky and others, 2008). Further, Indigenous, government, and nongovernment decision makers typically only consider their own information and analysis. Advances in web technology, platforms, and applications are supporting collaboration at expanded scales and making it easier to combine, visualize, and analyze data (Snaddon and others, 2013). The integration of shared standards, new collaboration tools, and data governance is providing expanded opportunities to build trust, close gaps, and increase the quality and usefulness of the data and information underpinning conservation assessments, plans, models, and implementation of sustainable natural resource management and practices.

HISTORICAL AND CURRENT STATE

Field notes collected by forest rangers on horseback were the original sources of natural resource data in the Northwest Boreal (NWB) region for agency and government use. Later, canvas maps were used to capture different resource values, such as timber and mineral potential. With the advent of the desktop computer and Geographic Information Systems (GIS), maps were digitized and managed locally. Over time, more sophisticated centralized data inventory systems emerged that disseminated data to users in the field supplemented by local cartographers and later by GIS specialists. Initially, due to cost and specialized training, government and large forest companies were the primary users of these systems. However, as the cost of the technology dropped and became more accessible, others started to use the systems, including First Nations, community colleges, nongovernment organizations, and private consultants.

OPPORTUNITIES FOR THE NORTHWEST BOREAL REGION

The recent and rapid increase in the amount of data being collected and disseminated, as well as society's growing awareness of its importance, provides both opportunities and challenges. Landscape conservation initiatives are increasingly using open data, tools, and applications to drive innovation, improve efficiency, and foster transparency and accountability. A range of existing and emerging resources can be leveraged to increase the pace and improve the quality of natural resources planning endeavors in the NWB region. These resources and technologies encourage data providers to share the data in ways that can build trust and support integration and analysis across organizational, cultural, and geographic boundaries. Successful data management and dissemination initiatives will continue to advance interests of contributors, carefully define target audiences, develop data standards, and address priority resource management and policy questions. The next generation of cataloging services, platforms, map and data viewers, tools, and social networks are being adapted and integrated to better support

BOX 7.4.1: DATA SHARING NETWORKS IN BRITISH COLUMBIA

As more people in northwestern British Columbia started using data, problems emerged with maintaining consistency and currency. For a time, the province increased local capacity, linking resource ministries—forests and environment—and field offices providing a high level of coordination for data. The regional government model was replaced by a more centralized version, and local government inventory and data management capacity declined. This coupled with more people outside of government manipulating data led again to multiple copies of data. The need for cross-sectorial collaboration on identifying authoritative datasets and setting standards became obvious to reduce costs and ensure the best available information was being used for decision-making. In the early 2000s, this lead to the establishment of groups which included the Provincial government, local industry, and First Nations, and allowed for an advanced coordinated and standardized geospatial datasets. These types of data-management governances require ongoing funding and organizational support to persist. To address these shortcomings, more formal governance mechanisms are being implemented in the northwest to coordinate across sectors to catalog data and information, and to encourage data standardization, such as the Skeena Knowledge Trust (http://skeenatrust.ca).

landscape conservation in the Northwest Boreal (NWB) Landscape Conservation Cooperative (LCC) and other regions (Table 7.4.1).

The NWB LCC makes priority biological, physical, and socioeconomic data and information widely available through a collaborative web-mapping application powered by Data Basin (Bachelet and others, 2011). The Alaska and Northwest Canada Regional Conservation Planning Atlas (https://aknwc.databasin.org/) contains curated datasets with associated metadata so the temporal and geographic scale of the data, along with methods used and units of measure, and any other relevant information, is stored along with the geospatial data. The Atlas allows users to create custom and exportable maps that can be used to engage stakeholders and will be integrated with a growing network of Landscape Conservation Cooperative (https://lccnetwork.org/) sponsored web-mapping resources. The Alaska and Northwest Canada Regional Conservation Planning Atlas is linked with a growing resource of data management and dissemination tools being used to improve conservation planning, resource management, and community engagement.

Table 7.4.1. Sample of online resources used in the Northwest Boreal Landscape Conservation Cooperative and broader conservation community.

Name	Description
LCC Integrated Data Management Network (https://fws.sciencebase.gov/idmn/LCC_IDMN_Group_ Report.pdf)	Project to develop a network and guidance for data management across Landscape Conservation Cooperatives (LCCs).
ScienceBase (https://www.sciencebase.gov/catalog/)	U.S. Geological Survey data catalog and platform. Provides web services and collaborative tools.
Alaska and Northwest Canada Conservation Planning Atlas (https://aknwc.databasin.org/)	Platform powered by Data Basin with open data access, mapping, analysis, and collaboration tools, and integration with ScienceBase. Joint effort of the Arctic, Western Alaska, Aleutian and Bering Sea Islands, and Northwest Boreal LCCs.
North Pacific Conservation Planning Atlas (https://nplcc.databasin.org/)	Platform powered by Data Basin with open data access, mapping, analysis, and collaboration tools, and integration with ScienceBase. Built and maintained by North Pacific LCC.
Boreal Avian Modeling Project Web Mapping Portal (https://borealbirds.databasin.org/)	Platform powered by Data Basin with open data access, mapping, analysis, and collaboration tools. Information is associated with abundance, distribution, and habitats of boreal birds and effects of human activity.
ArcGIS Online (http://www.esri.com/software/arcgis/arcgisonline)	Platform built using supporting maps, application, analytics, and collaboration.
Google Earth (https://www.google.com/earth/)	Desktop, mobile, or web application for viewing 3-dimensional Earth views.
CKAN (https://ckan.org/)	Platform to streamline publishing, sharing, finding, and using data. Used by data publishers (national and regional governments, companies, and organizations).
Southeast Alaska GIS Library (http://seakgis.alaska.edu/)	A geospatial data library with associated applications and services specific to southeast Alaska.
Griffin Groups (https://griffingroups.com/)	Application to support collaboration and community building in conservation. Each community is a private or public group that can manage various types of content.
West Coast Ocean Data Portal (http://portal.westcoastoceans.org/)	Network and data catalog containing information on sources and patterns of marine debris, adaptation to sea-level rise, impacts of ocean acidification on coasts, and marine planning.
National Snow and Ice Data Center (https://nsidc.org/)	The National Snow and Ice Data Center (NSIDC) supports research into our world's frozen realms: the snow, ice, glaciers, frozen ground, and climate interactions that make up Earth's cryosphere.
Scenarios Network for Alaska and Arctic Planning Data Portal (SNAP; http://ckan.snap.uaf.edu/dataset)	SNAP produces downscaled, historical, and projected climate data for subarctic and Arctic regions of Alaska and Canada and other project-specific data that cover larger regions.

BOX 7.4.2: PARTICIPATORY MAPPING PROJECT BUILDS AWARENESS OF LOCAL ECOSYSTEMS

The Discovery Islands, located between Vancouver Island and the mainland, are a distinctive ecological region in British Columbia. Until recently, there was no comprehensive ecosystem inventory or other resource management plan needed to inform land-use planning decisions. The Discovery Islands Ecosystem Mapping (DIEM) project (http://www.diemproject.org), a citizen-initiated project, used mapping and collaboration tools in Data Basin[1] to support their compilation and development of information about the terrestrial and freshwater ecosystems and enduring features of the Discovery Islands and nearby mainland inlets. The DIEM project staff developed rigorous standards to ensure the mapped results would be accepted as provincial government data. The digital format was designed to support additional map layers that document community knowledge, values, and interests. Local government and other decision makers will use this information to balance ecosystem values and effects when resource extraction and development is proposed. The communities of Discovery Islands are gaining new access to data and tools that allow shared recognition and evaluation of surrounding ecosystems and landscapes. Community mapping is the ongoing second stage of the DIEM project. Discovery Islands communities are leveraging the Data Basin technology to access new data and tools to visualize, integrate, and participate in land use planning decisions and development.

1 Data Basin (https://databasin.org/) is an open access platform supporting discovery, access, and visualization of spatial data. The Alaska and Northwest Canada Regional Conservation Planning Atlas (https://aknwc.databasin.org/) is a customized version or specific portal of Data Basin.

IMPLICATIONS FOR RESOURCE MANAGEMENT AND COMMUNITIES

As the need to collaborate at a landscape scale has increased, especially when dealing with environmental issues such as habitat connectivity and climate change, the focus has progressively turned to methods of gathering and sharing necessary data among project partners and stakeholders. Although government agencies may have the broadest, most organized data, they often lag behind in provision of the most current data. Thus, the data at the most appropriate scale may not have the most appropriate content for a given environmental issue. For this reason, science-based organizations of all scales should be included in the development of data sharing solutions.

The pace of resource management planning is tied directly to the efficiency of accessing the necessary data for decision-making. For regional, transboundary planning it is often difficult to find analogous data that can be compared across the entire geographic extent of a project. Finding the most accurate and up-to-date data can take significant staff time if no previous effort has been made. Anticipating these challenges before they occur and making a plan ahead of time can save significant resources of time and money.

The NWB region spans two countries with different standard metrics for measurement, along with different laws, regulations, and plans for data collection. Rather than continuing the large effort to stitch together disparate datasets into the future, it is important to come to an agreement among resource managers and engaged communities on data standardization across the NWB area of interest. This could include standard units of measure, best practices for the geographic scale and temporal sampling methodology for various species and habitats, and metadata standards.

INFORMATION GAPS

- A shared understanding or agreement is needed among resource managers and engaged communities on data standardization across the NWB region.
- A comprehensive inventory of regionally relevant data catalogs, platforms, map viewers, and associated social networks is needed.
- An understanding of trends in usage and demand for online datasets and tools among different audiences is important.
- For specific audiences or tools, an assessment of which variables, threats, and management strategies are most relevant to current conservation challenges would be valuable.

7.5
SCIENCE COMMUNICATION AS A DRIVER OF LANDSCAPE CHANGE

By Sarah F. Trainor[1]*, Kristin Timm*[2]*, Tina Buxbaum*[3]*, Alison York*[4]*, and Alison Perrin*[5]

KEY FINDINGS

- Communication is an ongoing, iterative, self-reflexive process. Ideally, a researcher learns about their stakeholders and audience, makes efforts to engage in a dialogue with them, evaluates the outcomes, revises the methods and model, and then tries again. In this way, science communication can evolve in step with changing landscapes, changing information needs, and scientific advances.
- Communication efforts will be most effective when the messenger and the information provided are perceived by the stakeholder or audience to be salient, legitimate, and credible. This can be increased through scientist–stakeholder collaborations and interactions that facilitate convening dialogue, translating information into usable forms, and mediating diverse viewpoints.
- The one-way flow of information is failing to meet the information needs of stakeholders in the rapidly changing boreal and northern Arctic communities. There is a need for more iterative two-way information exchanges between information producers and stakeholders.
- Full partnership and engagement between scientists and stakeholders requires investment in time, relationships, and trust building. Communication and

1 Alaska Center for Climate Assessment and Policy, University of Alaska Fairbanks, Fairbanks, Alaska, USA.

2 George Mason University, Fairfax, Virginia, USA.

3 Alaska Fire Science Consortium, University of Alaska Fairbanks, Fairbanks, Alaska, USA.

4 Yukon College, Whitehorse, Yukon, Canada.

5 Northern Climate ExChange, Whitehorse, Yukon, Canada.

relationship development can be enhanced by building nested networks of information exchange.

- Emerging social media and advancing technology have the potential to reach a large and diverse audience, but still require constant evaluation and revision. Face-to-face contact is still vital for deeper and more sustained engagement.

COMMUNICATION AS A DRIVER OF LANDSCAPE CHANGE

Landscape change in complex social-ecological systems involves linked processes of physical, ecological, and social drivers of change including parameters such as climate change, hydrologic change, wildfire, invasive species, land-clearing, policy, regulations, and economic incentives (Fig. 7.5.1). Scholarship in sustainability science and knowledge coproduction indicates that the ways in which science is *conducted* and *communicated* to land and resource managers, policy makers, and publics can be a driver of landscape change.

In this chapter we explore the ways in which communication can be a driver of landscape change. We offer a series of suggestions for how to achieve meaningful communication between scientists and resource managers, or stakeholders, and we provide examples of three organizations in Alaska and neighboring Canada that are working at the interface of science and management.

COMMUNICATION

Communication has been defined in a variety of ways, but Schirato and Yell (1997, p. 1) define communication as "the practice of producing meanings." Defined in this way, communication involves at least two or more individuals and is not focused on what is said by a speaker or heard by a listener, but rather by the meaning produced in that process or interaction. Meanings are further shaped by the context in which they occur, because communication is, "always informed by and produced within cultural contexts" (Schirato and Yell, 1997, p. 17).

Within the context of a given social ecological system, a variety of forms of communication can occur—each with their own histories, meanings, and expected outcomes. Concepts such as science communication, outreach, informal learning, environmental communication, public participation, science literacy, social learning, engagement, social marketing, coproduction, and others are used to describe the plethora of ways science can interact with society. While several authors have attempted to define and organize some of these concepts (see Bauer, Allum, and Miller, 2007; Burns, O'Connor, and Stocklmayer, 2003; Kurath and Gisler, 2009), a widely agreed upon taxonomy or organizing framework does not yet exist.

As an organizing concept that can overcome some of this ambiguity in terminology, recent research has turned to a focus on the desired effects of science communication and related activities (Besley, Dudo, and Yuan, 2017). Communication about science and the environment can have a range of effects on audiences, including but not limited to generating excitement about science, informing decision makers, educating, prompting a change in behavior and fostering trust (National Academies of Sciences, 2017; Maibach, 2017). Each of these desired outcomes of communication could be a unique goal or objective, with distinct means by which it is to be achieved and evaluated, and which different outcomes in the social ecological system.

For example, as a driver of landscape change, communication could involve the sharing of information among scientists and resource managers (Fig. 7.5.1). Scientists study the interactions and feedbacks between physical, biological, and human drivers of change and landscape change. As the landscape changes, there is ongoing scientific research about what is changing, the rate of change, and feedbacks between drivers of change. Results of scientific research are communicated through a range of processes, including peer-reviewed journals, scientific assessments, fact sheets, research briefs, webinars, workshops, and presentations at professional meetings and other non-academic venues. State, federal, tribal, and private resource managers, landowners, and other stakeholders use information related to the results of scientific research in their management decisions. Those decisions set an agenda for resource management, and thus a trajectory of landscape change. A classic example of the link between resource management and landscape change is strict fire suppression policy. Images of Smokey Bear and forest destruction communicated that fire was bad on the landscape and precipitated polices of wildfire exclusion from western landscapes, resulting in fuel build-up and successional change in forest ecosystems (Keane and others, 2002). On the whole, this cycle describes how the communication of scientific information can potentially influence landscape change. However, other models are also possible, and scientists can examine the extent to which communication influences management decisions and related landscape change as compared to other drivers of landscape change (Calef and others, 2015).

COMMUNICATION AS A DRIVER OF LANDSCAPE CHANGE

As previously described, communication is often conceptualized as an end product of science, but communication is also an integral part of the process of conducting science. Scientific research encompasses both the process of how science is conducted as well as the results of this process (Knapp and Trainor, 2013). Use-inspired science, including the coproduction of knowledge, is a process of scientific inquiry that directly integrates management concerns and information needs with research design and inquiry. In this context, communication must be practiced as a two-way exchange of information whereby scientists and potential information users collaborate to set the information agenda (Dilling and Lemos, 2011). Boundary organizations, such as the Northwest Boreal Landscape Conservation Cooperative, can assist with this, by

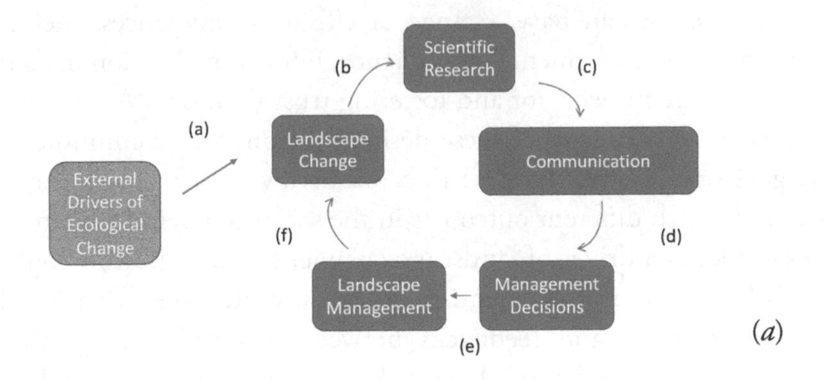

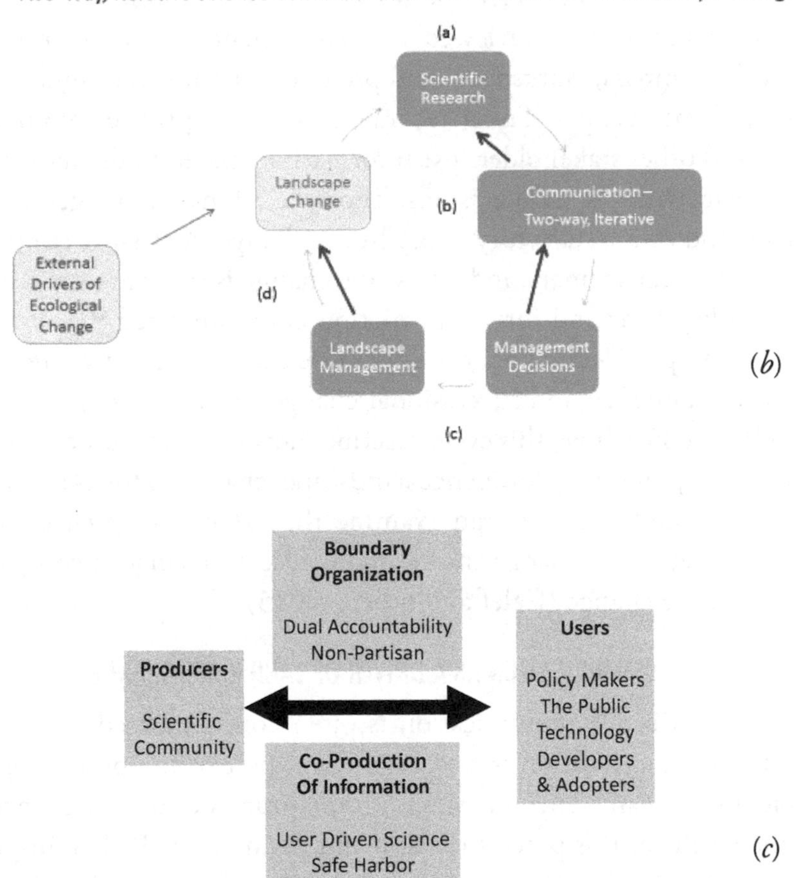

Figure 7.5.1. Science communication as a driver of landscape change. (*a*) Cycle of impacts. (*b*) Two-way iterative communication between scientists and managers can increase the magnitude of communication as a driver of landscape change. (*c*) Boundary organizations can play a central role in bridging scientists and decision makers. (Based on Figure 2-1 in Clark and Holiday 2006.)

spanning the divide between scientists and decision makers, or stakeholders, and have proven themselves as effective mechanisms for facilitating two-way communication and the coproduction of knowledge (Clark and Holliday, 2006; Dilling and Lemos, 2011; Kocher and others 2012).

With collaboration and iteration over time, the gap between research needs and scientific questions narrows (Ferguson and others, 2014). This two-way, iterative communication between scientists and managers enhances the usability, relevance, and applicability of science in land and resource management. This, in turn, creates a more potent link from science communication through management decision-making to landscape change. Thus, the ways in which science is conducted and communication channels are built between scientists and managers become drivers of landscape change.

The engagement of comanagement boards in Canada is an example of how two-way communication can affect management decisions. Comanagement boards are increasingly being used for wildlife management in the North and are entities that can promote two-way dialogue between scientists, policymakers, and community stakeholders. Comanagement boards also provide an avenue for considering and incorporating Traditional Ecological Knowledge into the decision-making process. Two-way information exchange can improve decisions concerning harvest management, particularly in situations where population health data is questioned, as was the case with beluga whales in the Beaufort Sea. Dialogue between members of the Canada/Inuvialuit Fisheries Joint Management Committee, which include the federal Department of Fisheries and Oceans, the Inuvialuit Game Council, and the Hunters and Trappers Committees of the Inuvialuit Settlement Region, has led to the development and evolution of an iterative beluga monitoring program that directly affects management decisions (Manseau and others, 2005).

The work being done by the Joint Fire Science Program's Fire Exchange Network, of which the Alaska Fire Science Consortium is a member (see Box 7.5.2, "Alaska Fire Science Consortium: Integrating Science and Humanities for Public Outreach"), is an example of this link between science communication and landscape change. These regionally based boundary organizations in the United States work at the interface of wildfire science and management (Kocher and others, 2012).[6] Although the Consortium members are not advocates for specific management policies or practices, they strive to build communication and integration of wildfire science and management to increase the resilience of both landscapes and communities to wildfire risk. Program evaluation is an integral part of this network, the results of which are actively used in project programming and strategic planning (Singletary and others, 2015; Maletsky and others 2018). Research is currently underway to document specific examples of how

6 See also: http://www.firescience.gov/JFSP_consortia_elements.cfm, and http://www.firescience.gov/JFSP_consortia_vision.cfm.

the work of these boundary organizations impacts wildfire and resource management and, in turn, landscape change.

TIPS FOR MEANINGFUL SCIENCE COMMUNICATION AND KNOWLEDGE COPRODUCTION

Know Your Audience, Implement, Evaluate, and Adapt

Strategic communication can provide a useful framework to understand different considerations in the communication process, which is ongoing, evolving, inherently self-reflexive, and can vary in terms of structure and fluidity. The first step of any outreach effort is to identify desired effects, through goals and objectives. Goals are defined as the "desired outcome[s] of a plan of action" (Kendall, 1992, p. 248) and communication objectives are incremental steps that "contribute toward achieving the goal[s]" (Kendall, 1992, p. 248). Audiences, tactics, and evaluation are identified from goals and objectives, and each possesses its own range of considerations.

One of the most important parts of the communication process is understanding the audience or the stakeholders that will be targeted (Bostrom, Bohm, and O'Connor, 2013; Hulme, 2014): What kind of knowledge do they already have? What information do they need? What are their decision needs? What are their limitations? Is there any bias? What is the best media for reaching the audience (printed matter, web-based tool, workshop, video, or other forms of communication)? These questions can be answered through formal or informal needs assessments.

Specific efforts can then be implemented to match these needs, and communication approaches can be tailored to the audience or stakeholder (Wholey and others, 2004; Bostrom, Bohm, and O'Connor, 2013). Communication efforts that aim to reach a diverse group of stakeholders, such as decision makers, scientists, the public, community members, and others, will therefore likely require a range of methods to reach each different audience.

Additionally, to make outreach efforts as effective as possible, they should be evaluated and adapted to meet evolving stakeholder needs and to match evolving modes of communication (Wall and others, 2017). No one model of communication is perfect for all instances, and meeting stakeholder needs can be a dynamic, ongoing process. Communication efforts are most successful when they are tailored to the specific audience, information needs, and information at hand and then adapted as time goes on to continue to meet the needs of stakeholders and the resources and capacity of the information providers (Dilling and Lemos, 2011; Knapp and Trainor, 2015).

The goal of knowledge translation is to apply knowledge into practice. Translating knowledge into action involves synthesizing, disseminating, exchanging, and applying information. Effective knowledge translation relies on understanding both the audience and the content—not all information needs to be translated—and how it is translated can make a big difference in the outcomes. Including stakeholders in knowledge creation and promoting the conditions necessary for desired actions to take place can improve the effectiveness of translation efforts. Knowledge translation is an iterative process that requires monitoring and adjusting both the content and the means of communication based on outcomes (Graham and others, 2006; Strauss and others, 2009).

As the ecological and biophysical landscape changes through a range of drivers, so too do the management, social, economic, and policy landscapes change over time in response to various drivers. Ecological change coupled with a range of changes in social and management spheres leads to ever-evolving stakeholder information needs, the continual generation of new scientific findings, and ongoing technological advances. Thus, communication efforts, which exist at the nexus of scientific advances and resource decision-making, also must be nimble, reflexive, and adaptive in order to remain relevant in this dynamic social-ecological system and to meet changing information needs (Salmon, Priestley, and Goven, 2015; Fischhoff, 1995).

Salience, Credibility, and Legitimacy

For information to be useful it must be perceived by the stakeholder as salient, credible, and legitimate, that is, relevant, to come from a trustworthy source, transparent in how it was derived, and unbiased (Cash and others, 2002; Druckman, 2015). It can be challenging to achieve all three attributes simultaneously. For example, increasing public participation can increase salience, but diminish credibility (Cash and others, 2006; Buizer and others, 2010).

Additionally, information must be presented on a scale that is matched temporally and spatially to the information needs of the user. Information that is too broad will not be relevant for stakeholders seeking to make decisions on a finer scale (Jacobs and others, 2005). At the same time information that is extremely detailed and at a fine resolution might present stakeholders with "information overload" and thus lose saliency (Cash and others, 2006; McNie, 2012). The timing of providing information for stakeholder use also is important. Information that comes too early or too late for proper use by stakeholders will inherently lack saliency.

A range of organizational structures can facilitate two-way, iterative information exchange, help create information that is salient, credible, and legitimate, and allow for evaluation and learning by both users and producers of knowledge. Useful models include knowledge networks, collaborative processes, and embedding information specialists within existing organizations. Boundary organizations and information brokers also can be useful in delivering information that meets stakeholder needs (Dilling and Lemos, 2011). Information providers can increase their salience, credibility, and legitimacy by facilitating scientist–stakeholder collaborations and interactions that facilitate convening dialogue, translate information into usable forms, and mediate diverse viewpoints (Cash and others, 2006).

Facilitating Information Exchange

There is a continuum of science communication models, in which one end is exemplified by the "loading dock" or "deficit model" approach in which information is created in a traditional curiosity-driven science approach and simply made available to stakeholders on a website, a newsletter, or a similar venue in a one-way flow of information (Cash and others, 2006; Bauer, Allum, and Miller, 2007). In contrast, full partnership or dialogic models involve scientists and stakeholders or other audiences engaging with each other directly (Fischhoff, 1995). In this partnership, there is a two-way, iterative flow of needs and information, resulting in research that is driven by stakeholder information needs (Cash and others, 2003; Brossard and Lewenstein, 2010; Fischhoff, 1995). Knowledge coproduction occurs when scientists partner with managers or stakeholders throughout various phases of the research process, including setting the research objective, identifying the research question, designing the project, collecting, analyzing, and interpreting data and disseminating results. Meadow and others (2015) have identified a range of modes of engagement between scientists and stakeholders using the parameters of research objective, origin of research question, type of relationship, and degree of stakeholder involvement.

Research on climate change information needs in Alaska shows that the traditional one-way flow of knowledge between information producers and stakeholders is not fully satisfying the information needs of stakeholders in the rapidly changing North. There is a need for more iterative two-way information exchange between information producers and stakeholders, as well as a need to make scientific data more transparent and accessible (Knapp and Trainor, 2013; 2015).

Ongoing, iterative partnership and engagement between scientists and stakeholders requires investment in time, relationship, and trust building, and is an audience-centered process. As explained by the National Academies of Sciences (2017, p. 89), "In science communication, as with other types of communication, the audience decides whether the sources of information or the institutions they represent are trustworthy." People use this assessment in deciding what information is worth their attention, and often what to think about that information. Developing trust is a long-term process (Tribbia and Moser, 2008; Lemos and others, 2012). Communication efforts can be enhanced by building nested networks of information exchange that leverage pre-existing trust and relationships (Lemos and others, 2014; Kettle and Trainor, 2015). Social scientists can enhance the use of science in management by increasing the understanding of decision contexts and the role of social capital, and establishing processes for the evaluation of outcomes (Wall et al. 2017; Bednarek et al. 2018). Recent literature emphasizes the need for strong interpersonal skills as well as for explicit training, career development and support for individuals and organizations who work at the interface of science and management, as these types of activities are not well incentivized in traditional scientific careers (Brugger et al. 2016; Bednarek et al. 2018).

Three examples of boundary organizations are the Alaska Center for Climate Assessment and Policy (see Box 7.5.1, "ACCAP Alaska Climate Webinar Series: Building a Nested Network of Information Exchange"), the Alaska Fire Science Consortium (Box 7.5.2, "Alaska Fire Science Consortium: Integrating Science and Humanities for Public Outreach"), and the Northern Climate ExChange (Box 7.5.3, "Northern Climate ExChange: Mainstreaming Climate Change into Government Decision-Making"). These examples describe successful outreach efforts in boreal Alaska and Canada, including nested networks, innovative interdisciplinary collaborations, and training workshops.

New and Emerging Forms of Communication

Novel forms of communication, including social media, can be vital for reaching certain populations and can be effective in bridging large distances and connecting otherwise remote stakeholders in the North. In boreal Alaska and Canada, this includes a growing regional audience as well as a distant yet potentially important audience in more southern populated regions (Trainor and others, 2016).

The proliferation of online and web-based outreach platforms—such as Twitter, Facebook, Vimeo, Flickr, Instagram, email listservs, and online webinars—has created new avenues and opportunities to engage in two-way communication with a larger, dispersed stakeholder base (Moloney and Unger, 2014). For example, a climate researcher at the Alaska Center for Climate Assessment and Policy has a closed Facebook group where residents of a region of Alaska have a safe, trusting environment to have conversations about weather and climate issues (personal conversation, R. Thoman).

CONCLUSION

As new modes of communication and technology emerge regularly, it is particularly important for outreach providers to keep in mind the basics: what are their goals and intentions in engaging in communication efforts, how are they developing a better understanding of and continually monitoring stakeholder and audience needs, are they responding and tailoring communications appropriately, and how are they evaluating the efficacy of their work and being adaptive to stakeholder feedback (Jacobs and others, 2005). Although the virtual interactions made possible by new technologies can simulate face-to-face interactions, in-person contact is still vital for deeper and more sustained outreach and engagement. There are a range of communication approaches and tactics, and the case studies illustrate some of the ways that they influence landscape change. Communication is rarely considered in the studies of linked social-ecological systems. This chapter presents an avenue for how communication can be a driver of landscape change and offers different dimensions of communication and information flow that could vary with respect to landscape change—providing an interesting avenue for future research.

INFORMATION GAPS

- More information is needed on stakeholder preference and applicability of social media in boreal Alaska and Canada.
- A better understanding of the formal and informal processes that deliver information and inform decision-making would be valuable.
- Documentation and evaluation of how outreach and scientific information have been directly applied in decision-making and resource management is needed.
- There is a need for additional case studies and research comparing different communication approaches (i.e. strategic, ad hoc) and factors (i.e. trust, credibility) and their influence on landscape change

BOX 7.5.1: ACCAP ALASKA CLIMATE WEBINAR SERIES: BUILDING A NESTED NETWORK OF INFORMATION EXCHANGE

The Alaska Center for Climate Assessment and Policy (ACCAP) is one of 11 Regional Integrated Sciences and Assessments (RISA) programs nationwide funded by the Climate Program Office of the National Oceanic and Atmospheric Administration. The RISA program supports research to build and expand the capacity to prepare for and adapt to climate variability and change. Paramount to the RISA mission is a commitment to building trust and partnerships with relevant stakeholders. Established in 2006, ACCAP engages stakeholders using several outreach mechanisms, including an email listserv, a quarterly newsletter, a website, and multiple decision-support tools. One of ACCAP's showcase outreach tools is its monthly climate webinar series.

In Alaska, webinars, or web-based seminars, have proven to be an especially useful technology for bridging vast distances, communicating science, and connecting scientists and stakeholders. ACCAP webinars have a diverse and expanding audience, including tribal administrators and environmental coordinators in remote Native villages, scientists from universities around the country, state and federal resource managers, non-profit organizations, and industry.

ACCAP has been hosting monthly webinars since June 2007 on various topics, including terrestrial, marine, and climate science; demonstration and showcase of decision-support tools; and presentation and discussion of national-level reports such as the National Climate Assessment. The goals of the webinar series are to (1) provide a forum for discussion of different aspects of climate change in Alaska, (2) identify information gaps and ways to fill them, and (3) improve the ability of Alaskans to prepare for and respond to climate change (Trainor and others, 2016).

The ACCAP webinar series platform has evolved with advancing technology in response to user feedback. Initially presented exclusively through telephone and online presentation, the ACCAP now uses a web-based platform in conjunction with phone access to reach both urban participants with high-speed internet access and rural participants with lower-speed technology. The webinars are recorded, and videos, audio files, and presentation slides are available for asynchronous viewing at the ACCAP website and through the ACCAP podcast subscription on the iTunes store and the ACCAP website. The webinars usually occur once a month, on a Tuesday at 10:00 a.m. Alaska time, and last one hour. The consistent day of the week and time allows participants to plan for and expect upcoming webinars. Additionally, by holding the webinars at 10:00 a.m. Alaska time it is possible to easily have workday participation from all the time zones within the United States and Canada.

A recent evaluation of the webinar series documents how participants and speakers benefit from participation, as well as how evolution of the webinar series over time has created and supported an expanding network of boundary organizations (Trainor and others, 2016). Notably, ACCAP webinars provide opportunities for additional networking between participants and organizations and thus allow for network growth.

One way this has been accomplished is through webinar satellite viewing hubs. These remote-viewing sites are hosted by ACCAP partner organizations and bring participants together in multiple locations throughout the state, which enables the webinar series to reach an expanded audience and facilitates local networking. Many of these sites report additional local discussion about webinar topics and follow-up with speakers. By convening the webinar series and leveraging the trusted networks of these other boundary organizations through the satellite viewing hubs, ACCAP has created a nested network of information exchange, which promotes two-way dialogue while reducing internal transaction costs (Kettle and Trainor, 2015).

BOX 7.5.2: ALASKA FIRE SCIENCE CONSORTIUM: INTEGRATING SCIENCE AND HUMANITIES FOR PUBLIC OUTREACH

Funded by the U.S. Joint Fire Science Program (JFSP) starting in 2008, the Alaska Fire Science Consortium (AFSC) is a member of the JFSP Fire Science Exchange Network, which consists of regionally focused consortia in the United States, designed to accelerate the awareness, understanding, and use of wildland fire science (Kocher and others, 2012). Wildland fire and resource managers are important stakeholders for all 15 consortia and, depending on a region's ecosystems, geography, and demography, secondary stakeholders may include large private landowners, non-profit organizations, and the public. Each consortium's outreach work ranges along a continuum from "loading dock" to full partnership, varying with the target audience.

The AFSC's major audience is the statewide interagency wildland fire-management community. Although historically most wildfires occurred in the boreal forest, increasing wildfires in the northern tundra and southeast Kenai Peninsula has expanded the range of interest in fire science application for management in Alaska. Based on initial needs assessments, the AFSC's early work focused on organizing and consolidating existing fire science information through its website and sharing researchers' work through webinars and workshops. More recently, the consortium increasingly emphasizes facilitating coproduction of science relevant to managers' concerns. This involves strengthening relationships between scientists and managers to identify common research interests and opportunities, and working directly with interagency management groups to co-lead workshops and seminars. Both of these require long-term relationship building and organizational longevity to make progress.

Based on more than five years of experience, AFSC has identified two main aspects of fire science outreach in Alaska. First, strategic knowledge, or tacit knowledge of what is politically and administratively feasible, plays a large role in the use of science and development of research priorities by management personnel (Hulme, 2014). Scientists are not trained or accustomed to identifying this type of knowledge and incorporating it into their research process. AFSC as a boundary organization has been able to bridge the management and scientific realms by explicitly communicating tacit management knowledge to scientists to increase the management application of their research. Second, due to the political sensitivity of wildland fire suppression, outreach to the public regarding fire science issues must be carefully coordinated with agency public information processes.

In a Time of Change: The Art of Fire

In 2012, the AFSC embarked on an innovative interdisciplinary visual art project designed to generate excitement, facilitate mutual understanding, and promote meaningful dialogue on issues related to fire science and society. In a deliberate effort to expand outreach to the public audience, this project partnered with the long-standing science and humanities organization in Fairbanks, Alaska, known as "In a Time of Change" and the Bonanza Creek Long Term Ecological Research Station.

Nine local artists were invited to embrace the inspiration of wildfire, fire science, and fire management to create a unique art exhibit, "In a Time of Change: The Art of Fire." Four field trips were organized for the artists to provide interactive opportunities with fire scientists and fire managers, giving them a behind-the-scenes look at what happens when a fire occurs, how scientific information is used in management decisions, and the many facets behind long-term ecological disturbance studies. The field trip element was structured for two-way exchange, allowing all participants to become immersed in the environment. The culmination of this project was a formal art exhibit at the Bear Gallery, Alaska Centennial Center for the Arts, Pioneer Park in Fairbanks, Alaska. The exhibit received more than 700 visitors in a three-week period with a notable 450 visitors at the opening alone. Simultaneous with the exhibit, AFSC hosted public lectures by the artists to discuss what they gained from the field trips, their creative process, and how they incorporated fire science into their art.

Evaluation surveys of artists, scientists, managers, and public exhibit visitors revealed that the exhibit motivated visitors to learn more about fire science, management, and wildfire protection in Alaska as well as an overall agreement that art exhibits such as this can be an effective mechanism for building public awareness and understanding of important issues (Trainor and Leigh, 2013).

The "Art of Fire" project was a deliberate expansion of AFSC's main target audience of fire managers to the public. It demonstrates how the interaction between artists, fire managers, and scientists can promote understanding and awareness of the scientific basis behind fire management practices in the context of changing ecosystems in Alaska. The project facilitated a sense of place and helped the public to understand the functionality of fire in the ecosystems of Interior Alaska. Understanding that role is an important part of effective land management.

BOX 7.5.3: NORTHERN CLIMATE EXCHANGE: MAINSTREAMING CLIMATE CHANGE INTO GOVERNMENT DECISION-MAKING

The Northern Climate ExChange (NCE), part of the Yukon Research Centre at Yukon College, focuses on the study of climate change in Yukon and other northern areas. The NCE provides a credible, independent source of information, develops shared understanding, promotes action, and coordinates research on climate change in Yukon and across northern Canada. Based in Whitehorse, Yukon, the NCE has been leading climate change research in the Yukon since 2000.

The NCE promotes and coordinates research on effects and adaptations, including risk and vulnerability assessments, coordinates the exchange of scientific and local knowledge and expertise, and provides mainstreaming and decision-making support, policy alternatives, and climate change education for a wide range of partners and audiences.

The NCE works with federal, territorial, municipal, and First Nation governments, academic institutions, industry, and communities to respond to northern needs. The NCE recognizes that community participation in research and decision-making is essential to long-term resource, environmental, and cultural sustainability in the North. Their work emphasizes the collection of local and scientific knowledge and expertise in order to reflect northern perspectives, priorities, and needs.

The "Decision-Making for Climate Change" course was created in 2010 to assist with mainstreaming climate change into government decision-making. The NCE runs a Climate Change Information and Mainstreaming Program through a partnership with Yukon government Climate Change Secretariat; the course was developed through that partnership to achieve the program goals.

The target audience for the course is Yukon government employees, in particular decision makers, managers, and policymakers. The course was designed to create a shared understanding of climate change issues and to provide decision makers with strategies to address climate change. The course content is delivered during one full day and provides students with an understanding of the science and policy of climate change. Participants leave with information about climate change in Yukon and tools to help them incorporate climate change considerations in their work.

The course gathers together decision makers from various fields and departments, creating an opportunity for interdisciplinary discussions and problem solving. One of the outcomes of the course is the formation of bridges between departments, which can lead to future collaborations and help to break down government silos. The course also builds an informal network of people working to address climate change, which leads to information sharing and new partnerships.

The success of the course comes from the opportunity to engage face to face in a two-way flow of information with a group of interested stakeholders. Although the course is aimed at government employees, course delivery takes place at Yukon College; and the NCE, as a part of the Yukon Research Centre and an academic institution, has credibility with participants as a neutral source of climate change science and expertise.

8
SUMMARY AND SYNTHESIS

By Aynslie Ogden[1] and F Stuart Chapin, III[2]

INTRODUCTION

The Northwest Boreal (NWB) region includes south-central and Interior Alaska, most of Yukon, the northern part of British Columbia, and a small part of Northwest Territories. Change is a constant in this vast region, which has been, and continues to be, highly dynamic. This book addresses the drivers of these changes and describes whether the rate and magnitude of ecosystem dynamics are changing.

This region is globally significant, in part because climate change is unfolding faster here than anywhere else on Earth (Clement and others, 2013), resulting in thawing of permafrost, decomposition of long-frozen organic matter, changes to lakes and rivers, and alterations of ecosystem structure and function. Local and international experts agree: the impacts of the changes that are happening within this region will not be limited to this region; they will feed back and influence the planet as a whole (Chapin and others 2014).

Resource management and conservation in the NWB region is done by a large number of agencies, both within and across international borders. This adds to the complexity of developing and implementing landscape-scale responses to changing conditions. The Northwest Boreal Landscape Conservation Cooperative (NWB LCC) was established in 2012 to bring together these agencies to identify shared interests and pool resources to address landscape-scale stressors.

The NWB LCC recognized that an important first step in addressing landscape-scale change is to understand what is changing and what is driving these changes. For this reason, the NWB LCC assembled an international team of experts to summarize the state of knowledge on drivers of landscape change in this region. The collaborative and voluntary nature of this effort is in itself significant and indicative of the willingness and capacity that exists in the NWB region to cooperate on understanding and addressing the drivers of landscape-scale change.

1 Government of Yukon, Whitehorse, Yukon, Canada.
2 University of Alaska Fairbanks, Fairbanks, Alaska, USA.

The objectives of this chapter are to (1) synthesize and summarize drivers of landscape change in the NWB region, (2) discuss the potential implications of these changes for resource management and conservation, and (3) suggest measures that can be taken as next steps to bring the various agencies together in understanding and addressing change in this region.

MAIN FINDINGS

This chapter summarizes the status and trends of drivers of landscape change in the NWB region. Tables are also provided that summarize the judgment of each chapter team on: the degree of confidence in the findings, and assessment of the state of knowledge and current monitoring capacity (Table 8.1); information gaps and the relative priority or need to address each of these gaps (Table 8.2); and emerging stewardship considerations and the relative priority or need to address each considerations (Table 8.3).

Landscape change in the NWB region is driven partly by external drivers. In addition to climate change, many other global influences place pressures on the environment of this region. Globalization, the long-range transport of pollutants, global population growth, and associated demands on natural resource consumption are examples of external drivers with implications for landscape-scale change in the NWB region.

Disturbance

Natural and human-caused disturbance affects forest and plant community structure and ecosystem function. Wildfire, forest insects, and pathogen outbreaks and the effects of introduced invasive species are important drivers that act independently, or in concert, across the landscape and over time, to shape the forests and the cultural landscape of the boreal region. Whether the climate becomes warmer/cooler and wetter/drier in any given area within the region, there will be changes in the nature and effect of the natural disturbance drivers on successional pathways and the resultant structure of forest within the boreal region.

Physical Drivers

During the late 20th and early 21st centuries, mean annual temperatures within the NWB region have increased by approximately 2°C (3.6°F). Precipitation also has increased, although the observed trends are small compared to year-to-year variability. Future scenarios indicate temperature and precipitation in the NWB LCC region likely will continue to increase. Changing climate is driving changes in other physical drivers, including the length of the growing season, degradation of permafrost, and changes in hydrological regimes.

Biological Drivers

Modern vegetation, the basis for all terrestrial subsistence resources, has been relatively stable across this region for the past 5,000 years. However, climate change is expected to result in fairly rapid shifts in species composition, which in turn will result in novel community and trophic assemblages. Changes in climate and ecological conditions could lead to an emergence or establishment of diseases that affect animal and human health, as well as influence the susceptibility of hosts to disease, rates of disease transmission, and facilitate invasion of novel pathogens and parasites. Changes in the viability of keystone species can result in cascading trophic effects on other species. Understanding ecological functions and processes is the most comprehensive means of assessing and dealing with change.

Socioeconomic Drivers

Landscape dynamics in the northwest boreal region are driven by human and natural factors. Many of the triggers to environmental impacts and landscape change in boreal Alaska and Canada are a complex result of regional demographics and traditions, local resource concentrations, and global economic and political incentives. Transparent documentation of these drivers and their effects is identified as a common need for understanding these pressures and agents of change, and is much needed in order to capture beneficial opportunities and manage or mitigate their negative consequences.

Interactions among Drivers

Drivers of landscape-scale change in the NWB LCC region do not occur in isolation. Physical, biological, anthropogenic, and institutional drivers interact. Many of the drivers, and the interactions between drivers, are novel. Current understanding of these interactions is hampered by complexity inherent to these social-ecological systems, creating considerable uncertainty about future conditions. Consequently, management and conservation efforts must plan for a range of possible future conditions.

Practices of Coproduction

Researchers can no longer merely "deliver" information and products to local peoples. Scientists are increasingly expected to engage in iterative collaborations with managers, communities, and other stakeholders to coproduce scientific knowledge that is relevant to the problem context and bridge traditional and scientific approaches to knowledge of ecosystem management. Decision-making, particularly across jurisdictions, requires a solid foundation of data and collaborative data management systems.

Learning by doing, in an experimental framework (that is, adaptive management), is the most useful approach to dealing with complexity. Vulnerability assessment, scenario planning, and appropriately scaled projection models can be used to capture existing knowledge and frame alternative futures. Management is mandated and delimited by institutional structures, and is best if it is flexible and responsive to new knowledge.

The institutions responsible for resource management and conservation also drive changes in the NWB LCC region. Institutions influence the pace and scope of responses to ongoing change. Because many natural resources (for example, salmon, caribou, air, water) cross jurisdictional boundaries, international agreements are another important driver of resource management and conservation.

IMPLICATIONS OF LANDSCAPE-SCALE CHANGE FOR RESOURCE MANAGEMENT AND CONSERVATION: THE NEED FOR STEWARDSHIP

Given the speed and complexity of recent and projected changes in the NWB LCC region, it is not likely that the ecology and communities of the region will return to the patterns that were typical of the last 5,000 years. Instead, plans for expected—but uncertain—future changes can be established.

Stewardship is a framework for actively shaping trajectories of ecological and social change to foster a more sustainable future for species, ecosystems, and society, sometimes promulgated in the context of public law. Its main features are active intervention, shaping of change, and recognition that people and the rest of nature are integral components of the same regional system. It advocates proactive management of changes to support ecosystem and human well-being rather than reacting to problems after they have become serious. Stewardship is based on respect for nature (meeting conservation needs) and for cultural integrity and societal well-being (meeting community needs; Chapin and others, 2015).

Stewardship is challenging to implement. Although many of the recent and likely future trends are clear in broad outline, there is substantial uncertainty about how these changes will play out over years to decades—a critical timeframe for management. Additionally, the interactions among studied drivers of change will lead to responses that are difficult to project, and many future changes are currently unknown or unanticipated. Even when actions seem clear, they often are difficult to implement for institutional reasons. Treaties, laws, and regulations, for example, may not allow these actions to be taken, or people may decide not to take needed actions for political, economic, or cultural reasons.

Despite these challenges, there never has been a better time to implement a stewardship framework in the NWB LCC region. Recent and projected changes never have been so clearly documented or so well understood. Moreover, the collaboration of Indigenous, government, and university partners that has emerged within the NWB LCC demonstrates a commitment among many stakeholders to find a suite of integrated solutions.

A Stewardship Approach

The goals of stewardship are to:

1. Reduce the magnitude and rate of preventable changes;
2. Adapt to and shape changes that cannot be prevented; and
3. Avoid or escape unsustainable traps by fostering transformation to an alternative state that might provide better opportunities for ecosystems and society.

Stewardship is not trying to prevent inevitable changes or treating people as separate from nature or taking actions without consultation. There are several practical steps to implementing stewardship goals.

Stewardship must be based on traditional knowledge and scientific information that are needed for sound management of ecosystems and society. This includes information that documents recent and projected changes in ecosystem properties and processes, their effects on ecosystem services (the benefits that people receive from ecosystems), and the communities that depend on these services, as well as the effects of people on ecosystems (Fig. 8.1). Sound data management requires accurate information that is accessible and useful to various potential users, including communities, managers, and scientists.

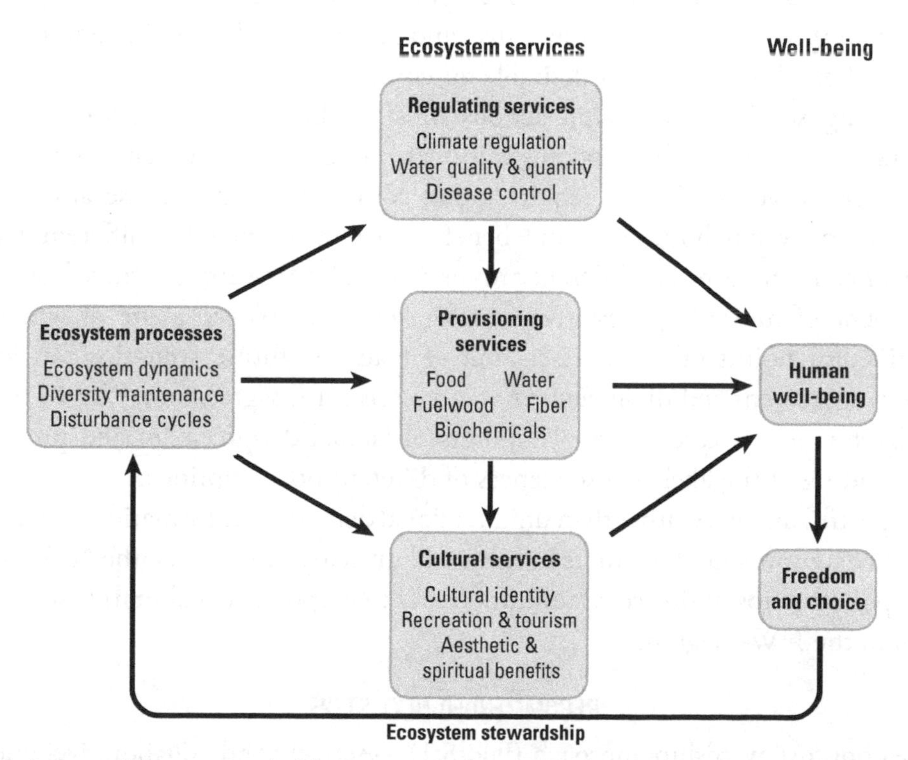

Figure 8.1. A stewardship approach. This approach does not seek to prevent inevitable changes, treat people as separate from nature, nor take actions without broad engagement and collaboration; it actively, and collaboratively, plans for expected—but uncertain—future changes.

Stewardship that seeks to shape changes should be based on cautious small-scale experimentation that recognizes uncertainties and gaps in human understanding. For example, adaptive management monitors the effects of management, learns from these observations, and adjusts actions to improve outcomes. Traditional knowledge often provides guidelines, based on generations of small experiments, of ways that people can respectfully interact with the rest of nature for their mutual benefit. Stewardship also benefits from observations of "natural experiments" such as the ecological changes that result from climate-driven changes in wildfire and permafrost.

When substantial interventions seem necessary to escape from unsustainable traps (for example, unsustainable rates of erosion near communities or dependence on diesel fuel), it is important to plan carefully for potential transformations through careful preparation, actions, and follow-up. Preparation might include the identification of possible futures (both good and bad) and of pathways that favor the likelihood of good outcomes. Implementation requires flexible strategies that are responsive to expected and unforeseen events. Follow-up requires building the resilience of those aspects of the new system that are favorable to ecosystems and society.

Policies strongly influence the opportunities for stewardship. It is important to be aware of policies that either foster or inhibit stewardship opportunities or to imagine new policies that might improve stewardship. Policies are most likely to foster stewardship if they are flexible, responsive, and adaptive to changes in conditions, and robust in their ability to deliver desirable outcomes.

Planning specific stewardship actions must be built on sound science and an understanding of the policy and institutional frameworks that shape opportunities and barriers to action. It also requires good communication because any potential action or policy has both costs and benefits that may differ for different types of species or ecosystems or for different groups of people. Stewardship therefore requires engagement of multiple perspectives among key stakeholders, some of whom may have different points of view. Discussion of plausible future scenarios can identify points of agreement and disagreement and reasons for disagreement. This in turn sets the stage for new research to resolve points of factual disagreement and provides an understanding of the likely consequences of different policy options.

Respectful dialogue rather than unidirectional delivery of information is an essential element of stewardship. Due to its breadth and diversity of partners, the NWB LCC is ideally poised to foster this communication and to empower collaborative stewardship actions in the NWB region.

RECOMMENDED NEXT STEPS

1. Adopt a stewardship approach (Fig. 8.1)—Actively and collaboratively plan for expected—but uncertain—future changes.

2. Enhance monitoring capacity (Table 8.1)—Long-term monitoring networks are needed to improve the understanding of drivers of landscape-scale change and to assess the efficacy of alternative management responses. The authors of this chapter have assessed the current monitoring capacity for each driver of landscape change and have provided suggestions as to where additional capacity is needed.
3. Address key information gaps (Table 8.2)—Because of the complex nature of forest systems and the uncertainty associated with how they may respond to changing climatic conditions, it is important that resource managers have access to information to guide management decisions. The authors of this report have identified a number of key information gaps, and have judged each according to the level of priority (or need) to be addressed.
4. Consider stewardship implications (Table 8.3)—The authors of this chapter have identified a number of considerations for resource managers that may facilitate the achievement of stewardship objectives for this region in light of the findings in this report. Each stewardship implication has been judged according to the level of priority (or need) to be addressed. Engaging in a deliberate process to consider these implications, and assess how best to pool limited resources to address landscape-scale stressors, is recommended.
5. Continue collaboration across disciplines, agencies, jurisdictions, and borders through the NWB LCC—Collaboration is an essential component of responding to landscape-scale change. The NWB LCC already has made significant progress in bringing together agencies across jurisdictions to identify shared interests and pool resources to better understand and address landscape-scale stressors.
6. Enhance data management and dissemination—The NWB LCC is making biological, physical, and socioeconomic data available through a collaborative web-mapping application. Future work on data standardization across the region is needed to improve the ability to bring together data from different sources.
7. Enhance communication—Communication is an important part of assessing and adapting to change. Communication is needed to enhance dialogue between those with varying interests in the future of the NWB LCC region in order to balance and effectively integrate these interests.
8. Increase the coproduction of knowledge—Participatory and collaborative approaches to research can enhance dialogue and understanding between scientists, policy makers, aboriginal groups, and industry, both within and across jurisdictions. Coproduction of knowledge also improves the understanding of the likely consequences of different policy options.

Table 8.1. Drivers of landscape change as assessed for the NWB region.

Column 1—Driver of landscape change: Indicates the driver of landscape change, as assessed in this book.

Agreement ↑			
	High agreement Limited evidence (M)	High agreement Medium evidence (H)	High agreement Robust evidence (VH)
	High agreement Limited evidence (L)	Medium agreement Medium evidence (M)	Medium agreement Robust evidence (H)
	Low agreement Limited evidence (VL)	Low agreement Medium evidence (L)	High agreement Robust evidence (M)
	Evidence (type, amount, quality, consistency) →		

Confidence Scale

Confidence: A depiction of evidence and agreement statements and their relationship to confidence. Confidence increases towards the top-right corner as suggested by the increasing strength of the shading. Generally evidence is most robust when there are multiple, consistent, independent lines of high-quality evidence. Source: Masterandrea and others, 2010.

Column 2—Status and trend (confidence): Summarizes the status and trend of the driver of landscape change, as well as the degree of confidence in the main findings. Confidence in the validity of a finding is based on the type, amount, quality, and consistency of evidence (for example, mechanistic understanding, theory, data, models, expert judgement). Confidence is expressed qualitatively using a five-point scale using the method described by Masterandrea and others (2010)1: *Very high confidence*—The author team has very high confidence in the finding(s) status and trend due to the availability of multiple, consistent, independent lines of high-quality evidence that are in high agreement; *High confidence*—The author team has high confidence in the finding(s) of status and trend due to the existence of either high agreement and medium evidence, or medium agreement and robust evidence. Medium confidence—The author team has medium confidence in the finding(s) of status and trend due to the existence of one of the following conditions: high agreement and limited evidence, medium agreement and medium evidence, or low agreement and robust evidence; *Low confidence*—The author team has low confidence in the finding(s) of status and trend due to the existence of either medium agreement and limited evidence, or low agreement and medium evidence; *Very low confidence*—The author team has very low confidence in the finding(s) of status and trend due to low agreement and limited evidence.

Column 3—Why it is important: Describes the importance of the status/trend of the driver of landscape change.

Column 4—State of knowledge: Expresses the judgment of the author team regarding the state of knowledge of drivers using the following qualitative five point scale: *Excellent*—The author team considers the state of knowledge about the driver to be excellent. There are no substantive knowledge gaps; *Very good*—The author team considers the state of knowledge about the driver to be very good. A minor knowledge gap has been noted; *Good*—The author team considers the state of knowledge about the driver to be good. Several minor knowledge gaps or one substantive knowledge gap has been noted; *Fair*—The author team considers the state of knowledge about the driver to be fair. Numerous minor knowledge gaps or several substantive knowledge gaps have been noted; *Poor*—The author team considers the state of knowledge about the driver to be poor. Numerous substantive knowledge gaps have been noted.

Column 5—Current monitoring capacity: Expresses the judgment of the author team regarding current monitoring capacity using the following qualitative five point scale: *Excellent*—The author team considers current monitoring capacity to be excellent. No enhancements to current monitoring programs are needed at present; *Very good*—The author team considers current monitoring capacity to be very good. Minor enhancements to current monitoring programs are needed in some regions; *Good*—The author team considers current monitoring capacity to be good. Minor enhancements to current monitoring programs are needed across the region; *Fair*—The author team considers current monitoring capacity to be fair. Major enhancements to current monitoring programs are needed in some regions; *Poor*—The author team considers current monitoring capacity to be poor. Major enhancements to current monitoring programs are needed across the entire region.]

[1]The confidence figure, used with permission, from Intergovernmental Panel on Climate Change (IPCC) from Mastrandrea and others (2010).

Driver of landscape change	Status and trend (confidence)	Why is it important?	State of knowledge	Current monitoring capacity
Disturbance—Wildfire	Increasing fire size and frequency (High confidence), and changing seasonality and severity (Medium confidence).	Changing fire conditions can affect patterns of forest succession.	Fair—There is poor historical or paleoecological data from which to infer trends in seasonality and severity.	Good—Remote sensing of fire activity across broad areas, but challenges in assessing fire severity and post-fire recovery.
Disturbance—Wildfire/ permafrost interactions	There is a trend toward increased wildfire in permafrost areas (Medium confidence).	There is the potential for fire to stimulate rapid permafrost degradation, increasing erosion and stimulating carbon release.	Fair—Both fire characteristics and permafrost response depend strongly on local conditions, and despite a few compelling examples, how this will affect future conditions is not yet understood.	Poor—Many of the important interactions require multiyear, field based observations in remote areas.
Disturbance—Wildfire/ climate change interactions	Disturbances may catalyze a rapid shift in biotic communities experiencing directional trends in climate (Low confidence).	This could dramatically alter the pace and direction of ecosystem responses to climate.	Fair—Responses to climate following disturbance may be qualitatively different from responses in undisturbed ecosystems.	Poor—Biotic monitoring often preferentially focuses on undisturbed areas.
Disturbance—Forest insect and pathogen epidemics	Climate oscillations are driving some insect outbreaks, increasing severity and extent (Medium confidence); other irruptive insect cycles appear to be driven mainly by interspecies interactions (Medium confidence). Temperature and precipitation increases are influencing the behavior of some forest pathogens (Medium confidence).	Insect and pathogen epidemics alter forest composition and structure.	Fair—Drivers of insect outbreaks not understood for many irruptive species, similar situation for pathogens though trend towards warmer and wetter conditions favor many fungi.	Good for insects as aerial surveys monitor status and trends for insects, but not causes. Fair for pathogens as most pathogens require ground based monitoring and the NWB region is vast.
Disturbance—Invasive species	Exotic and potentially invasive species are being introduced to the region at an increasing rate (Medium confidence).	Once introduced, it is expensive to remove an invasive species and the chances of full removal are very low.	Fair—There is little information on the ecological effects of invasive species.	Fair—Current monitoring capacity varies between jurisdictions.

Driver of landscape change	Status and trend (confidence)	Why is it important?	State of knowledge	Current monitoring capacity
Physical—Climate and climate change	Temperature is increasing and will continue to increase at time scales of 30-year climate normals (High confidence). Precipitation is increasing and will continue to increase at time scales of 30-year climate normals (Medium confidence).	Temperature and precipitation affect the physical and ecological template of the entire region, and exert strong influence on the physical processes, biology, and hydrology of the region.	Good—Subregional variation in historical climatology is less well understood than in comparably sized regions to the south due to the sparseness of stations with long-term data and poor representation of high-elevation stations; however, regional trends are evident.	Fair—Major enhancements to current monitoring programs are needed in some regions. Specifically, higher elevations are underrepresented, and the number of long-term climate stations is insufficient to adequately characterize the diversity of climatic variations and future climate downscaling at scales required for some applications.
Physical—Hydrologic response	An earlier onset of, and more rapid snowmelt, has been observed as well as more frequent midwinter rainfall periods. Continuous permafrost zones: mean flows have increased, annual peak flows have decreased, and winter low flows have had significant increases. Discontinuous permafrost zones: mean flows have increased, annual peak flows have slightly decreased, and winter low flows have had significant increases. Sporadic permafrost zones: variability in mean flows and winter low flows have been observed, annual peak flows have slightly decreased. (High confidence)	Resource development and climate change have already affected water resources in the region. Changing hydrological conditions have wide-ranging effects on ecosystems, disturbance regimes, transportation and hydroelectricity production, among other factors.	Fair—Additional baseline information is required for inventory purposes, economic development, public safety, and transboundary monitoring.	Fair—An enhanced water quantity and quality monitoring network is required to respond to additional pressures on water resource regimes.

Driver of landscape change	Status and trend (confidence)	Why is it important?	State of knowledge	Current monitoring capacity
Physical—Permafrost	Ground temperatures in boreal regions have remained stable or have decreased slightly during last decade. Permafrost degradation has already delivered significant consequences to ecosystems, land use, and infrastructure stability. (High confidence)	Permafrost, ground that remains below 0°C for 2 or more years, is a fundamental factor affecting the patterns and processes of boreal ecosystems and the human use of northern lands.	Fair—Ground ice amount and distribution is poorly quantified across the region.	Good—Long-term monitoring of active-layer thickness by a large international collaborative team (CALM) has documented trends over the last several decades.
Physical—Growing season length	The length of the growing season has increased (High confidence).	Growing season affects ungulate behavior, winter transportation, species composition, drought (longer growing season, less effective moisture) and the length of the fire season.	Fair—A better understanding of temperature, precipitation, and soil moisture projections will improve future scenarios of growing season productivity.	Good—The main source of information on growing season length is satellite observations.
Physical—Enduring features	Enduring features (abiotic components of landscapes) are generally stable over time, and have long-term influence. Many are well mapped. Hydrological and cryospheric elements are trending in some parameters, but most features are stable. (High confidence)	Many organisms rely on particular enduring features as habitat, or for dispersal and colonization. Other enduring features may block movements. Shifts in distribution, a primary response to climate change, need understanding of these facilitating or blocking actions.	Good—Many enduring features are well mapped; a considerable number still need inventory and mapping. Gaps in interpreting their relevance to various taxa remain.	Good—Many enduring features are stable enough that monitoring can be intermittent. Others (notably water and ice) require more attention.
Biological—Vegetation composition change	Modern vegetation in the northwest boreal forest has remained fairly stable for the last 5,000 years. Direct and indirect effects of climate warming are strong drivers of vegetation change in Alaska and northwest Canada. (High confidence)	Low species diversity contributes to high vulnerability to the effects of climate change. Vegetation is the basis for all terrestrial subsistence resources and therefore understanding drivers of vegetation change is critical for management.	Fair—Improvements in the understanding of the relation between climate and vegetation change from paleorecords along with enhanced species composition map coverage and experimental studies will increase the ability to accurately estimate vegetation change into the future.	Fair—Additional monitoring is needed to improve ability to understand and estimate vegetation change.

Driver of landscape change	Status and trend (confidence)	Why is it important?	State of knowledge	Current monitoring capacity
Biological—Novel community and trophic assemblages	Novel ecosystems have repeatedly emerged in the NWB region since the last glacial maximum, and current climate change will result in fairly rapid establishment of novel communities. (Medium confidence)	In a world of directional change, novel futures should be expected. Intrinsically, change is neither bad nor good, but in this context is simply the reshuffling of species distributions and ecological processes in response to directional perturbations.	Fair—A better understanding of dominant ecological processes and functional interrelationships is needed to improve projection models.	Fair—Additional monitoring of changes in species distributions and in regimes of disturbance and land use is needed.
Biological—Wildlife parasite and pathogen lifecycles	Climate warming and anthropogenic activity have been linked to emerging infectious diseases in wildlife. (High confidence)	The rapid rate of ecological change in the NWB region may lead to emergence or establishment of diseases that affect animal and human health.	Fair—Baseline data and understanding of relation between environmental variables and lifecycles of pathogens and parasites about wildlife disease in many parts of northern Canada and the United States are lacking.	Fair—Major enhancements to current monitoring programs are needed in some regions. Baseline data are being collected and wildlife disease is being researched in some areas of northern Canada and the United States on a range of species, but large knowledge gaps remain regarding pathogen distribution and lifecycles, and effects on wildlife populations.
Biological—Marine derived nutrients (MDNs)	The available nutrients and biomass from returning anadromous fishes strongly affect freshwater and riparian communities and have trophic consequences that affect the abundance of other species. The structure and abundance of other species within the riparian system feeds back to freshwater through nutrients and prey affecting riverine productivity. (High confidence)	Salmon and other anadromous species interact synergistically with bears, wolves, other carnivores, and scavengers to provide a keystone ecosystem service, maintaining nutrient flow and sustaining biodiversity in interior boreal riparian systems.	Fair—Although marine derived nutrients strongly affect freshwater and terrestrial communities, how many species are affected, the extent to which they are affected, and the response of an ecosystem to varying salmon escapement numbers (for example, what escapement supports bear or wolf densities, or stream food web productivity levels) are not known.	Fair—Difficulties in reliable escapement monitoring prevent widespread and accurate adult return counts.

Driver of landscape change	Status and trend (confidence)	Why is it important?	State of knowledge	Current monitoring capacity
Socioeconomic—Land use change and resource extraction	The NWB region remains largely wilderness, dominated by natural ecological processes and a low human impact. (High confidence)	The impacts of land use change and resource extraction result in more immediate and localized effects.	Good—The geographic footprint of resource development and land use change is moderately well documented. Poor—The overall effects on ecological processes and off-site impacts remain poorly understood.	Fair—There is no central depository of information on resource development approvals and progress.
Socioeconomic—Rural and Indigenous livelihoods	Many aboriginal peoples of the NWB region continue to depend on the harvesting of wild mammals, birds, fish, berries and fuel wood to meet their basic needs. (High confidence)	Although specific effects and industrial development factors are recognized as drivers of change, the dispersed use of biological resources by many individual households also could be having significant effects on fish, wildlife, and other resources.	Fair—Although specific effects of hunting and foraging on broader ecosystem processes are largely unknown, work elsewhere suggests they are greatest near villages and transportation corridors.	Fair—Consistent harvest data are lacking for the Northwest Boreal region, limiting understanding of the effects of hunting and fishing on ecosystems.
Socioeconomic—Contaminants	The NWB region contains hundreds to thousands of sites with organic and inorganic contamination from military, state and federal government, and commercial uses and is subject to contaminant loading from long-range transport. (High confidence)	Contaminated sites across the NWB region pose risks to human and environmental health through a number of exposure pathways.	Fair—Not all point and non-point sources of contamination have been documented.	Fair—The diverse diet and geography of the region has prevented full characterization of contaminants in subsistence foods.
Socioeconomic—Law and policy	Most land in the NWB region is publicly owned, and is managed in trust by democratically elected governments. Governance of citizens' actions regarding land uses and land management in the NWB region is complicated by multiple layers of government (federal, state, provincial or territorial; Indigenous), often with overlapping spatial and topical interests and responsibilities.	Decisions made by government institutions frequently require referral and consultation with numerous agencies and the public in order to achieve social license. Despite short-term lengthening of decision-making processes, acquiring social license increases effectiveness of decisions and their subsequent actions.	Fair to Good—In Canada, the courts, along with future land claim agreements, need to provide greater clarity as to the powers First Nations have over decision-making on public lands.	Fair—Additional work is needed to monitor the effectiveness of government and institutions in achieving resource management objectives.

Driver of landscape change	Status and trend (confidence)	Why is it important?	State of knowledge	Current monitoring capacity
Socioeconomic—Values and ethics	Values and ethics drive policy and management actions. The emergence of conservation social science highlights the demand for multidisciplinary, complex social-ecological analysis techniques. (High confidence)	Adaptation requires selecting among trade-offs, informed by human beliefs and ethics just as much as by facts. Values determine how the public receives and responds to scientific knowledge and policy-making.	Poor—Current adaptive management approaches rarely incorporate values and ethics. Making these more explicit will allow practitioners to constructively harness such knowledge.	Fair to Good—Theory and techniques to understand and incorporate values and ethics into study designs are well developed; however, their application in environmental management is limited.
Interactions—Interactions among drivers	Complex interactions and feedback loops occur between disturbance, physical drivers, biological drivers, and socioeconomic drivers. These feedback loops can be positive or negative, and either immediate or delayed.	Feedback and interactions can radically alter the ultimate effects of any one driver, either exacerbating or ameliorating its impacts.	Very good to poor—Complex modeling efforts are increasing the ability to simulate interactions. However, lack of historical data and lack of fine-scale temporal or geographic resolution in what data do exist make model calibration difficult. Threshold values are particularly hard to pinpoint.	Fair—Monitoring of interactions and feedback is dependent on monitoring of multiple variables, and thus can only be as good as the weakest of these efforts.
Interactions—Cumulative effects	Cumulative effects and resulting impacts are variable across the NWB region. In areas of increasing industrial development or human use, greater changes in the landscape are expected, and thus greater resulting impacts to important ecological values. Cumulative effects give rise to important uncertainties that must be considered during assessment and monitoring. (High confidence)	Cumulative effects represent the additive, synergistic, non-linear or unpredictable changes to ecological systems and values. Thus, both natural and anthropogenic drivers of change can interact and be integrated as impacts with relevance to ecological processes and the many consumptive and nonconsumptive values across the region.	Very good to fair—The extent of cumulative effects are likely well quantified in some areas of the NWB region, depending on past planning and assessment efforts. Likewise, the impacts to some values are well known or can be predicted. Parts of the NWB region with limited industrial activity and ecological values with no direct relevance to human populations are not considered or well understood. The additive or interactive effects of natural and anthropogenic disturbance sources are particularly uncertain.	Fair—Modeling tools are available for quantifying cumulative effects and impacts. However, identifying, measuring, predicting and then monitoring impacts will require considerable investment in strategic and inclusive assessment processes.

Driver of landscape change	Status and trend (confidence)	Why is it important?	State of knowledge	Current monitoring capacity
Practices of Coproduction—Meaningfully engaging communities	Recent research from both social and natural sciences has shown that to meaningfully engage Indigenous communities, researchers can no longer merely "deliver" information and products to local peoples (High confidence).	Research relationships between scientists and communities historically have been premised on colonial power relationships that have often caused Indigenous communities to mistrust researchers and research products. Indigenous communities know different kinds of information about the ecosystem than that produced through western sciences.	Fair—More work is needed to bridge traditional and scientific approaches to ecological knowledge and ecosystem management.	Fair—Additional efforts to monitor and evaluate the effectiveness of various processes to meaningfully engage communities in research are warranted.
Practices of Coproduction—Evolving roles of scientists and managers	Scientists are increasingly expected to engage in iterative collaborations with managers, communities, and other stakeholders to coproduce scientific knowledge that is relevant to the problem context. (High confidence). Managers are increasingly required to engage in participatory processes that help to balance and integrate multiple interests to effectively respond to and shape ecological change (High confidence).	Professional ideals predominant in natural resource management (for example, historical condition, naturalness, the "scientific management" approach, and ecological integrity) may not be viable in directionally changing social and environmental conditions.	Fair—Little is published about how management regimes or agencies cope with/adapt to biophysical changes in the NWB region; comparative studies of management regimes and collaboration are few and limited in geographic scope.	Poor—There is no mechanism for monitoring and assessing the effectiveness or adaptive capacity of management regimes.
Practices of Coproduction—Assessing and projecting change	The development and application of scenarios and simulation tools in natural resource management is increasing. Confidence in the effectiveness or usefulness of these approaches compared to others in decision-making is medium to low, given the recent and limited number of applications.	Effective conservation action must be grounded in an understanding of change caused by biophysical and anthropogenic drivers. As we create scenarios/projections of future conditions, more informed and collaborative decisions can be made and possible management intervention points to influence and adapt to agents of change can be identified.	Fair to Good—Scenario planning has a long history of use outside of natural resource management (business and military) and approaches and methods are fairly well established. Developing scenarios for natural resource management applications is more recent and limited by knowledge gaps such as approaches for integrated assessment of climate and land-use effects.	Fair to Good—Some status and trends monitoring systems are in place and focusing on climate change and understanding future trajectories and biophysical response (for example, National Park Service, Inventory and Monitoring Program (I&M), and U.S. Fish and Wildlife Service Refuge system). The monitoring around management action effectiveness in the face of climate change and other drivers is poor to fair.

Driver of landscape change	Status and trend (confidence)	Why is it important?	State of knowledge	Current monitoring capacity
Practices of Coproduction—Data management and dissemination	A solid foundation of data management is required to support analysis, monitoring and decision-making. Lack of coordination in data-management systems can lead to duplication of costs and effort. (High confidence).	As the need to collaborate at a landscape scale increases, especially when dealing with environmental issues such as habitat connectivity and climate change, methods of gathering and sharing necessary data among project partners and stakeholders increases in importance.	Good—The next generation of databases, tools, platforms, and social networks are being adapted and integrated to better support landscape conservation in the NWB region and other regions.	Good—The NWB LCC is making priority biological, physical, and socioeconomic data and information widely available through a collaborative web-mapping application—the Alaska and Northwest Canada Conservation Planning Atlas (https://aknwc.databasin.org/).
Practices of Coproduction—Outreach	Practitioners are increasingly realizing that effective outreach is an ongoing, iterative, self-reflexive process leading to a two-way flow of information (High confidence).	This flow creates knowledge-to-action networks that can be important drivers in the process of assessing and adapting to change.	Fair—Emerging social media and advancing technology have the potential to reach a large and diverse audience, but require constant evaluation and revision of efforts.	Fair—Ongoing monitoring of the effectiveness of outreach activities and experimentation with new approaches is warranted.

Table 8.2. Summary of main information gaps identified for the NWB LCC.

Column 1—Driver of landscape change: Indicates the drivers of landscape change, as assessed in this book.

Column 2—Information gap: Highlights an information gap associated with the drivers of landscape change, as identified in this book.

Column 3—Priority: Describes the judgment of the author team regarding the priority (or need) to address each information gap using the following qualitative three point scale. *High*—Addressing this information gap is considered to be high priority by the author team. Knowledge-base enhancements in this area are considered to be critical (because, for example, they are likely to affect decision-making AND (OR) the understanding of important ecosystem processes); *Medium*—Addressing this information gap is considered to be medium priority by the author team. Knowledge-base enhancements in this area are considered to be important because (for example, they are likely to affect decision-making AND (OR) the understanding of important ecosystem processes); *Low*—Addressing this information gap is considered to be low priority by the author team. Knowledge-base enhancements in this area are considered to of lower priority because (for example, they are unlikely to affect decision-making AND (OR) the understanding of important ecosystem processes).

Driver of landscape change	Information gap	Priority (high/medium/low)
Disturbance—Wildfire	Little is known about how fire and vegetation succession are likely to interact with permafrost in lowland forests and peatlands or in tundra environments.	High—These interactions may be very important to forest carbon dynamics over the coming century.
	Fire-initiated changes in forest vegetation may affect large mammals such as moose, caribou, elk, and bison. However, there is no good indicator of how a shifting forest mosaic may lead to increases in some mammal populations and decreases in others. Additionally, how will changes in the distribution and abundance of large mammals affect patterns of forest succession?	High—Large mammals such as moose, caribou, elk, and bison are essential to food security and cultural identity in many aboriginal communities.
	Although it is known that fire may initiate the conversion of conifer forests to deciduous broadleaf forests, the degree to which these changes are likely to occur at landscape and regional scales is not known. What are the implications of such a conversion for landscape and regional processes of water, energy, and carbon cycling?	High—A greater ability to understand and anticipate landscape responses to changing wildfire characteristics, that builds on the understanding of ecosystem succession and fire behavior, is essential.
	Further research is needed on how shifts in forest succession patterns and the mosaic of forest types will influence future fire regime and direct climate effects on boreal forests.	Medium—Research is needed to understand the extent to which shifts from conifer to deciduous forest cover may compensate for an intensified fire weather regime by altering landscape flammability.

Driver of landscape change	Information gap	Priority (high/medium/low)
Disturbance—Forest insect and pathogen epidemics	Drivers of insect outbreak cycles for most insect and pathogen cycles are unknown. Reconstruction of past climate combined with dendrochronology and life-history studies have enabled the understanding of the relationship of climate and outbreaks in the case of the spruce beetle; however these kinds of data are lacking for the most irruptive species. More information is needed on the relation between climatic drivers such as El Niño and foliar and rust pathogens.	Medium—Understanding drivers of irruptive species would enable estimation of outbreak likelihood and inform selection of appropriate management options.
	Although conspicuous insect and pathogen outbreaks are currently monitored through aerial surveys, the kinds of data needed to describe irruptive cycles—long-term, quantitative data on insect densities and outbreak incidence—are lacking for most of our irruptive species.	Medium—Identifying cyclic events is a prerequisite to learning and understanding the causes of cyclicity.
Disturbance—Invasive species	Baseline information is still needed on native species' distributions within the NWB region, especially for insects and pathogens.	Medium—In some cases it has been difficult to determine whether a newly detected species is an overlooked native or a new invader.
	Risk assessment of potentially invasive species that considers likelihood of introduction, potential vectors, likelihood of persistence, and ecological/economic/social impacts.	High—Invasive species can have profound effects on native ecosystems, causing displacement or mortality of endemics, outcompeting native species for limited resources, and altering food webs.
	More information on the ecological effects of the most aggressive invaders already in the region (sweetclover, bird vetch, Elodea, and green alder sawfly) would help land managers determine appropriate responses.	High—There is little information on the ecological impacts of invasive species in the NWB region.
	Because pesticide degradation rates are significantly slower in cold climates than in the climates where the pesticides were developed and tested, information on pesticide fate in the climate of the NWB region is needed.	Medium—Information on the fate of pesticides in this region is needed to help managers evaluate their utility in managing infestations.
Physical—Climate and climate change	Subregional variation in historical climatology is less well understood than in comparably sized regions of southern Canada and the conterminous United States due to the sparseness of stations with long-term data.	Medium—An improved understanding of subregional variation in climate will improve projections of climate and climatically derived variables, which is important for adaptation planning.
	High-elevation stations are poorly represented in weather monitoring networks.	High—Climate at higher elevations may be disproportionally important for understanding the future utility of refugia.
	Projecting derived climatically related variables, such as drought or runoff, depends greatly on interactions with other factors, such as permafrost, glaciers, and vegetation, which may also change with climate change.	Medium—Refinement of projections of climate-related variables will support adaptation planning.
Physical—Hydrologic response	Enhanced water-quality and -quantity monitoring networks and baseline information are required to respond to additional pressures on water resource regimes and to support economic development, public safety, and transboundary monitoring.	High—Improved and expanded systematic and comprehensive monitoring will improve the ability to respond to changes in the hydrologic system.

Driver of landscape change	Information gap	Priority (high/medium/low)
Physical—Permafrost	Quantification of ground ice amount and distribution across the region.	High—The magnitude of societal and ecological consequences of permafrost degradation will depend on climate and ground-ice conditions.
Physical—Growing season length	Future changes in effective moisture.	Medium—Effective moisture scenarios are needed to indicate whether browning or greening can be expected.
	Snow cover changes as related to soil moisture and storage.	Medium—Soil moisture and storage scenarios will inform expectations on growing season productivity.
	Future scenarios of growing season length.	Medium—Scenarios of growing season length are needed to inform and educate adaptation planning.
Physical—Enduring features	Conservation planners have applied mapping of enduring features relatively rarely in even static conservation planning, and less so in the context of shifting distributions, so the approach needs more testing.	Medium—The value of applying enduring features to projections of future biodiversity distributions needs to be better understood.
	The approach depends on a complete inventory of each selected enduring feature in the planning region, and such inventories may still have to be developed (or completely overlooked) especially for some of the spatially limited features such as glacial landforms, springs, and karst features.	Low—Inventories of all potential enduring features are difficult and expensive to accomplish. Considerable advances can be made by dealing with other information gaps first.
	Meaningful application of enduring features or land facets to planning in the context of climate change will require a systematic cataloging of the potential role of these features in the distribution and colonization ability of diverse species or species guilds, and careful selection of connectivity models and their parameterization, to derive meaningful results. Some such information may be lacking, and application may require an iterative and exploratory approach in most regions.	High—Understanding the potential value of enduring features depends on a better matching of particular features with habitat and movement characteristics of individual species. Exploring these options is a high priority in any assessment and application of enduring features as a conservation planning tool.
Biological—Vegetation composition change	Spatial resolution of paleorecords	Low—Although having a larger-scale understanding of historical vegetation change, due to the nature of data collection, this information gap will continually be altered and improved.
	Vegetation mapping based on species composition	High—Plant communities will change in response to climate at the species level first, vegetation mapping efforts that are species-based will be of critical importance as changes in plant community and function are identified.
	Mapping that links the interior of Alaska and western Yukon	High—The arbitrary "line" that separates Canada from Alaska has limited the ability of researchers to observe and study large-scale vegetation patterns.
	Experimental vegetation studies that manipulate both the direct and indirect effects of climate	Medium—Disentangling the direct and indirect effects of climate can be challenging; however, ecosystem-scale experiments can increase knowledge about rate and mechanisms of change.
	Drivers of vegetation change at both the northern extent and southern boundary	Medium—Factors affecting species migration will provide important information for modeling efforts.

Driver of landscape change	Information gap	Priority (high/medium/low)
Biological—Novel community and trophic assemblages	Fine-scale spatially explicit species richness data; that is, extant species assemblages	Medium—Need to know and monitor which species exist where, but a select focal subset might suffice given the enormity of the task.
	Rates of ecosystem drivers and thresholds (for example, fire, spruce bark beetle, extreme weather events)	Low—Rates are crucial for building projection models of ecosystems, but accurate projections of timing will not be possible.
	Measures, tool, and management approaches (for example, assessing the need to intervene)	High—New monitoring datasets, combined with modeling tools and experimental approaches to management, will be the foundation for assessing options and investigating effects of changing species distributions.
	Measuring ecological difference in monitored metrics to determine what is novel	Medium—Novelty can occur in species distributions, ecological processes and functions, and human actions and values. The potential for novelty is derivative of monitoring and projecting change in distributions, and processes and values inside/adjacent to the NWB region.
	Capacity to detect ecological surprises	Low—Change creates surprise when it is adverse and unexpected to people. Detecting surprise is derivative of understanding values and projecting change, and thus contingent on other priorities (values assessment; projection tools).
	Articulating management goals in a directional world	High—Management goals may need to change given the long-term directional change underway. Assessing those goals in light of changes and expressing them publicly are crucial steps prior to, or in tandem with, monitoring, scenario development, and experimental actions.
	Underlying values that shape social, ethical, and policy dimensions of embracing and even designing novel ecosystems	High—Investment in understanding and projecting change depends on social licenses and political will, which are fundamentally rooted in social values and attitudes, which need to be periodically articulated and critiqued with reference to how change is progressing.

Driver of landscape change	Information gap	Priority (high/medium/low)
Biological—Wildlife parasite and pathogen life cycles	Baseline information about the current distribution of pathogens and parasites in the NWB region is lacking, limiting the ability to identify and track emergence or expansion of infectious diseases.	High—Baseline information about wildlife disease provides a basis for monitoring changes in geographical and host distribution and prevalence. Novel wildlife pathogens and parasites also may be detected as part of long-term surveillance programs.
	Relationships between environmental variables, such as temperature and precipitation, and life cycles of many pathogens and parasites that infect northern wildlife are not well understood.	High—More field data are required to produce and validate models of projected pathogen lifecycle and distribution changes associated with climate warming. Results will inform decision-making by wildlife managers and provide guidance about human health risks associated with zoonotic diseases.
	Important aspects of the ecology of many wildlife diseases, such as transmission routes, reservoirs, and carryover effects, have not been established in northern ecosystems.	Medium—Although many diseases are well described in domestic animals and (or) in wildlife in southern regions of North America, research and monitoring are required to understand the ecology of these diseases in northern wildlife.
	Demographic and fitness consequences of disease on wildlife populations are not well known, making it difficult to project the outcomes of changing pathogen and parasite communities.	High—It is critical for wildlife managers to have a better understanding of the effects of disease on wildlife populations. Morbidity, mortality, and reproductive loss may have enormous impacts on the viability and maintenance of wildlife populations and, consequently, on the people who rely on wildlife resources for subsistence.
Biological—Marine derived nutrients (MDNs)	Information is sparse on the feedbacks from salmon between streams and riparian habitats, and the indirect effects from spawning salmon and other fishes. For instance, does the nutrient input from salmon feed back to streams via more and better quality prey for fishes?	High—Trophic consequences of reduction in the available MDN to riparian ecosystems may result in reduced biodiversity and (or) species loss. Clearer understanding of how MDNs are transported through and feedback into ecosystems can provide information to manage or mitigate species loss.
	Research is needed to understand density-dependent mortality in salmon and the implications of increased escapement on this mortality (for example, corresponding nutrients needed to maintain productive rearing salmonid populations). Do salmon returns of higher or lower escapement levels mean higher or lower returns (from trophic benefits of salmon biomass) of their own and other species?	High—MDNs can have profound implications to salmon stock size, yet the science of fisheries stock assessment currently does not account for MDN density-dependent mechanisms. Without an understanding of the effect of MDN feedback on salmon returns, estimating future of harvest levels is challenging.
	Further research is needed to understand nutrient levels and carcass biomass needed to maintain both freshwater and terrestrial ecosystem components. Little is known about whether ecosystems are operating on the legacy effects of past salmon runs. These riparian systems may be functioning at a higher level of inherent productivity from hundreds of years of salmon returns.	High—If salmon and food-web productivity is dependent on legacy effects of past salmon runs, eventual depletion of these long-term stores could lead to long-term salmon population and freshwater productivity declines.
Interactions—Interactions among drivers	Many interactions are unknown, not just in magnitude, but also in direction. Even in cases where the directionality of interactions is clear, thresholds have not been precisely pinpointed. Unknown thresholds may exist that cause rapid and unprecedented shifts.	High—Without a clear understanding of feedback between drivers, the socioeconomic and biophysical effects of changes to specific drivers will be systematically under- or overestimated, and poor management choices will result.

Driver of landscape change	Information gap	Priority (high/medium/low)
Interactions—Cumulative effects	Assessment and management of cumulative effects will require a full spatial accounting of past, present, and reasonably foreseeable future effects (that is, landscape change). These data are likely complete for some drivers of change in some areas of the NWB region. An accounting of effects should be followed by a process to identify important values and the relationships between the effects and the 'health' or persistence of those values (that is, negative or positive effects).	High—Cumulative effects represent all drivers of change and their potential interactions. Thus, a full accounting of present and future landscape change is achieved through an integrative approach such as a cumulative effects assessment and planning process.
Socioeconomic—Land use change and resource extraction	Centralized compilation and regular updating of current land use designations and resource extraction activities is required for quick referral.	Low—Information on land-use designations is largely already available, but needs to be centralized and made more easily accessible.
Socioeconomic—Pollution and contaminants	Existing contaminated sites and non-point sources of contamination have not yet been fully documented for the region. The quantity of contaminants sequestered in the environment and subject to liberation due to climate change and other factors is not yet known.	High—A basic inventory of contaminated sites is needed before prioritization for control or remediation can be undertaken.
	The rates and rate-limiting factors for contaminant transformation and biodegradation need to be more fully understood in order to better project contaminant fate and develop technologies for mitigation in the face of climate change.	Medium—Contaminant translocation and transformation trajectories need to be understood in the long term.
	Contaminants in subsistence foods are not well characterized due to the varied diet and vast geography of the NWB region.	High—High levels of contaminants in subsistence foods can pose significant risks to human health and longevity.
Socioeconomic—Rural and Indigenous livelihoods	Although recreational hunting and fishing may be well regulated, the degree to which wild forest resources are used by northern residents for subsistence and household use is poorly documented.	High—Subsistence use of wild resources has both ecological and socioeconomic implications, and represents a significant gap in our overall understanding of the NWB social-ecological system.
	The social, economic, and cultural value of subsistence uses of boreal resources is poorly documented, especially in Canada.	High—The lands and resources of the NWB have long been fundamental to the subsistence of many Indigenous and rural people in Canada and the United States.
	The relationship of subsistence resource use to economic factors (for example, access to market products and the wage economy) and environmental factors (for example, resource abundance and reliability) has been little investigated in boreal regions.	High—Today and in the past, fluctuations in harvested wildlife and botanical resources probably represent a complex conflation of yearly weather events and human influences.
	Effects of modern hunting and foraging, including strategies such as prey switching, on ecosystem processes and the sustainability of wild resources are largely unknown.	High—The effects of prey switching strategies have proven difficult to monitor with existing data collection practices in Alaska.
	The ecosystem- and landscape-level impacts of fish and game enhancement policies (for example, by means of predator control and prescribed burns) are understudied.	High—The active and intensive management of wild fish and game for use by local peoples is an important driver in the structure and function of boreal forest ecosystems

Driver of landscape change	Information gap	Priority (high/medium/low)
Socioeconomic—Law and policy drivers	In Canada, the courts, along with future land claim agreements, need to provide greater clarity as to the powers First Nations have over decision-making on public lands.	Medium—In Canada, First Nations' rights are constitutionally entrenched; however, despite recent land claims agreements with many First Nations, and because of a lack of agreements with many others, the division of jurisdictions, influence, and responsibility between governments is still unclear and frequently being contested in court. The resulting uncertainty is slowing down governance decisions.
Socioeconomic—Values and Ethics	Adaptive management research designs need to incorporate a model of the assumed ethics of human-environment relationships.	High—Different publics, institutions, and cultures practice their everyday lives from specific ethical standpoints reflecting values about human-environment relationships, yet rarely are these made explicit in research designs or management actions. When divergent, these ethics can prevent learning within adaptive management programs, and even exacerbate political inequalities across the social system.
	Assessment of whether and how usefully the different analytical values schema developed by non-Indigenous researchers correspond with northern Indigenous peoples' values.	High—The natural and social sciences have played a large role in the political subjugation of Indigenous peoples in the modern era; in many places globally, researchers have connected Indigenous dispossession with subsequent degradation of ecosystems and (or) maladaptive social-ecological systems. Understanding the ways in which northern Indigenous peoples' values differ from those of non Indigenous researchers will contribute to more effective methodological designs to 'coproduce' knowledge and learn with other constituencies.
	Application and testing of institutional design principles for commons resources in northern social-ecological systems.	High—Ostrom's institutional design principles were developed from decades of empirical research and provide the strongest, most-generalizable theoretical framework for understanding and improving the management of common-pool resources. That categorization includes most northern resources and ecosystems, yet there has been no systematic effort to test this framework in northern contexts, especially under conditions of rapid social and (or) ecological change.
	Reconciling value conflicts across scales, for example, when a global public criticizes local communities' traditional resource management practices.	High—Political conflicts can greatly impede effective multistakeholder resource management and erode the trust necessary to institute and follow regulations.
	Studies should more frequently integrate methodological tools such as Q-Method into their research designs to systematically understand the values sought and shaped by research participants.	High—How social systems are currently and most frequently measured offers only blunt tools to understand how different publics might make tradeoffs, or even their preferences amongst difficult policy choices. Yet there are numerous quantitative, qualitative, and mixed-methods research designs from the social sciences that can produce reliable measurements of values, as well as change in values over time. In these ways, elements of an adaptive system such as social learning can be quantitatively monitored over time.

Driver of landscape change	Information gap	Priority (high/medium/low)
Practices of Coproduction—Meaningfully engaging communities	Traditional management practices and Traditional Ecological Knowledge (TEK) in the different areas of the NWB region need to be understood prior to developing research questions.	High—Allowing space in the research procedure for community input to shape the research questions and agenda is an important element of a meaningful research relationship.
	More work is needed to bridge scientific and traditional approaches to knowledge of ecosystem management.	High—TEK contributes information and insights about ecosystems and ecosystem change at scales and ways of knowing that are different from approaches in western scientific disciplines. Meaningfully including these insights in the process of research will increase trust and make for more effective decision-making based on this collaborative research.
Practices of Coproduction—Evolving roles of scientists and managers	Appraisals of management performance across or within jurisdictions, over time for the NWB region (including comanagement regimes). Appraisals of how management regimes or agencies cope with/adapt to biophysical changes in the NWB region. Comparative studies of management regimes, with increased geographic scope.	High—Empirical understanding of the functioning of management programs and systems is necessary in order to avoid assuming that they actually work when they do not, and the reverse (assumption: information about management will actually be used in decision-making processes).
Practices of Coproduction—Assessing and projecting change	Projection and assessment of the integrated effects of climate change and land use.	Medium—A holistic perspective is needed to assess risks posed by the cumulative effects of climate change and multiple land uses operating over large spatial and temporal scales. Scenario analysis and simulation modeling can provide this perspective to guide identification of mitigation and adaptation opportunities.
Practices of Coproduction—Data management and dissemination	It has been observed that although the governments at different levels may have the broadest and most well-organized data, it often lags behind in provision of the most current data. Thus, a challenge lies in that the data at the most appropriate scale may not have the most appropriate content for a given environmental issue.	High—The pace of resource management planning is directly tied to the efficiency of accessing the necessary data for decision-making. For regional, transboundary planning, it is often difficult to find analogous data that can be compared across the entire geographic extent of the project. Finding the most accurate and most up-to-date data can take significant staff time if no previous effort has been made. Anticipating these challenges before they occur and making a plan ahead of time can save significant resources of time and money.

Driver of landscape change	Information gap	Priority (high/medium/low)
Practices of Coproduction—Outreach	Stakeholder/management preference and applicability of social media in boreal Alaska and Canada.	Medium/low—Emerging social media and advancing technology efforts have the potential to reach a large and diverse audience, but still require constant evaluation and revision of efforts. Face-to-face contact is still vital for deeper and more sustained engagement.
	Variable information needs across a vast distance, varied land managers, and different habitat types.	High—For information to be useful, it must be perceived by the stakeholder to be salient, credible, and legitimate; for example, to be relevant, come from a trustworthy source, and be unbiased. Understanding the different information needs across such a vast and variable region is paramount and challenging.
	Understanding the dynamics between formal and informal processes that deliver information and those that are useful in decision-making	High—Understanding the mechanisms behind decision-making will inform the manner and the information communicated and a firm understanding is integral in the two-way iterative flow of information between producers and consumers.
	Documentation and assessment of how outreach and scientific information have been directly applied in decision-making and resource management.	Medium—Outreach is an ongoing, iterative, self-reflexive process. Learn about stakeholders/audience, implement outreach efforts to reach them, evaluate the outcomes and then revise methods/model and try again. Documenting how information is applied not only helps justify the efforts but informs future interactions.

Table 8.3. Summary of emerging stewardship considerations as identified for the NWB LCC.

Column 1—Driver of landscape change: Indicates the drivers of landscape change, as assessed in this book.

Column 2—Emerging stewardship considerations: The Northwest Boreal Landscape Conservation Cooperative (NWB LCC) is promoting a stewardship concept, defined as the active shaping of pathways of social and ecological change for benefit of ecosystems and society to reduce the rate and magnitude of preventable changes, adapt to and shape changes that cannot be prevented, avoid or escape unsustainable traps by fostering transformation to an alternative state that might provide better opportunities for ecosystems and society.

Column 3—Priority: This table communicates the judgment of the author team regarding the priority (or need) to address each stewardship consideration noted using a qualitative three point scale. Concisely state the rationale for the assessment (one or two sentence description): *High*—Addressing this emerging stewardship consideration is considered to be high priority by the author team. These considerations are considered to be critical (for example, drivers of landscape change are likely to affect important ecosystem values AND (OR) rapid response is likely to have significant benefit); *Medium*—Addressing this emerging stewardship consideration is considered to be medium priority by the author team. These considerations are considered to be important (for example, drivers of landscape change are likely to affect important ecosystem values AND (OR) rapid response is likely to show some benefit); *Low*—Addressing this emerging stewardship consideration is considered to be lower priority by the author team. These considerations are considered to be important (for example, drivers of landscape change are unlikely to affect important ecosystem values AND (OR) rapid response is not likely to show some benefit)]

Driver of landscape change	Emerging stewardship considerations	Priority (high/medium/low)
Disturbance—Wildfire	Climate-driven changes in the fire regime are likely to overwhelm human capacity to manage fire in sparsely populated areas, and landscape-scale conservation strategies likely will need to accommodate changing fire and vegetation mosaics.	High—Wildfire is the primary agent of natural disturbance to boreal forests in Alaska and northwest Canada, and fires play a central role in organizing the physical and biological attributes of the biome.
Disturbance—Forest insect and pathogen epidemics	Managers should work with the understanding that climate change and introductions of new forest pests are likely to alter forest insect and pathogen relationships in ways that are difficult to project. As a consequence, management toward historical regimes may be unrealistic.	Medium—Generally, changes in insect and pathogen irruptiveness brought about by climate or introductions cannot be controlled at a regional scale, but they can be mitigated and worked with through management practices.
Disturbance—Invasive species	Overall, there is a need to build awareness of invasive species and their potential impacts in the NWB region. The likelihood of invasive species reaching new areas in the NWB LCC region is increasing with resource development and climate change. Once invasive species become established, removal is expensive and the chances of eradication are low.	Medium—All sectors of society can play a role in preventing the introduction of exotic and potentially invasive species to the NWB region. Proactive planning and risk assessment by land managers will enable them to respond rapidly and effectively when new infestations occur.
Physical—Climate and climate change	Climate system may be approaching a point where lower emission scenarios are unrealistic to represent future climatic conditions. Scenario planning exercises should consider medium to high emission scenarios as more likely than lower scenarios in characterizing possible future climate conditions and associated impacts.	High—Interactions of precipitation and temperature determine the consequences of climate change for variables with important thresholds, such as the amount of precipitation expected to fall as snow and climatic moisture deficit.

Driver of landscape change	Emerging stewardship considerations	Priority (high/medium/low)
Physical—Hydrologic response	Changes in ice-cover dynamics in lakes and rivers will alter the frequency and magnitude of extreme ice-jam floods, affecting ecology as well as posing hazards to communities and infrastructure, and may decrease the availability of river and lake-ice transportation routes.	High—Changing hydrological regimes are already having widespread human, ecological, and economic impacts, which will probably intensify in the future.
Physical—Permafrost	The societal and ecological consequences of permafrost degradation in the boreal regions include changes to the hydrologic regime and freshwater discharge, changes to habitat for fish and wildlife, increased emissions of methane and changes in soil carbon, impairment of water quality from sediment input from thaw slumping on slopes, and damage to human infrastructure.	High—Adaptations in the design specifications of infrastructure projects may prevent or minimize risks to infrastructure from permafrost degradation.
Physical—Growing season length	The potential social effects of a longer growing season on rural native populations and subsistence lifestyles include: mismatches in timing of the natural system events and the human reliance of those systems, such as hunting and ungulate behavior; unreliable transportation corridors used in winter for traveling and hunting; new species moving north and old species in decline; and increased wildfires. Efforts to inform and educate communities living in the region and to support adaptation planning are needed.	Medium—Current monitoring capacity is considered to be good; future efforts should be directed at improving scenarios of growing season length along with enhanced education and adaptation efforts.
Physical—Enduring features	Many elements of biodiversity will better adapt to climate change when connectivity between current and future potential distributions is maintained or enhanced. When enduring features assist or interfere with connectivity, human assistance may be required to achieve distribution change. Additionally, landscape-scale diversity of enduring features is likely to enhance regional biodiversity, at some scales.	Medium—The subset of species whose distributions are highly dependent on enduring features is not likely to include many with high economic or cultural value (except fish), so managing them is not as likely to receive societal priority (except hydrological connectivity for fish).
Biological—Vegetation composition change	Changes in vegetation are often associated with physical drivers (for example, changes in permafrost) and disturbance (for example, wildfire), therefore vegetation change often is one of many cascading effects. For example, changes in wildfire could change distribution of deciduous trees, which in turn will affect moose distribution. Because vegetation composition is often linked to variations in mammals and birds, any changes in vegetation will have large societal effects. Direct effects of vegetation change include redistricting of subsistence harvesting areas, and loss of important non-forest timber products.	High—Vegetation and associated habitat change will alter the harvest of subsistence resources.
Biological—Novel community and trophic assemblages	A suite of appropriately scaled and connected protected areas to act as reference areas and perhaps climate refugia needs to be maintained.	High—Reference conditions are crucial controls for assessing the effects of shifting climate in relative isolation of other causal agents of landscape change. Climate refugia offer best hope for long-term conservation, if appropriately scaled.

Driver of landscape change	Emerging stewardship considerations	Priority (high/medium/low)
Biological—Wildlife parasite and pathogen lifecycles	Greater understanding of current and historical distribution and prevalence of known wildlife pathogens in the NWB allows for detection of recent changes, and can be used to simulate future conditions. Wildlife disease monitoring and research, including surveillance for novel pathogens and assessment of demographic impacts, within the context of climate and environmental change is important for making management decisions in the NWB region.	Medium—These considerations are important because pathogens and parasites have the potential to affect wildlife at both the individual and population level; wildlife species of concern may be especially vulnerable and emerging diseases may have significant consequences for animal and human health. Climate change and anthropogenic influences can affect occurrence and distribution of established and emerging diseases and distribution of hosts, leading to effects on wildlife health, and, potentially, domestic animals and human health.
Biological—Marine derived nutrients (MDNs)	Maintaining adequate salmon returns, populations of large carnivores that disperse MDNs, retaining riparian and freshwater habitat complexity, and protecting riparian zones from human development will help ensure nutrient delivery and retention, and help sustain species and ecosystem health into the future.	High—Naturally functioning ecosystems and habitat complexity are essential for productivity and the sustainability of salmon and the host of other species they support, including humans.
Socioeconomic—Land use change and resource extraction	Exploration for oil and natural gas is extending into shale beds and coal reserves that previously were not being developed. High diesel costs and undeveloped timber resources that are far from markets are prompting many northern communities to consider the use of wood fiber (bioenergy) for local electricity generation. These pressures, combined with emerging technologies such as hydraulic fracturing and use of bioenergy, can have important environmental impacts at the landscape level.	Medium—Watch for developments, surges in industrial/exploitative activity. It is important to monitor locations and methods by which new resource developments proceed.
Socioeconomic—Rural and Indigenous livelihoods	Changes in climate, biotic, industrial, economic, and (or) political drivers can readily tip the balance regarding the degree of direct utilization of natural resources by households, especially in rural and Aboriginal communities.	High—Many unknowns and potentially great social-ecological sensitivity. The added and cumulative effects of natural resource exploitation and climate change have infringed upon the adaptive capacity of many communities.
Socioeconomic—Contaminants	Careful attention to regulations and compliance for ongoing and future resource development and transport projects is critical to protecting the environmental and human health of the NWB region.	High—Clean-up of existing hot spots and prevention of future contamination are priorities for the protection of both human and ecosystem health.
Socioeconomic—Law and policy	Each country would benefit from more coherent governmental responses to the challenges that climate change poses for landscape conservation.	High—Priorities are to address (1) lack of recognition of the pervasive effects of climate change on mandates of virtually all government agencies and consequent need for cross-government integration, and (2) lack of adequate public profile for climate change issues when assessing policy and governmental action.

Driver of landscape change	Emerging stewardship considerations	Priority (high/medium/low)
Socioeconomic—Values and ethics	Managers can work toward selecting methods to make explicit the values and ethics of their constituencies and all coordinating institutions. Developing expertise in the methodological tools of conservation social science can assist managers in creating rigorous, effective, and societally supported adaptive management research programs, as well as fostering learning across the system and across politically contentious constituencies.	High—Because values and ethics shape all individual and institutional learning within a changing social-ecological system, it is essential to improve access to and use of the methodological tools to make these explicit, and to monitor divergences as well as changes over time as new insights are achieved.
Interactions—Interactions among drivers	Points of interaction among drivers are the logical junctures at which many of the above stewardship concerns coalesce.	High—Prioritization of stewardship goals must take interactions into account, or resulting rules, standards, decisions, and directives may prove counterproductive.
Interactions—Cumulative effects	Cumulative effects are a complex multiscale problem that requires strategic considerations. Tiered assessment and decision-making should be considered for the NWB region. This involves the assessment of large projects within conventional regulatory processes while placing those decisions in the context of broader regional targets for cumulative effects set through strategic and broadly inclusive processes (for example, Regional Strategic Environmental Assessment).	High—Project-based environmental assessment, the current process, is not adequate for assessing and managing the full range of small to large drivers of cumulative effects and impacts. Considering that this NWB region is relatively undisturbed, there is the opportunity to set and achieve a strategic direction that will maintain important values.
Practices of Coproduction—Meaningfully engaging communities	Research relationships between scientists and communities have historically been premised on colonial power relationships that have often caused Indigenous communities to mistrust researchers and research products. All research partners need the fiscal capacity to work together. Research monies need to help build capacity within tribes and organizations to administer research relationships. Funding cycles and grant-writing procedures need to be reformulated to encourage iterative research designs, knowledge coproduction, and to reflect the time it takes to conduct rigorous community-based work. Research budgets should ensure that communities fiscally benefit from the process of research, at both individual and tribal/First Nation scales.	High—Trust is the fundamental condition of meaningful collaboration, and trust is accumulated through actual participation at multiple stages of a research and management program. This includes being equitable about the fiscal benefits to completing such studies, and recognizing the labor and knowledge required for all parties to participate.

Driver of landscape change	Emerging stewardship considerations	Priority (high/medium/low)
Practices of Coproduction—Evolving roles of scientists and managers	The transition toward ecosystem stewardship is challenging for managers trained in scientific management, both as individuals and within institutions that were organized around older approaches. The NWB LCC could serve these agencies and affected communities by building capacity and opportunities to engage in adaptive governance practices such as fostering iterative learning environments and participatory group processes. Managers are generally trained in areas outside the NWB region and come to this region without sufficient understanding of its social context to be fully effective. The NWB region could enhance the capacity of managers and scientists to understand regional context, including identifying conflicting problem definitions and solutions across multiple scales and stakeholders.	High—Scientific management of natural resources is an approach based on institutional requirements for control and assumptions about sufficiency of information that have proven difficult to meet in northern North America. In a directionally changing social-ecological system, meeting such requirements will become increasingly challenging. Ecological and social outcomes have often not been those sought. Fundamental change in the approaches of scientists and managers (within an increasingly supportive policy context) will be required in order to invent, adapt, and diffuse viable alternative approaches for clarifying and achieving ecological and social goals in the common interest.
Practices of Coproduction—Assessing and projecting change	Take the time to learn about and engage with methods and tools designed to embrace uncertainty linked to system complexities and uncontrollable drivers in decision-making.	High—Although climate- and ecological-response sciences are improving understanding and future projections for some system elements, there have been and always will be uncertainties for which assessment and approaches like adaptive management will be relevant.
Practices of Coproduction—Data management and dissemination	The NWB region spans two countries with different standard metrics for measurement, along with different laws, regulations, and plans for data collection. It will be important to come to an agreement among resource managers and engaged communities on data standardization across the NWB target area.	High—Cross-jurisdictional data management agreements should consider standard units of measure, best practices for the geographic scale, a temporal sampling methodology for various species and habitats, and metadata standards.
Practices of Coproduction—Outreach	As new modes of communication and technology emerge regularly, it is particularly important for outreach efforts to monitor preferred stakeholder communication methods, evaluate the efficacy of their work, and revise efforts based on stakeholder feedback. Likewise, although the virtual interactions made possible by new technologies can simulate face-to-face interactions, in person contact is still vital for deeper and more sustained outreach and engagement.	High—Outreach efforts will be most effective when the information provided is perceived by the stakeholder to be salient, legitimate, and credible. This can be increased through scientist—stakeholder collaborations and interactions that facilitate convening dialog, translate information into usable forms, and mediate diverse viewpoints. Science outreach can evolve in sync with changing landscapes, changing information needs, and scientific advances if practitioners grow and nurture the more iterative two-way information exchange between information producers and stakeholders.

REFERENCES CITED

Acevedo-Whitehouse, K., and Duffus, A.L., 2009, Effects of environmental change on wildlife health: Philosophical Transactions of the Royal Society of London B, v. 364, p. 3,429–3,438.

Arctic Council and the International Arctic Science Committee (ACIA), 2005, Arctic Climate Impact Assessment: Cambridge University Press, 1,042 p., accessed November 11, 2017, at https://www.amap.no/documents/doc/arctic-arctic-climate-impact-assessment/796.

Adageirsdottir, G., Echelmeyer, K.A., and Harrison, W.D., 1998, Elevation and volume changes on the Harding Icefield, Alaska: Journal of Glaciology, v. 44, p. 570–582.

Adams, L.G., Farley, S.D., Stricker, C.A., Demma, D.J., Roffler, G.H., Miller, D.C., and Rye, R.O., 2010, Are inland wolf–ungulate systems influenced by marine subsidies of Pacific salmon?: Ecological Applications, v. 20, p. 251–262.

Adger, W.N., Brown, K., Nelson, D.R., Berkes, F., Eakin, H., Folke, C., Galvin, K., Gunderson, L., Goulden, M., O'Brien, K., Ruitenbeek, J., and Tompkins, E.L., 2011, Resilience implications of policy responses to climate change: Wiley Interdisciplinary Reviews—Climate Change, v. 2, no. 5, p. 757–766.

Ahlenius, Hugo, 2006, Pathways of contaminants to the Arctic: UNEP/GRID-Arendal website, accessed September 4, 2014, at http://www.grida.no/graphicslib/detail/pathways-of-contaminants-to-the-arctic_54e4.

Akcil, A., and Koldas, S., 2006, Acid mine drainage (AMD)—Causes, treatment and case studies: Journal of Cleaner Production, v. 14, no. 12, p. 1,139–1,145.

Alaska Center for Conservation Service, 2016, Alaska exotic plant information clearinghouse: Anchorage, University of Alaska database, accessed November 6, 2017, at http://aknhp.uaa.alaska.edu/botany/akepic/.

Alaska Climate Impact Assessment Commission, 2008, Final commission report: Juneau, Alaska State Legislature, 124 p.

Alaska Climate Research Center, 2013, Temperature changes in Alaska: Alaska Climate Research Center website, accessed July 2, 2014, at http://climate.gi.alaska.edu/ClimTrends/Change/TempChange.html.

Alaska Committee for Noxious and Invasive Plant Management, 2016, Strategic plan: University of Alaska, Fairbanks, Cooperative Extension Service website, accessed March 29, 2016, at https://www.uaf.edu/ces/cnipm/.

Alaska Department of Fish and Game, 2016, Subsistence in Alaska—A year 2014 update: Anchorage, Alaska Department of Fish and Game, 4 p., accessed November 11, 2017, at http://www.adfg.alaska.gov/static/home/subsistence/pdfs/subsistence_update_2014.pdf.

Alaska Division of Geology and Geophysical Surveys, 1989–2015, Alaska's mineral industry: Fairbanks, Alaska Department of Natural Resources, Special Reports 44, 49, 55, 62 and 70 for the years 1989, 1994, 2000, 2007 and 2014, respectively, accessed December 9, 2015, at http://dggs.alaska.gov/pubs/minerals.

Alaska Department of Health and Social Services, 2008, Health risks in Alaska among adults— Alaska behavioral risk factor survey: Alaska Department of Health and Social Services, 2007 Annual Report, 84 p.

Alaska Industrial Development and Export Authority, 2015, Ambler Mining District industrial access road: Ambler Access website, accessed December 16, 2015, at http://www.ambleraccess.org/projects/ambler/index.html.

Alaska Oil and Gas Association, 2015, AOGA fact sheet—Cook Inlet oil & gas production: Anchorage, Alaska Oil and Gas Association, 3 p., accessed November 27, 2016, at http://www.aoga.org/sites/default/files/news/cook_inlet_fact_sheet_final.pdf.

Allan, J.D., Wipfli, M.S., Caouette, J.P., Prussian, A., and Rodgers, J., 2003, Influence of streamside vegetation on terrestrial invertebrate inputs to salmonid food webs: Canadian Journal of Fisheries and Aquatic Sciences, v. 60, p. 309–320.

Allen, C.R., Fontaine, J.J., Pope, K.L., and Garmestani, A.S., 2011, Adaptive management for a turbulent future: Journal of Environmental Management, v. 92, no. 5, p. 1,339–1,345, doi:10.1016/j.jenvman.2010.11.019.

Allison, S.D., and Treseder, K.K., 2011, Climate change feedbacks to microbial decomposition in boreal soils: Fungal Ecology, v. 4, no. 6, p. 362–374, doi:10.1016/j.funeco.2011.01.003.

Altizer, S., Dobson, A., Hosseini, P., Hudson, P., Pascual, M., and Rohani, P., 2006, Seasonality and the dynamics of infectious diseases: Ecology Letters, v. 9, no. 4, p. 467–484.

Altizer, S., Ostfeld, R.S., Johnson, P.T., Kutz, S., and Harvell, C.D., 2013, Climate change and infectious diseases—From evidence to a predictive framework: Science, v. 341, p. 514–519.

Anderson, D., and Nuttall, M., eds., 2004, Cultivating Arctic landscapes—Knowing and managing animals in the circumpolar North: New York, Berghahn Books, 256 p.

Anderson, J.E., 1991, A conceptual framework for evaluating and quantifying naturalness: Conservation Biology, v. 5, p. 347–352.

Anderson, M.G., and Ferree, C.E., 2010, Conserving the stage—Climate change and the geophysical underpinnings of species diversity: PLoS ONE, v. 5, no. 7, p. e11554, doi:10.1371/journal.pone.0011554.

Angell, A.C., and Kielland, K., 2009, Establishment and growth of white spruce on a boreal forest floodplain—Interactions between microclimate and mammalian herbivory: Forest Ecology and Management, v. 258, p. 2,475–2,480.

Anielski, M., and Wilson, S., 2009, Counting Canada's natural capital—Assessing the real value of Canada's boreal ecosystems: Ottawa, Ontario, Canadian Boreal Initiative, 76 p., accessed October 24, 2017, at https://www.cbd.int/financial/values/canada-countcapital.pdf.

Appenzeller, T., 2015, The new North: Science, v. 349, n. 6,250, p. 806–809, doi:10.1126/science.349.6250.806.

Arctic Council, 2013, Arctic resilience interim report: Stockholm, Stockholm Environment Institute and Stockholm Resilience Centre, accessed November 10, 2017, at http://arctic-council.org/arr/resources/project-publications/.

Arctic Monitoring and Assessment Programme, 2003, AMAP assessment 2002—The Influence of global change on contaminant pathways to, within, and from the Arctic: Oslo, Norway, Arctic Monitoring and Assessment Programme (AMAP), 65 p.

Arctic Monitoring and Assessment Programme, 2010a, AMAP assessment 2009—Persistent organic pollutants (POPs) in the Arctic: Science of the Total Environment, Special Issue 408, p. 2,851–3,051.

Arctic Monitoring and Assessment Programme, 2010b. AMAP assessment 2009—Radioactivity in the Arctic: Oslo, Norway, Arctic Monitoring and Assessment Programme, 92 p.

Arctic Monitoring and Assessment Programme, 2011, Snow, water, ice and permafrost in the Arctic (SWIPA)—Climate change and the cryosphere: Oslo, Norway, Arctic Monitoring and Assessment Programme (AMAP), 538 p.

Arctic Monitoring and Assessment Program, 2016, AMAP Assessment 2015—Radioactivity in the Arctic: Oslo, Norway, Arctic Monitoring and Assessment Programme (AMAP), 89 p.

Ascher, W., Steelman, T., and Healy, R., 2010, Knowledge and environmental policy—Re-imagining the boundaries of science and politics: Cambridge, Massachusetts, MIT Press, 280 p.

Aumann, C., Farr, D.R., and Boutin, S., 2007, Multiple use, overlapping tenures, and the challenge of sustainable forestry in Alberta: Forestry Chronicle, v. 83, no. 5, p. 642–650.

Axys, 2001, Thresholds for addressing cumulative effects on terrestrial and avian wildlife in the Yukon: Report for the Department of Indian and Northern Affairs Environmental Directorate by Axys Environmental Consulting, Calgary, Alberta, accessed November 14, 2017, at http://emrlibrary.gov.yk.ca/ylupc/thresholds_for_addressing_cumulative_effects_on_terrestrial_and_avian_wildlife_2001.pdf.

Bachelet, D., Comendant, T., and Strittholt, J., 2011, Web platform for sharing spatial data and manipulating them online: EOS, Transactions American Geophysical Union, v. 92, no. 14, p. 118–119.

Baird, R.A., Verbyla, D., and Hollingsworth, T.N., 2012, Browning of the landscape of interior Alaska based on 1986–2009 Landsat sensor NDVI: Canadian Journal of Forest Research, v. 42, p. 1,371–1,382.

Balshi, M.S., McGuire, A.D., Duffy, P., Flannigan, M., Walsh, J., and Melillo, J., 2009, Assessing the response of area burned to changing climate in western boreal North America using a Multivariate Adaptive Regression Splines (MARS) approach: Global Change Biology, v. 15, no. 3, p. 578–600, accessed November 10, 2017, at doi:10.1111/j.1365- 2486.2008.01679.x.

Barber, V.A., Juday, G.P., and Finney, B.P., 2000, Reduced growth of Alaskan white spruce in the twentieth century from temperature-induced drought stress: Nature, v. 405, p. 668–763.

Barrand N.E., and Sharp, M.J., 2010 Sustained rapid shrinkage of Yukon glaciers since the 1957–1958 International Geophysical Year: Geophysical Research Letters, v. 37, no. 7, p. L07501.

Baxter, C.V., Fausch, K.D., and Saunders, C.W., 2005, Tangled webs—Reciprocal flows of invertebrate prey link streams and riparian zones: Freshwater Biology, v. 50, p. 201–220. Beach, D., 2014, Lessons from scenario planning for wildlife management in the southwest Yukon: Saskatoon, Canada, University of Saskatchewan, M.E.S. thesis.

Beck, P.S.A., Juday, G.P., Alix, C., Barber, V.A., Winslow, S.E., Sousa, E.E., Heiser, P., Herriges, J.D., and Goetz, S.J., 2011, Changes in forest productivity across Alaska consistent with biome shift: Ecology Letters, v. 14, no. 4, p. 373–379.

Bednarek, A. T., Wyborn, C., Cvitanovic, C., Meyer, R., Colvin, R.M., Addison, P.F.E., Close, S.L., et al., 2018, Boundary spanning at the science–policy interface: the practitioners' perspectives: Sustainability Science, v. 13, no. 4, p. 1175–1183.

Beever, E.A., O'Leary, J., Mengelt, C., West, J.M., Julius, S., Green, N., Magness, D., Petes, L., Stein, B., Nicotra, A.B., Hellmann, J.J., Robertson, A.L., Staudinger, M.D., Rosenberg, A.A., Babij, E., Brennan, J., Schuurman, G.W., and Hofmann, G.E., 2015, Improving conservation outcomes with a new paradigm for understanding species' fundamental and realized adaptive capacity: Conservation Letters, May/June, 7 p., doi: 10.1111/conl.12190.

Beget, J.E., Stone, D., and Verbyla, D.L., 2006. Regional overview of interior Alaska, part 1, chap. 2 of Chapin, F.S., and others, eds., Alaska's changing boreal forest: New York, Oxford University Press, p. 12–20.

Beier, C., 2011, Factors influencing adaptive capacity in the reorganization of forest management in Alaska: Ecology and Society, v. 16, no. 1, accessed November 14, 2015, at http://www.ecologyandsociety.org/vol16/iss1/art40/.

Beier, P., and Brost, B.M., 2010, Use of land facets to plan for climate change—Conserving the arenas, not the actors: Conservation Biology, v. 24, no. 1, p. 701–710.

Beier, P., Hunter, M.L., and Anderson, M., 2015, Special section—Conserving nature's stage: Conservation Biology, v. 29, no. 3, p. 613–617, doi:10.1111/cobi.12511.

Bell, M., Hill, T., and Whelan, V., 2012, Yukon agriculture state of the industry report 2010–2011–2012: Whitehorse, Government of Yukon, Energy, Mines and Resources, Agriculture Branch and Agriculture Canada, 46 p., accessed November 7, 2015, at http://www.emr.gov.yk.ca/agriculture/pdf/20102012_Agriculture_StateofIndustry_InterimReport.pdf.

Bella, E.M., 2011, Invasion prediction on Alaska trails—Distribution, habitat, and trail use: Invasive Plant Science and Management, v. 4, p. 296–305.

Ben-David, M., Hanley, T.A., and Schell, S.M., 1998, Fertilization of terrestrial vegetation by spawning Pacific salmon—The role of flooding and predator activity: Oikos, v. 83, p. 47–55.

Bengtsson, J., Angelstam, P., Elmqvist, T., Emanuelsson, U., Folke, C., Ihse, M. and Nyström, M., 2003, Reserves, resilience and dynamic landscapes: AMBIO—A Journal of the Human Environment, v. 32, no. 6, p. 389–396.

Bennett, B.A., and Mulder, R.S., 2009, Natives gone wild—Climate change and a history of a Yukon invasion, in Darbyshire, S.J., and Prasad, R., eds., The view from the North: Proceedings of the Weeds Across Borders 2008 Conference, May 27–30, 2008, Banff, Alberta, Canada, accessed April 7, 2016, at http://cfs.nrcan.gc.ca/pubwarehouse/pdfs/31658.pdf.

Bennett K.E., Cannon, A.J., and Hinzman, L., 2015, Historical trends and extremes in regional scale interior and western Alaska river basins: Journal of Hydrology, v. 527, p. 590–607.

Bennett, K.E., and Walsh, J.E., 2014, Spatial and temporal changes in indices of extreme precipitation and temperature for Alaska: International Journal of Climatology, v. 35, no. 7, p. 1,434–1,452.

Bennett, N.J., and Roth, R., eds., 2015, The conservation social sciences—What? How? and Why?: Vancouver, University of British Columbia, Canadian Wildlife Federation and Institute for Resources, Environment, and Sustainability, 77 p.

Bennett, N.J., Roth, Robin, Klain, S.C., Chan, K.M.A., Clark, D.A., Cullman, Georgina, Epstein, Graham, Nelson, M.P., Stedman, Richard, Teel, T.L., Thomas, R.E.W., Wyborn, Carina, Curran, Deborah, Greenberg, Alison, Sandlos, John, and Verissimo, Diogo, 2016, Mainstreaming the social sciences in conservation: Conservation Biology, v. 31, no. 1, p. 56– 66, accessed November 10, 2017, at doi:10.1111/cobi.12788.

Berg, E.E., 2000, Studies in the wilderness areas of the Kenai National Wildlife Refuge—Fire, bark beetles, human development, and climate change in McCool, S.F., Cole, D.N., Borrie, W.T., and O'Loughlin, Jennifer, eds., Proceedings of the Wilderness Science in a Time of Change, May 23–27, 1999: Missoula, Montana. U.S. Forest Service, RMRS-P-15-VOL-3., p. 63–67.

Berg, E.E., and Anderson, R.S., 2006, Fire history of white and Lutz spruce forests on the Kenai Peninsula, Alaska, over the last two millennia as determined from soil charcoal: Forest Ecology and Management, v. 227, p. 275–283.

Berg, E.E., Henry, J.D., Fastie, C.L., De Volder, A.D., and Matsuoka, S.M., 2006, Spruce beetle outbreaks on the Kenai Peninsula, Alaska, and Kluane National Park and Reserve, Yukon Territory—Relationship to summer temperatures and regional differences in disturbance regimes: Forest Ecology and Management, v. 227, p. 219–232.

Berg, E.E., Hillman, K.M., Dial, R., and DeRuwe, A., 2009, Recent woody invasion of wetland on the Kenai Peninsula lowlands, south-central Alaska—A major regime shift after 18,000 years of wet Sphagnum-sedge peat recruitment: Canadian Journal of Forest Research, v. 39, p. 2,033–2,046.

Berkes, Fikret, 1999, Sacred ecology—Traditional ecological knowledge and resource management: New York, Taylor and Francis, 232 p.

Berkes, Fikret, 2012, Sacred ecology (3rd ed.): New York, Routledge Press, 392 p.

Berkes, Fikret, and Berkes, M.K., 2009, Ecological complexity, fuzzy logic, and holism in Indigenous knowledge: Futures, v. 41, no. 1, p. 6–12.

Berkes, Fikret, Reid, W.V., Wilbanks, T.J., and Capistrano, D., 2006, Conclusions—Bridging scales and knowledge systems, chap. 17 of Bridging scales and knowledge systems—Concepts and applications in ecosystem assessment: Washington, D.C., Island Press, p. 315.

Bersamin, A., Sidenberg-Cherr, S., Stern, J.S., and Luick, B.R., 2007, Nutrient intakes are associated with adherence to a traditional diet among Yup'ik Eskimos living in remote Alaska Native communities: The CANHR Study, International Journal of Circumpolar Health, v. 66, no. 1, p. 62–70.

Besley, J.C., Dudo, A., Yuan, S., 2017, Scientists' views about communication objectives: Public Understanding of Science, v. 27, no. 6, p. 708–730.

Boertje, R.D., Keech, M.A., and Paragi, T.F., 2010, Science and values influencing predator control for Alaska moose management: The Journal of Wildlife Management, v. 74, no. 5, p. 917–928.

Berryman, A.A., 1996, What causes population cycles of forest Lepidoptera?: Trends in Ecology & Evolution, v. 11, no. 1, p. 28–32.

Bettinetti, R., Quadroni, S., Galassi, S., Bacchetta, R., Bonardi, L., and Vailati, G., 2008, Is meltwater from alpine glaciers a secondary DDT source for lakes?: Chemosphere, v. 73, p. 1,027–1,031.

Bigelow, N.H., Brubaker, L.B., Edwards, M.E., Harrison, S.P., Prentice, I.C., Anderson, P.M., Andreev, A.A., Bartlein, P.J., Christensen, T.R., Cramer, W., Kaplan, J.O., Lozhkin, A.V., Matveyeva, N.V., Murray, D.F., McGuire, A.D., Razzhivin, V.Y., Ritchie, J.C., Smith, B., Walker, D.A., Gajewski, K., Wolf, V., Holmqvist, B.H., Igarashi, Y., Kremenetskii, K., Paus, A., Pisaric, M.F.J., and Volkova, V.S., 2003, Climate change and Arctic ecosystems—1— Vegetation changes north of 55°N between the last glacial maximum, mid-Holocene, and present: Journal of Geophysical Research, Atmospheres, v. 108, no. D19, doi:10.1029/2002JD002558.

Biggs, R.T., Blenckner, C., Gordon, L., Norström, A., Nyström, M., and Peterson, G.D., 2012, Regime shifts, in Hastings, A., and Gross, L., eds., Encyclopedia of theoretical ecology: Berkeley, California, University of California Press, p. 609–617.

Binford, L.R., 1978, Nunamiut ethnoarchaeology: New York, Academic Press, 509 p. Blais, J.M., Schindler, D.W., Muir, D.C.G., Kimpe, L.E., Donald, D.B., and Rosenberg, B., 1998, Accumulation of persistent organochlorine compounds in mountains of western Canada: Nature, v. 395, p. 585–588.

Blais, J.M., Schindler, D.W., Muir, D.C.G., Sharp, M., Donald, D., Lafrenière, M., Braekevelt, E., and Strachan, W.M.J., 2001, Melting glaciers—A major source of persistent organochlorines to subalpine Bow Lake in Banff National Park, Canada: Ambio, v. 30, p. 410– 415.

Blancher, P., 2003, Importance of Canada's boreal forest to landbirds: Ottawa, Ontario, Canadian Boreal Initiative and the Boreal Songbird Initiative, 42 p.

Boertje, R.D., Keech, M.A., and Paragi, T.F., 2010, Science and values influencing predator control for Alaska moose management: The Journal of Wildlife Management, v. 74, no. 5, p. 917–928, doi:10.2193/2009–261.

Boggs, K., Sturdy, M., Rinella, D.J., and Rinella, M.J., 2008, White spruce regeneration following a major spruce beetle outbreak in forests on the Kenai Peninsula, Alaska: Forest Ecology and Management, v. 255, no. 10, p. 3,571–3,579

Bolton, W.R., Hinzman, L., and Yoshikawa, K., 2000, Stream flow studies in a watershed underlain by discontinuous permafrost, in Kane, D.L., ed., American Water Resources Association's 2000 Spring Specialty Conference, —Water Resources in Extreme Environments, Proceedings: Middleburg, Virginia, American Water Resources Association, p. 31–36.

Bolton, W.R., Hinzman, L.D. and Yoshikawa, K., 2004, Water balance dynamics of three small catchments in a sub-Arctic boreal forest, in Kane, D.L., and Yang, D., eds., Northern research basins water balance: International Association of Hydrological Sciences Publication 290, p. 213–223.

Borthwick, I., 1997, Environmental management in oil and gas exploration and production—An overview of issues and management approaches: The United Nations Environment Programme, Industry and Environment Centre, Paris, France, UNEP IE/PAC Technical Report 37, 68 p., accessed November 7, 2015, at http://www.ogp.org.uk/pubs/254.pdf.

Bostrom A., Böhm, G. and O'Connor, R.E., 2013,Targeting and tailoring climate change communications, Wiley Interdisciplinary Reviews: Climate Change, vol 4, no. 5, p. 447–455.

Brackley, A.M., Barber, V.A., and Pinkel, C., 2010, Developing estimates of potential demand for renewable wood energy products in Alaska: U.S. Forest Service, Portland, Oregon, General Technical Report PNW-GTR-827, 31 p.

Bradley, M.J., Kutz, S.J., Jenkins, E., and O'Hara, T.M., 2005, The potential impact of climate change on infectious diseases of Arctic fauna: International Journal of Circumpolar Health, v. 64, p. 468–477.

Bravo, M., and Sörlin, S., eds., 2002. Narrating the Arctic—A cultural history of Nordic scientific practices: Canton, Massachusetts, Watson Publishing, 373 p.

Brossard, D., and Lewenstein, B.V., 2010, A critical appraisal of models of public understanding of science—Using practice to inform theory, in Kahlor, LeeAnn, and Stout, Patricia, eds., Communicating science—New agendas in communication: New York, Routledge, p. 11–39.

Brost, B., 2010, Use of land facets to design conservation corridors—Conserving the arenas, not the actors: Flagstaff, Northern Arizona University, M.S. thesis.

Brost, B.M., and Beier, P., 2012, Use of land facets to design linkages for climate change: Ecological Applications, v. 22, no. 1, p. 87–103.

Brotton, J., and Wall, G., 1997, Climate change and the Bathurst caribou herd in the Northwest Territories, Canada: Climatic Change, v. 35, no. 1, p. 35–52.

Brown, S.R., 1980, Political subjectivity—Applications of Q methodology in political science: New Haven, Connecticut, Yale University Press.

Brown, C., and Johnstone, J.F., 2012, One burned, twice shy—Repeat fires reduce seed availability and alter substrate constraints on Picea mariana: Forest Ecology and Management, v. 266, p. 36–41.

Brown, R., Derksen, C., and Wang, L., 2010, A multi-dataset analysis of variability and change in Arctic spring snow cover extent: Journal of Geophysical Research, v. 115, no. D16111, doi:10.1029/2010JD013975.

Brown, R.D., and Robinson, D.A., 2011, Northern hemisphere spring snow cover variability and change over 1922–2010 including an assessment of uncertainty: The Cryosphere, v. 5, p. 219–229, doi:10.5194/tc-5-219-2011.

Brubaker, L.B., Anderson, P.M., Edwards, M.E., and Lozhkin, A.V., 2004, Beringia as a glacial refugium for boreal trees and shrubs—New perspectives from mapped pollen data: Journal of Biogeography, v. 31, p. 1–16.

Brugger, J., Meadow, A., and Horangic, A., 2015, Lessons from first-generation climate science integrators: Bulletin of the American Meteorological Society, v. 97, no. 3, p. 355–365.

Brunner, R.D., and Lynch, A.H., 2010, Adaptive governance and climate change: Chicago, American Meteorological Society, 404 p, 10.1007/978-1-935704-01-0.

Brunner, R.D., and others, 2005, Adaptive governance—Integrating science, policy, and decision-making: New York, Columbia University Press, 368 p.

Buizer, J., Jacobs, K., and Cash, D., 2010, Making short-term climate forecasts useful—Linking science and action: Proceedings of the National Academy of Sciences, v. 113, no. 17, p. 4,597–4,602.

Buma, Brian, Brown, C.D., Donato, D.C., Fontaine, J.B., and Johnstone, J.F., 2013 The impacts of changing disturbance regimes on serotinous plant populations and communities: BioScience, v. 63, no. 11, p. 866–876, accessed November 10, 2017, at https://doi.org/10.1525/bio.2013.63.11.5.

Bureau of Land Management, 2015, Yukon River Lowlands—Kuskokwim Mountains—Lime Hills rapid ecoregional assessment (REA)—Final report: Fairbanks, Alaska, Bureau of Land Management, accessed November 23, 2015, at http://www.blm.gov/style/medialib/blm/wo/Communications_Directorate/public_affairs/landscape_approach/landscape_3.Par.62649.File.dat/YKL_Final_Report_09012015.pdf.

Burek, K.A., Gulland, F.M., and O'Hara, T.M., 2008, Effects of climate change on Arctic marine mammal health: Ecological Applications, v. 18, p. 126–134.

Burns, T.W., O'Connor, D.J., and Stocklmayer, S.M., 2003, Science communication: A contemporary definition: Public Understanding of Science, v. 12, no. 2, p. 183–202.

Burnside, R., Schultz, M., Lisuzzo, N., and Kruse, J., 2010, Assessing mortality and regeneration of larch (Larix laricina) after a 1999–2004 landscape level outbreak of the larch sawfly (Pristiphora erichsonii) in Alaska, chap. 14 of Potter, K.M., and Conkling, B.L., eds., Forest health monitoring—National status, trends, and analysis: U.S. Forest Service, General Technical Report SRS-176, p. 143–150.

Burton, A.C., Huggard, D., Bayne, E., Schieck, J., Sólymos, P., Muhly, T., Farr, D., and Boutin, S., 2014, A framework for adaptive monitoring of the cumulative effects of human footprint on biodiversity: Environmental Monitoring and Assessment, v. 186, no. 6, p. 3,605–3,617.

Burton, P.J., Messier, C., Weetman, G.F., Prepas, E.E., Adamowicz, W.L., and Tittler, R., 2003, The current state of boreal forestry and the drive for change, in Burton, P.J., Messier, C., Smith, D.W., and Adamowicz, W.L., eds., Towards sustainable management of the boreal forest: Ottawa, Ontario, Canada, NRC Research Press, p. 1–40.

Butler, L.G., Kielland, K., Rupp, T.S., and Hanley, T.A., 2007, Interactive controls of herbivory and fluvial dynamics over vegetation patterns along the Tanana River, interior Alaska: Journal of Biogeography, v. 34, p. 1,622–1,631.

Buttrick, S., Popper, K., Schindel, M., McRae, B., Unnasch, B., Jones, A., and Platt, J., 2015, Conserving nature's stage—Identifying resilient terrestrial landscapes in the Pacific Northwest: Portland, Oregon, The Nature Conservancy, 104 p., accessed November 14, 2015, at http://nature.ly/resilienceNW.

Cajete, G., 2000, Native science—Natural laws of interdependence: Santa Fe, New Mexico, Clear Light Publishers, 315 p.

Calef, M.P., Varvak, A., McGuire, A.D., Chapin, F.S., and Reinhold, K.B., 2015, Recent changes in annual area burned in Interior Alaska—The impact of fire management: Earth Interactions, v. 19, p. 1–17.

Callaghan, T.V., Johansson, M., Brown, R.D., Groisman, P.A., Labba, N., Radionov, V., Barry, R.G., Bradley, R.S., Blangy, S., Bulygina, O.N., Christensen, T.R., Colman, J., Essery, R.L.H., Forbes, B.C., Forchhammer, M.C., Frolov, D.M., Golubev, V.N., Grenfell, T.C., Honrath, R.E., Juday, G.P., Rae Melloh, R., Meshcherskaya, A.V., Petrushina, M.N., Phoenix, G.K., Rautio, A., Razuvaev, V.N., Robinson, D.A., Rodionov, V., Romanov, P., Schmidt, N.M., Serreze, M.C., Shevchenko, V., Shiklomanov, A.I., Shindell, D., Shmakin, A.B., Sköld, P., Sokratov, S.A., Sturm, M., Warren, S., Woo, M., Wood, E.F., and Yang, D., 2011a, Changing snow cover and its impacts, in Arctic Monitoring and Assessment Programme (AMAP), Snow, Water, Ice, and Permafrost (SWIPA): Oslo, Norway, Arctic Council, p. 4–1–4–59.

Callaghan, T.V., Johansson, M., Brown, R.D., Groisman, P.Y., Labba, N., Radionov, V., Bradley, R.S., Blangy, S., Bulygina, O.N., Christensen, T.R., Colman, J., Essery, R.L.H., Forbes, B.C., Forchhammer, M.C., Golubev, V.N., Honrath, R.E., Juday, G.P., Meshcherskaya, A.V., Phoenix, G.K., Pomeroy, J., Rautio, A., Robinson, D.A., Schmidt, N.M., Serreze, M.C., Shevchenko, V., Shiklomanov, A.I., Shmakin, A.B., Sköld, P., Sturm, M., Woo, M., and Wood, E.F., 2011b, Multiple effects of changes in Arctic snow cover: Ambio, v. 40, no. 1, p. 32–45.

Camacho, A.E., 2009, Adapting governance to climate change—Managing uncertainty through a learning infrastructure: Emory Law Journal, v. 59, p. 1–77.

Cameron, E.K., and Bayne, E.M., 2009, Road age and its importance in earthworm invasion of northern boreal forests: Journal of Applied Ecology, v. 46, p. 28–36.

Canadian Council of Ministers of the Environment, 2009, Regional strategic environmental assessment in Canada—Principles and guidance: Winnipeg, Manitoba, Canadian Council of Ministers of the Environment, accessed November 25, 2015, at http://www.ccme.ca/files/Resources/enviro_assessment/rsea_principles_guidance_e.pdf.

Carlson, M., Antoniuk, T., Farr, D., Francis, S., Manuel, K., Nishi, J., Stelfox, B., Sutherland, M., Yarmoloy, C., Aumann, C., and Pan, D., 2010, Informing regional planning in Alberta's Oilsands Region with a land-use simulation model, in Swayne, D.A., Yang, W., Voinov, A.A., Rizzoli, A., and Filatova, T., eds., Proceedings of the 2010 International Congress on Environmental Modeling and Software, July 2010, Ottawa, Ontario, Canada: International Congress on Environmental Modeling and Software website, accessed November 14, 2015, at http://www.iemss.org/iemss2010/proceedings.html.

Carlson, M., and Browne, D., 2015, The future of wildlife conservation and resource development in the western boreal forest—A technical report on cumulative effects modeling of future land use scenarios: Kanata, Ontario, Canadian Wildlife Federation, 100 p.

Carlson, M., and Schmiegelow, F., 2002, Cost-effective sampling design applied to large-scale monitoring of boreal birds: Conservation Ecology, v. 6, no. 2, article 11, accessed November 14, 2015, at http://www.consecol.org/vol6/iss2/art11/.

Carlson, M., Stelfox, B., Purves-Smith, N., Straker, J., Berryman, S., Barker T., and Wilson B., 2014, ALCES Online—Web-delivered scenario analysis to inform sustainable land-use decisions, in Ames, D.P., Quinn, N.W.T., and Rizzoli, A.E.,, eds., Proceedings of the 2014 International Congress on Environmental Modeling and Software, June 2014, San Diego, California: International Congress on Environmental Modeling and Software website, accessed November 14, 2015, at http://www.iemss.org/sites/iemss2014/proceedings.php.

Carlson, M.L., and Shephard, M., 2007, Is the spread of non-native plants in Alaska accelerating?: U.S. Forest Service, General Technical Report PNW GTR-694, p. 111–127.

Carter, S., 1999, Aboriginal people and colonizers of western Canada to 1900 (vol. 5): Toronto, Ontario, Canada, University of Toronto Press, 195 p.

Case, David, and Voluck, David, 1984, Alaska natives and American laws (2d ed.): Fairbanks, University of Alaska Press, 586 p.

Cash, D.W., Borck, J.C., and Patt, A.G., 2006, Countering the loading-dock approach to linking science and decision-making: Science, Technology and Human Values, v. 3, no. 21, p. 465– 494.

Cash, D.W., Clark, W.C., Alcock, F., Dickson, N., Eckley, N., and Jäger, J., 2002, Salience, credibility, legitimacy and boundaries—Linking research, assessment and decision-making: Cambridge, Massachusetts, Harvard University, John F. Kennedy School of Government, KSG Working Papers Series RWP02–046, 25 p.

Cash, D.W., Clark, W.C., Alcock, F., Dickson, N.M., Eckley, N., Guston, D.H., Jäger, J., and others, 2003, Knowledge systems for sustainable development: Proceedings of the National Academy of Sciences, v. 100, no. 14, p. 8,086–8,091.

Cattadori, I.M., Haydon, D.T., and Hudson, P.J., 2005, Parasites and climate synchronize red grouse populations: Nature, v. 433, p. 737–741.

Caves, J.K., Bodner, G.S., Simms, K., Fisher, L.A., and Robertson, T., 2013, Integrating collaboration, adaptive management, and scenario-planning—Experiences at Las Cienegas National Conservation Area: Ecology and Society, v. 18, no. 3, article 43, doi:10.5751/ES-05749-180343.

Cederholm, C.J., Houston, D.B., Cole, D.L., and Scarlett, W.J., 1989, Fate of coho salmon (Oncorhynchus kisutch) carcasses in spawning streams: Canadian Journal of Fisheries and Aquatic Sciences, v. 46, p. 1,347–1,355.

Chamberlain, Adam, and Haile, Webnesh, 2016, Environmental assessment and environmental impact review in the post-devolution Northwest Territories, in Hanna, K.S., ed., Environmental impact assessment—Practice and participation (3rd ed.): Toronto, Ontario, Canada, Oxford University Press, p. 241–266.

Chamberlain, E.C., Rutherford, M.B., and Gibeau, M.L., 2012. Human perspectives and conservation of grizzly bears in Banff National Park, Canada: Conservation Biology, v. 26, no. 3, p. 420–431.

Chapin F.S., III, Carpenter, S.R., Kofinas, G.P., Folke, C., Abel, N., Clark, W.C., Olsson, P., Smith, D.M., Walker, B., Young, O.R., Berkes, F., Biggs, R., Grove, J.M., Naylor, R.L., Pinkerton, E., Steffen, W., and Swanson, F.J., 2010, Ecosystem stewardship—Sustainability strategies for a rapidly changing planet: Trends in Ecology & Evolution, v. 25, no. 4, p. 241– 249.

Chapin, F.S., III, Cochran, P., Huntington, O.H., Knapp, C.N., Brinkman, T.J., and Gadamus, L.R., 2013, Traditional knowledge and wisdom—A guide for understanding and shaping Alaskan social-ecological change, in Linking ecology and ethics for a changing world: Dordrecht, Netherlands, Springer, p. 49–62.

Chapin, F.S., III, Hollingsworth, T.N., Murray, D.F., Viereck, L.A., and Walker, M.D., 2006, Floristic diversity and distribution in Alaska's boreal forest in Chapin, F.S., III, Oswood, M., Van Cleve, K., Viereck, L.A., and Verbyla, D., eds., Alaska's changing boreal forest: New York, Oxford University Press, p. 81–99.

Chapin, F.S., III, Lovecraft, A.L., Zavaleta, E.S., Nelson, J., Robards, M.D., Kofinas, G.P., Trainor, S.F., Peterson, G.D., Huntington, H.P., and Naylor, R., 2006, Policy strategies to address sustainability of Alaskan boreal forests in response to a directionally changing climate: Proceedings of the National Academy of Sciences, v. 103, no. 45, p. 16,637–16,643.

Chapin, F.S., III, McGuire, A.D., Ruess, R.W., Hollingsworth, T.N., Mack, M.C., Johnstone, J.F., Kasischke, E.S., Euskirchen, E.S., Jones, J.B., Jorgenson, M.T., Kielland, K., Kofinas, G.P., Turetsky, M.R., Yarie, J., Lloyd, A.H., and Taylor., D.L., 2010, Resilience of Alaska's boreal forest to climatic change: Canadian Journal of Forest Research, v. 40, no. 7, p. 1,360– 1,370, doi:10.1139/X10–074.

Chapin, F.S., III, Robards, M.D., Huntington, H.P., Johnstone, J.F., Trainor, S.F., Kofinas, G.P., Ruess, R.W., Fresco, N., Natcher, D.C., and Naylor, R.L., 2006, Directional changes in ecological communities and social-ecological systems—A framework for prediction based on Alaskan examples: The American Naturalist, v. 168, no. S6, p. S36–S49, doi:10.1086/509047.

Chapin, F.S., III, Robards, M.D., Johnstone, J.F., Lantz, T.C., and Kokelj, S.V., 2013, Case study—Novel socio-ecological systems in the North—Potential pathways toward ecological and societal resilience, in Hobbs, R.J., Higgs, E.S., and Hall, C.M., eds., Novel ecosystems— Intervening in the new ecological world order: New York, John Wiley and Sons, p. 334–344.

Chapin, F.S., III, Sommerkorn, M., Robards, M.D., and Hiller-Pegram, K., 2015, Ecosystem stewardship—A resilience framework for Arctic conservation: Global Environmental Change, v. 34, p. 207–217, accessed November 10, 2017, at http://dx.doi.org/10.1016/j.gloenvcha.2015.07.003.

Chapin, F.S., III, Sturm, M., Serreze, M.C., McFadden, J.P., Key, J.R., Lloyd, A.H., McGuire, A.D., Rupp, T.S., Lynch, A.H., Schimel, J.P., Beringer, J., Chapman, W.L., Epstein, H.E., Euskirchen, E.S., Hinzman, L.D., Jia, G., Ping, C.L., Tape, K.D., Thompson, C.D.C., Walker, D.A., and Welker, J.M., 2005, Role of land-surface changes in Arctic summer warming: Science, v. 310, p. 657–660.

Chapin, F.S., III, Trainor, S.F., Cochran, P., Huntington, H., Markon, C., McCammon, M., McGuire, A.D., and Serreze, M., 2014, Alaska, chap. 22 of Melillo, J.M., Richmond, T.C., and Yohe, G.W. eds., Climate change impacts in the United States—The Third National Climate Assessment: U.S. Global Change Research Program, p. 514–536, doi:10.7930/J00Z7150.

Chetkiewicz, C., and Lintner, A.M., 2014, Getting it right in Ontario's north—The need for regional strategic environmental assessment in the Ring of Fire [Wawangajing]: Toronto, Ontario, Canada, Wildlife Conservation Society Canada and Ecojustice, accessed November 10, 2017, at http://wcscanada.org/Portals/96/Documents/RSEA_Report_WCSCanada_Ecojustice_FINAL.pdf.

Chisana Caribou Herd Working Group, 2012, Management plan for the Chisana caribou herd, 2010–2015: Whitehorse, Yukon, Government of Yukon Department of Environment, 48 p., accessed November 10, 2017, at https://ecos.fws.gov/ServCat/DownloadFile/49096?Reference=48817.

Chivers, M.R., Turetsky, M., Waddington, J.M., Harden, J., and McGuire, A.D., 2009, Effects of experimental water table and temperature manipulations on ecosystem CO2 fluxes in an Alaskan rich fen: Ecosystems, v. 12, p. 1,329–1,342.

Christensen, L., and Krogman, N., 2012, Social thresholds and their translation into social- ecological management practices: Ecology and Society, v. 17, no. 1, article 5, doi: 0.5751/ES- 04499–170105.

Churchill, A.C., Turetsky, M.R., McGuire, A.D., and Hollingsworth, T.N., 2015, Response of plant community structure and primary productivity to experimental drought and flooding in an Alaskan fen: Canadian Journal of Forest Research, v. 45, p. 185–193.

Clark D.A., Fluker, S., and Risby, L., 2008, Deconstructing ecological integrity policy in Canadian national parks. in Hanna, K.S., Clark, D.A., and Slocombe, D.S., eds., Transforming parks and protected areas—Policy and governance in a changing world: Abingdon, United Kingdom, Routledge, p. 154–186.

Clark, D.A., and Slocombe, D.S., 2009, Respect for grizzly bears—An aboriginal approach for co-existence and resilience: Ecology and Society, v. 14, no. 1, article 42, accessed November 10, 2017, at http://www.ecologyandsociety.org/vol14/iss1/art42.

Clark, D.A., Workman, L., and Jung, T., 2016, Impacts of reintroduced bison on First Nations people in Yukon, Canada—Finding common ground through participatory research and social learning: Conservation and Society, v. 14, no. 1, p. 1–12.

Clark, D.A., Workman, L., and Slocombe, D.S., 2014, Science-based grizzly bear conservation in a co-management environment—The Kluane region case, Yukon, in Clark, S.G., and Rutherford, M.B., eds., Large carnivore conservation—Integrating science and policy in the North American West: Chicago, University of Chicago Press, p. 108–139.

Clark, S.G., 2002, The policy process—A practical guide for natural resource professionals: New Haven, Connecticut, Yale University Press, 215 p.

Clark, S.G., 2011, The policy process—A practical guide for natural resource professionals (2d ed.): New Haven, Connecticut, Yale University Press.

Clark, W. and Holliday, L., 2006, Linking knowledge with action for sustainable development—The role of program management—Summary of a Workshop: Washington, D.C., The National Academies Press, 134 p., https://doi.org/10.17226/11652.

Clegg, B.F., and Hu, F.S., 2010, An oxygen-isotope record of Holocene climate change in the south-central Brooks Range, Alaska: Quaternary Science Reviews, v. 29, p. 928–939.

Clement, J.P., Bengtson, J.L., and Kelly, B.P., 2013, Managing for the future in a rapidly changing Arctic—A report to the President: Washington, D.C., Interagency Working Group on Coordination of Domestic Energy Development and Permitting in Alaska, D.J. Hayes, Chair, 59 p., accessed March 16, 2018, at https://www.afsc.noaa.gov/publications/misc_pdf/iamreport.pdf.

Climate Change in Alaska Adaptation Advisory Group, 2010, Public infrastructure, chap. 4 of Draft final report: Climate Change in Alaska Adaptation Advisory Group Web site, accessed September 4, 2014, at http://www.climatechange.alaska.gov/aag/docs/aag_Ch4_27Jan10.pdf.

Coakley, S.M., Scherm, H., and Chakraborty, S., 1999, Climate change and plant disease management: Annual Review of Phytopathology, v. 37, no. 1, p. 399–426.

Cochran, Patricia, Huntington, O.H., Pungowiyi, Caleb, Stanley, Tom, Chapin, F.S., III, Huntington, H.P., Maynard, N.G., and Trainor, Sarah, 2013, Indigenous frameworks for observing and responding to climate change in Alaska: Climatic Change, v. 12, no. 3, p. 557– 567.

Cole, D.N., and Yung, L., eds., 2010, Beyond naturalness—Rethinking park and wilderness stewardship in an era of rapid change: Washington, D.C., Island Press, 304 p.

Collen, P., and Gibson, R.J., 2001, The general ecology of beavers (Castor spp.), as related to their influence on stream ecosystems and riparian habitats, and the subsequent effects on fish—A review: Reviews in Fish Biology and Fisheries, v. 10, no. 4, p. 439–461.

Comeau, P.G., and Thomas, K.D., eds., 1996, Silviculture of temperate and boreal broadleaf-conifer mixtures: Victoria, Canada, British Columbia Ministry of Forests—Summary of papers presented at Silviculture of Temperate and Boreal Broadleaf-Conifer Mixtures Workshop: Province of British Columbia, Ministry of Forests Research Program, February 28 and March 1, 1995, Richmond, British Columbia, 34 p.

Conn, J., Beattie, K., Shephard, M., Carlson, M., Lapina, I., Hebert, M., Gronquist, R., Densmore, R., and Rasy, M., 2008a, Alaska Melilotus invasions—Distribution, origin, and susceptibility of plant communities: Arctic and Alpine Research, v. 40, no. 2, p. 298–308.

Conn, J., Stockdale, C., and Morgan, J.C., 2008b, Characterizing pathways of invasive plant spread to Alaska— I, Propagules from container-grown ornamentals: Invasive Plant Science and Management, v. 1, p. 331–336.

Conn, J., Stockdale, C., Werdin-Pfisterer, N., and Morgan, J., 2010, Characterizing pathways of invasive plant spread to Alaska— II, Propagules from imported hay and straw: Invasive Plant Science and Management, v. 3, p. 276–285.

Conn, J., Werdin-Pfisterer, N., Beattie, K., and Densmore, R., 2011, Ecology of invasive Melilotus albus on Alaskan glacial river floodplains: Arctic, Antarctic and Alpine Research, v. 43, p. 343–354.

Cortés-Burns, H., Lapina, I., Klein, S., Carlson, M., and Flagstad, L., 2008, Invasive plant species monitoring and control—Areas impacted by 2004 and 2005 fires in interior Alaska—A survey of Alaska BLM lands along the Dalton, Steese, and Taylor Highways: Prepared for the Bureau of Land Management, Alaska State Office by the Alaska Natural Heritage Program, University of Alaska, Anchorage, 162 p.

Cortner, H., and Moote, M.A., 1999, The politics of ecosystem management: Washington, D.C., Island Press, 224 p.

Council of Canadian Academies, 2014, Aboriginal food security in northern Canada—An assessment of the state of knowledge: Ottawa, Ontario, Council of Canadian Academies, 256 p.

Council of Yukon First Nations, 2016, The umbrella final agreement: Council of Yukon First Nations website, accessed November 10, 2017, at http://cyfn.ca/agreements/umbrella-final-agreement/.

Cronon, W., 1996, The trouble with wilderness; or, getting back to the wrong nature: Environmental History, v. 1, no. 1, p. 7–28.

Cross, M.S., Zavaleta, E.S., Bachelet, D., Brooks, M.L., Enquist, C.A.F., Fleishman, E., Graumlich, L.J., Groves, C.R., Hannah, L., Hansen, L., Hayward, G., Koopman, M., Lawler, J.J. Malcolm, J., Nordgren, J., Petersen, B., Rowland, E.L., Scott, D., Shafer, S.L., Shaw, M.R., and Taber, G.M., 2012, The adaptation for conservation targets (ACT) framework—A tool for incorporating climate change into natural resource management: Environmental Management, v. 50, no. 3, p. 341–351, doi:10.1007/s00267-012-9893-7.

Cruikshank, Julie, 2005, Do glaciers listen?—Local knowledge, colonial encounters, and social imagination: Vancouver, Canada, University of British Columbia Press, 312 p.

Csiszar, I., Justice, C.O., McGuire, A.D., Cochrane, M.A., Roy, D.P., Brown, F., Conard, S.G., Frost, P.G.H., Giglio, L., Elvidge, C., Flannigan, M.D., Kasischke, E., McRae, D.J., Rupp, T.S., Stocks, B.J., and Verbyla, D.L., 2004, Land use and fires, in Gutman, G., Janetos, A.C., Justice, C.O., Moran, E.F., Mustard, J.F., Rindfuss, R.R., Skole, D., Turner, B.L., and Cochrane, M.A., eds., Land change science: Dordrecht, Netherlands, Kluwer Academic Publishers, p. 329–350.

Cutler, S.J., Fooks, A.R., and van der Poel, W.H.M., 2010, Public health threat of new, reemerging, and neglected zoonoses in the industrialized world: Emerging Infectious Diseases, v. 16, p. 1–7.

Cwynar, L.C., and Ritchie, J.C., 1980, Arctic steppe-tundra—A Yukon perspective: Science, v. 208, p. 1,375–1,377.

Dale, Chelsea, and Natcher, D.C., 2014, What is old is new again—The reintroduction of Indigenous fishing technologies in British Columbia: Local Environment—The International Journal of Justice and Sustainability, v. 20, no. 11, p. 1,309–1,321, http://dx.doi.org/10.1080/13549839.2014.902371.

Danby, R.K., and Hik, D.S., 2007, Variability, contingency and rapid change in recent subarctic alpine tree line dynamics: Journal of Ecology, v. 95, no. 2, p. 352–363, doi:10.1111/j.1365-2745.2006.01200.x.

Danby, R.K., and Slocombe, D.S., 2002, Protected areas and intergovernmental cooperation in the St. Elias Region: Natural Resources Journal, v. 42, no. 2, p. 247–282.

Danby, R.K., and Slocombe, D.S., 2005, Regional ecology, ecosystem geography, and transboundary protected areas in the St. Elias Mountains: Ecological Applications, v. 15, p. 405–422, doi:10.1890/04-0043.

D'Arrigo, R.D., Kaufmann, R.K., Davi, N., Jacoby, G.C., Laskowski, C., Myneni, R.B., and Cherubini, P., 2004, Thresholds for warming-induced growth decline at elevational tree line in the Yukon Territory, Canada: Global Biogeochemical Cycles, v. 18, no. 3, doi:10.1029/2004GB002249.

Daszak, P., Cunningham, A.A., and Hyatt, A.D., 2001, Anthropogenic environmental change and the emergence of infectious diseases in wildlife: Acta Tropica, v. 78, p. 103–116.

Day, J.C., and Affum, J., 1995, Windy Craggy—Institutions and stakeholders: Resources Policy, v. 21, no. 1, p. 21–26.

Decker, D.J., Jacobson, C.A., and Brown, T.L., 2006, Situation-specific "impact dependency" as a determinant of management acceptability—Insights from wolf and grizzly bear management in Alaska: Wildlife Society Bulletin, v. 34, no. 2, p. 426–432, doi:10.2193/0091-7648(2006)34[426:SIDAAD]2.0.CO;2.

Denevan, W.M., 1992, The pristine myth—The landscape of the Americas in 1492: Annals of the Association of American Geographers, v. 82, no. 3, p. 369–385.

Derksen, C., and Brown, R., 2012, Spring snow cover extent reductions in the 2008–2012 period exceeding climate model projections: Geophysical Research Letters, v. 39, no. 19, L19504, doi:10.1029/2012GL053387.

DeVelice, R., 2003, Non-native plant inventory—Kenai trails: U.S. Forest Service, Chugach National Forest, Anchorage, Alaska, R10-TP-124, 24 p.

Dial, R.J., Berg, E.E., Timm, K., McMahon, A., and Geck, J., 2007, Changes in the alpine forest- tundra ecotone commensurate with recent warming in southcentral Alaska—Evidence from orthophotos and field plots: Journal of Geophysical Research, v. 112, p. G04015, doi:10.1029/2007JG000453.

Dial, Roman J., Smeltz, T.S., Sullivan, P.F., Rinas, C.L., Timm, Katriina, Geck, J.E., Tobin, S.C., Golden, T.S., and Berg, E.C., 2016, Shrubline but not treeline advance matches climate velocity in montane ecosystems of south-central Alaska: Global Change Biology, v. 22, no. 5, p. 1,841–1,856, doi:10.1111/gcb.13207.

Dilling, L., and Lemos, M.C., 2011, Creating usable science—Opportunities and constraints for climate knowledge use and their implications for science policy: Global Environmental Change, v. 21, no. 2, p. 680–689.

Dinkel, C.L., and Czapla, P.K., 2012, Alaska forage manual: State of Alaska, Department of Natural Resources, Division of Agriculture, Plant Materials Center, Palmer, 107 p., accessed November 7, 2015, at http://dnr.alaska.gov/ag/akpmc/forage/AFM-full_web_v2.pdf.

Dobson, A., and Foufopoulos, J., 2001, Emerging infectious pathogens of wildlife: Philosophical Transactions of the Royal Society of London B, v. 356, p. 1,001–1,012.

Donald, D.B., Syrgiannis, J., Crosley, R.W., Hodsworth, G., Muir, D.C.G., Rosenberg, B., Sole, A., and Schindler, D.W., 1999, Delayed deposition of organochlorine pesticides at a temperate glacier: Environmental Science and Technology, v. 33, p. 1,794–1,798.

Douglas, M., 2004 [reprint 1970], Natural symbols—Explorations in cosmology: New York, Routledge, 177 p.

Druckman, J. N., 2015, Communicating policy-relevant science: Political Science and Politics, v. 48, no. 1, p. 58–69.

Duinker, P.N., Burbidge, E.L., Boardley, S.R., and Greig, L.A., 2013, Scientific dimensions of cumulative effects assessment—Toward improvements in guidance for practice: Environmental Reviews, v. 21, no. 1, p. 40–52.

Duinker, P.N., and Greig, L.A., 2006, The impotence of cumulative effects assessment in Canada—Ailments and ideas for redeployment: Environmental Management, v. 37, no. 2, p. 153–161.

Duinker, P.N., and Greig, L.A., 2007, Scenario analysis in environmental impact assessment—Improving explorations of the future: Environmental Impact Assessment Review, v. 27, no. 3, p. 206–219, doi:10.1016/j.eiar.2006.11.001.

Dukes, J.S., Pontius, J., Orwig, D., Garnas, J.R., Rodgers, V.L., Brazee, N., Cooke, B., Theoharides, K.A., Stange, E.E., and Harrington, R., 2009, Responses of insect pests, pathogens, and invasive plant species to climate change in the forests of northeastern North America—What can we predict?: Canadian Journal of Forest Research, v. 39, no. 2, p. 231– 248.

Edwards, M.E., and Brubaker, L.B., 1986, Late Quaternary vegetation history of the Fishhook Bend area, Porcupine River, Alaska: Canadian Journal of Earth Science, v. 23, p. 1,765–1,773.

Edwards, M.E., Brubaker, L.B., Lozhkin, A.V., and Anderson, P.M., 2005, Structurally novel biomes—A response to past warming in Beringia: Ecology, v. 86, p. 1,696–1,703.

Ellis, E.C., Goldewijk, K.K., Siebert, S., Lightman, D., and Ramankutty, N., 2010, Anthropogenic transformation of the biomes, 1700 to 2000: Global Ecology and Biogeography, v. 19, no. 5, p. 589–606.

Elsner, W.K., and Jorgenson, J.C., 2009, White spruce seedling (Picea glauca) discovered north of the Brooks Range along Alaska's Dalton Highway: Arctic, v. 62, p. 342–344.

Environment Canada, 2013, Toxic substances list: Environment and Climate Change Canada website, accessed November 13, 2015, at http://www.ec.gc.ca/lcpe-cepa/default.asp?lang=En&n=0DA2924D-1.

Environment Canada, 2014a, Climate trends and variation bulletin: Environment and Climate Change Canada website, accessed June 25, 2015, accessed November 10, 2017 at https://www.canada.ca/en/environment-climate-change/services/climate-change/trends-variations.html .

Environment Canada, 2014b, National inventory report, 1990–2012— Greenhouse gas sources and sinks in Canada: Ottawa, Ontario, Environment Canada, Greenhouse Gas Division, accessed March 28, 2015, at http://www.ec.gc.ca/ges-ghg/.

Environment Yukon, 2010, Status of Yukon fisheries 2010: Whitehorse, Canada, Yukon Environment, Fish and Wildlife Branch, 95 p., accessed March 28, 2015, at http://www.env.gov.yk.ca/publications-maps/documents/status_yukon_fisheries2010.pdf.

Environment Yukon, 2014, Plans and reports—Moose: Whitehorse, Canada, Environment Yukon, accessed March 28, 2015, at http://www.env.gov.yk.ca/publications-maps/plansreports.php#moose.

Estes, J.A., Terborgh, J., Brashares, J.A., Power, M.E., Berger, J., Bond, W.J., Carpenter, S.R., Essington, T.E., Holt, R.D., Jackson, J.B.C., Marquis, R.J., Oksanen, L., Oksanen, T., Paine, R.T., Pikitch, E.K., Ripple, W.J., Sandin, S.A., Scheffer, M., Schoener, T.W., Shurin, J.B., Sinclair, A.R.E., Soulé, M.E., Virtanen, R., and Wardle, D.A., 2011, Trophic downgrading of planet Earth: Science, v. 333 p. 301–306.

Federal Energy Regulatory Commission, 2015. Alaska natural gas transportation projects: Federal Energy Regulatory Commission website, accessed December 16, 2015, at http://www.ferc.gov/industries/gas/indus-act/angtp.asp.

Ferguson, D.B., Rice, J.L., and Woodhouse, C.A., 2014, Linking environmental research and practice—Lessons from the integration of climate science and water management in the Western United States: Tucson, University of Arizona, Climate Assessment for the Southwest, 21 p, accessed on December 2, 2017, at http://www.climas.arizona.edu/publication/report/linking-environmental-research-and-practice.

Festa-Bianchet, M., Ray, J.C., Boutin, S., Côté, S.D., and Gunn, A., 2011, Conservation of caribou (Rangifer tarandus) in Canada—An uncertain future: Canadian Journal of Zoology, v. 89, no. 5, p. 419–434.

Fischer, F., 2009, Democracy and expertise—Reorienting policy inquiry: New York, Oxford University Press, 304 p.

Fischhoff, B., 1995, Risk perception and communication unplugged: twenty years of process: Risk Analysis, v. 15, no. 2, p. 137–145.

Flannigan, M.D., Logan, K.A., Amiro, B.D., Skinner, W.R., and Stocks, B.J., 2005, Future area burned in Canada: Climatic Change, v. 72, nos. 1–2, p. 1–16.

Fleming, S.W., 2005, Comparative analysis of glacial and nival streamflow regimes with implications for lotic habitat quantity and fish species richness: River Research and Applications, v. 21, p. 363–379.

Folke, C., Hahn, T., Olsson, P., and Norberg, J., 2005, Adaptive governance of social-ecological systems: Annual Review of Environment and Resources, v. 30, p. 441–473, doi:10.1146/annurev.energy.30.050504.144511.

Foote, L., Krogman, N., and Spence, J., 2009, Should academics advocate on environmental issues?: Society and Natural Resources, v. 22, no. 6, p. 579–589.

Forman, R.T.T., and Alexander, L.E., 1998, Roads and their major ecological effects: Annual Review of Ecology and Systematics, v. 29, p. 207–231.

Francis, S.R., and Hamm, J., 2011, Looking forward—Using scenario modeling to support regional land use planning in northern Yukon, Canada: Ecology and Society, v. 16, no. 4, p. 18, accessed November 25, 2015, at http://dx.doi.org/10.5751/ES-04532-160418.

French, H.M., and Shur, Y., 2010, The principles of cryostratigraphy: Earth-Science Reviews, v. 101, p. 190–206.

Frick, W.F., Pollock, J.F., Hicks, A.C., Langwig, K.E., Reynolds, D.S., and others, 2010, An emerging disease causes a regional population collapse of a common North American bat species: Science, v. 239, p. 679–682.

Gamache, I., Jaramillo-Correa, J.P., Payette, S., and Bousquet, J., 2003, Diverging patterns of mitochondrial and nuclear DNA diversity in subarctic black spruce—Imprint of a founder effect associated with postglacial colonization: Molecular Ecology, v. 12, p. 891–901.

Gamon, J.A., Huemmrich, K.F., Stone, R.S., and Tweedie, R.S., 2013, Spatial and temporal variation in primary productivity (NDVI) of coastal Alaskan tundra—Decreased vegetation growth following earlier snowmelt: Remote Sensing of Environment, v. 129, p. 144–153.

Garamszegi, L.Z., 2011, Climate change increases the risk of malaria in birds: Global Change. Biology, v. 17, p. 1,751–1,759.

Garcia, R.A., Cabeza, M., Rahbek, C., and Araújo, M.B., 2014, Multiple dimensions of climate change and their implications for biodiversity: Science, v. 344, no. 6183, doi:10.1126/science.1247579.

Garibaldi, A., and Turner, N., 2004, Cultural keystone species—Implications for ecological conservation and restoration: Ecology and Society, v. 9, no. 3, article 1, accessed November 25, 2015, at http://www.ecologyandsociety.org/vol9/iss3/art1.

Garmestani, A.S, and Allen, C.R., eds., 2014, Social-ecological resilience and law: New York, Columbia University Press, 416 p.

Garmestani, A.S., Allen, C.R., and Benson, M.H., 2013, Can law foster social-ecological resilience?: Ecology and Society, v. 18, no. 2, p. 37.

Gates, C.C., Freese, C.H., Gogan, P.J.P., and Kotzman, M., eds., 2010, American bison—Status survey and conservation guidelines 2010: Gland, Switzerland, International Union for Conservation of Nature, 154 p., accessed on November 14, 2017, at https://cmsdata.iucn.org/downloads/american_bison_report.pdf.

Gau, R.J., Case, R., Penner, D.F., and McLoughlin, P.D., 2002, Feeding patterns of barren-ground grizzly bears in the central Canadian Arctic: Arctic, v. 55, no. 4, p. 339–344.

Gauthier, S., Bernier, P., Kuuluvainen, T., Shvidenko, A.Z., and Schepaschenko, D.G., 2015, Boreal forest health and global change: Science, v. 349, p. 819–822.

Gende, S.M., Edwards, R.T., Willson, M.F., and Wipfli, M.S., 2002, Pacific salmon in aquatic and terrestrial ecosystems—Pacific salmon subsidize freshwater and terrestrial ecosystems through several pathways, which generate unique management and conservation issues but also provide valuable research opportunities: BioScience, v. 52, p. 917–928.

Gende, S.M., Quinn, T.P., and Willson, M.F., 2001, Consumption choice by bears feeding on salmon: Oecologia, v. 127, p. 372–382.

Gende, S.M., Quinn, T.P., Willson, M.F., Heintz, R., and Scott, T.M., 2004, Magnitude and fate of salmon-derived nutrients and energy in a coastal stream ecosystem: Journal of Freshwater Ecology, v. 19, p. 149–160.

Gerhardt, K.E., Huang, X.D., Glick, B.R., and Greenberg, B.M., 2009, Phytoremediation and rhizoremediation of organic soil contaminants—Potential and challenges: Plant Science, v. 176, p. 20–30.

Gerlach, S.C., Loring, P.A., Turner, A.M., and Atkinson, D.E., 2011, Food systems, climate change, and community needs, in Lovecraft, A.L., and Eicken, Hajo, eds., North by 2020: Fairbanks, Alaska, University of Alaska Press, p. 111–134.

Gleeson, J., Gray, P., Douglas, A., Lemieux, C.J., and Nielsen G., 2011, A practitioner's guide to climate change adaptation in Ontario's ecosystems: Sudbury, Canada, Ontario Centre for Climate Impacts and Adaptation Resources, 74 p.

Glick, P., Stein, B.A., and Edelson, N.A., 2011, Scanning the conservation horizon—A guide to climate change vulnerability assessment: Washington, D.C., National Wildlife Federation, 168 p.

Goetz, S.J., Bunn, A.G., Fiske, G.J., and Houghton, R.A., 2005, Satellite-observed photosynthetic trends across boreal North America associated with climate and fire disturbance: Proceedings of the National Academy of Sciences, v. 102, p. 13,521–13,525.

Gottesfeld, L.M.J., 1994a, Aboriginal burning for vegetation management in northwest British Columbia: Human Ecology, v. 22, no. 2, p. 171–188.

Gottesfeld, L.M.J., 1994b, Conservation, territory, and traditional beliefs—An analysis of Gitksan and Wet'suwet'en subsistence, northwest British Columbia, Canada: Human Ecology, v. 22, no. 4, p. 443–465.

Government of Canada, 2014, Federal Contaminated Sites Inventory: Treasury Board of Canada Secretariat, accessed on April 11, 2018, at http://www.tbs-sct.gc.ca/fcsi-rscf/home-accueil-eng.aspx and https://databasin.org/datasets/fea28a4dfc544eaebd466ef2f96e8bd8.

Graham, I., Logan, J., Harrison, M.B., Straus, S.E., Tetroe, J., Caswell, W. and Robinson, N., 2006, Lost in knowledge translation—Time for a map?: Journal of Continuing Education in the Health Professions, v. 26, p. 13–24.

Graves, K., 2003, Resilience and adaptation among Alaska Native men: Fairbanks, University of Alaska, Ph.D. dissertation, 217 p.

Groot, C., and Margolis, L., eds., 1991, Pacific salmon life histories: Vancouver, Canada, University of British Columbia Press 564 p.

Grosse, G., and others, 2011, Vulnerability of high-latitude soil organic carbon in North America to disturbance: Journal of Geophysical Research—Biogeosciences, v. 116, p. G00K06.

Groves, C.R., 2003, Drafting a conservation blueprint—A practitioner's guide to planning for biodiversity: Washington, D.C., The Nature Conservancy and Island Press, 404 p.

Groves, C.R., Game, E.T., Anderson, M.G., Cross, M., Enquist, C., Ferdaña, Z., Girvetz, E., Gondor, A., Hall, K.R., Higgins, J., Marshall, R., Popper, K., Schill, S., and Shafer, S.L., 2012, Incorporating climate change into systematic conservation planning: Biodiversity and Conservation, v. 21, p. 1,651–1,671, doi:10.1007/s10531-012-0269-3.

Groves, C.R., Jensen, D.B., Valutis, L.L., Redford, K.H., Shaffer, M.L., Scott, J.M., Baumgartner, J.V., Higgins, J.V., Beck, M.W., and Anderson, M.G., 2002, Planning for biodiversity conservation—Putting conservation science into practice—A seven-step framework for developing regional plans to conserve biological diversity, based upon principles of conservation biology and ecology, is being used extensively by the nature conservancy to identify priority areas for conservation: BioScience, v. 52, no. 6, p. 499–512, accessed November 10, 2017, at http://bioscience.oxfordjournals.org/content/52/6/499.full.

Groves, C.R., Valutis, L., Vosick, D., Neely, B., Wheaton, K., Touval, J., and Runnels, B., 2000, Volume II of Designing a geography of hope—A practitioner's handbook for ecoregional conservation planning: The Nature Conservancy, p. A6–1–A6–4, accessed November 14, 2015, at http://www.denix.osd.mil/nr/upload/Design_geo_hope.pdf.

Grünzweig, J.M., Sparrow, S.D., Yakir, D., and Chapin, F.S., III, 2004, Impact of agricultural land-use change on carbon storage in boreal Alaska: Global Change Biology, v. 10, no. 4, p. 452–472.

Guay, K.C., Beck, P.S.A., Berner, L.T., Goetz, S.J., Baccini, A., and Buermann, W., 2014, Vegetation productivity patterns at high northern latitudes—A multi-sensor satellite data assessment: Global Change Biology, v. 20, no. 10, p. 3,147–3,158.

Guisan, A., Tingley, R., Baumgartner, J.B., Naujokaitis-Lewis, I., Sutcliffe, P.R., Tulloch, A.I.T., Regan, T.J., Brotons, L., McDonald-Madden, E., Mantyka-Pringle, C., Martin, T.G., Rhodes, J.R., Maggini, R., Setterfield, S.A., Elith, J., Schwartz, M.W., Wintle, B.A, Broennimann, O., Austin, M., Ferrier, S., Kearney, M.R., Possingham, H.P., and Buckley, Y.M., 2013, Predicting species distributions for conservation decisions: Ecology Letters, v. 16, no. 12, p. 1,424–1,435, doi:10.1111/ele.12189.

Gunderson, L.H., and Holling, C.S., eds., 2002, Panarchy—Understanding transformations in human and natural systems: Washington, D.C., Island Press, 536 p.

Gunderson, L.H., Holling, C.S., and Light, S.S., 1995, Barriers and bridges to the renewal of regional ecosystems: New York, Columbia University Press, 593 p.

Gunn, J.H., and Noble, B.F., 2009a, A conceptual basis and methodological framework for regional strategic environmental assessment (R-SEA): Impact Assessment and Project Appraisal, v. 27, no. 4, p. 258–270.

Gunn, J.H., and Noble, B.F., 2009b, Integrating cumulative effects in regional strategic environmental assessment frameworks—Lessons learned from practice: Journal of Environmental Assessment Policy and Management, v. 11, no. 3, p. 267–290.

Gunn, J.H., and Noble, B.F., 2011, Conceptual and methodological challenges to integrating SEA and cumulative effects assessment: Environmental Impact Assessment Review, v. 31, no. 2, p. 154–160.

Gurnell, A.M., 1998, The hydrogeomorphological effects of beaver dam-building activity: Progress in Physical Geography, v. 22, no. 2, p. 167–189.

Hall, E.S., Jr., Gerlach, S.C., and Blackman, M.B., 1985, In the national interest—A geographically based study of Anaktuvik Pass Iñupiat subsistence through time: Barrow, Alaska, North Slope Borough, 97 p.

Hallett, L.M., Standish, R.J., Hulvey, K.B., Gardener, M.R., Suding, K.N., Starzomski, B.M., Murphy, S.D., and Harris, J.A., 2013, Towards a conceptual framework for novel ecosystems, in Hobbs, R.J., Higgs, E.S., and Hall, C.M., eds., Novel ecosystems—Intervening in the new ecological world order: Oxford, United Kingdom, Wiley-Blackwell, p. 16–28.

Hamann, A., and Aitken, S.N., 2013, Conservation planning under climate change—Accounting for adaptive potential and migration capacity in species distribution models: Diversity and Distributions, v. 19, no. 3, p. 268–280, doi:10.1111/j.1472-4642.2012.00945.x.

Hansen, W.D., Brinkman, T.J., Chapin, F.S., III, and Brown, C., 2013, Meeting Indigenous subsistence needs—The case for prey switching in rural Alaska: Human Dimensions of Wildlife, v. 18, no. 2, p. 109–123, doi:10.1080/10871209.2012.719172.

Haraway, Donna, 1988. Situated knowledges—The science question in feminism and the privilege of partial perspective: Feminist Studies, v. 14, no. 3, p. 575–596.

Harding, Sandra, 1998, Is science multicultural?—Postcolonialisms, feminisms, and epistemologies: Bloomington, Indiana University Press, 256 p.

Harris, J.A., Hobbs, R.J., Higgs, E., and Aronson, J., 2006, Ecological restoration and global climate change: Restoration Ecology, v. 14, p. 170–176.

Harrison, X.A., Blount, J.D., Inger, R., Norris, D.R., and Bearhop, S., 2011, Carry-over effects as drivers of fitness differences in animals: Journal of Animal Ecology, v. 80, p. 4–18.

Harte, J., 2001, Land use, biodiversity, and ecosystem integrity—The challenge of preserving Earth's life support system: Ecology Law Quarterly, v. 27, p. 929–965.

Harvell, C.D., Mitchell, C.E., Ward, J.R., Altizer, S., Dobson, A.P., Ostfeld, R.S., and Samuel, M.D., 2002, Climate warming and disease risks for terrestrial and marine biota: Science, v. 296, no. 5,576, p. 2,158–2,162

Harvest Management Coalition, 2012, Fortymile caribou herd harvest plan 2012–2018: Fairbanks, Alaska Department of Fish and Game, 24 p., accessed November 10, 2017, at https://www.adfg.alaska.gov/static/research/plans/pdfs/fortymile_harvest_plan_2012_2018.pdf.

Healy, R.G., and Ascher, W., 1995, Knowledge in the policy process—Incorporating new environmental information in natural resources policy making: Policy Sciences, v. 28, no. 1, p. 1–19.

Hebblewhite, M., 2017, Billion dollar boreal woodland caribou and the biodiversity impacts of the global oil and gas industry: Biological Conservation, v. 206, p. 102–111.

Helfield, J.H., and Naiman, R.J., 2002, Salmon and alder as nitrogen sources to riparian forests in a boreal Alaskan watershed: Oecologia, v. 133, p. 573–582.

Henry, D., Landry, A., Elliot, T., Gorecki, L., Gates, M., and Chow, C., 2008, State of the Park Report—Kluane National Park and Reserve of Canada: Parks Canada, Whitehorse, Canada, 72 p.

Hewitt, R.E., Bennett, A.P., Breen, A.L., Hollingsworth, T.N., Taylor, D.L., Chapin, F.S., III, and Rupp, T.S., 2016, Getting to the root of the matter—Landscape implications of plant- fungal interactions for tree migration in Alaska: Landscape Ecology, v. 31, no. 4, p. 895–911, doi:10.1007/s10980-015-0306-1.

Hezel, P.J., Zhang, X., Bitz, C.M., Kelly, B.P., and Massonnet, F., 2012, Projected decline in spring snow depth on Arctic sea ice caused by progressively later autumn open ocean freeze- up this century: Geophysical Research Letters, v. 39, no. L17505, doi:10.1029/2012GL052794.

Hilderbrand, G.V., Hanley, T.A., Robbins, C.T., and Schwartz, C.C., 1999, Role of brown bears (Ursus arctos) in the flow of marine nitrogen into a terrestrial ecosystem: Oecologia, v. 121, p. 546–550.

Hinzman, L.D., Bettez, N.D., Bolton, W.R., Chapin, F.S., Dyurgerov, M.B., Fastie, C.L., Griffith, B., Hollister, R.D., Hope, A., Huntington, H.P., Jensen, A.M., Jia, G.J., Jorgenson, T., Kane, D.L., Klein, D.R., Kofinas, G., Lynch, A.H., Lloyd, A.H., McGuire, A.D., Nelson, F.E., Nolan, M., Oechel, W.C., Osterkamp, T.E., Racine, C.H., Romanovsky, V.E., Stone, R.S., Stow, D.A., Sturm, M., Tweedie, C.E., Vourlitis, G.L., Walker, M.D., Walker, D.A., Webber, P.J., Welker, J., Winker, K.S., and Yoshikawa, K., 2005, Evidence and implications of recent climate change in northern Alaska and other Arctic regions: Climate Change, v. 72, no. 3, p. 251–298.

Hinzman, L.D., Deal, C., McGuire, A.D., Mernild, S.H., Polyakov, I.V., and Walsh, J.E., 2013, Trajectory of the Arctic as an integrated system: Ecological Applications, v. 23, no. 8, p. 837– 1,868.

Hinzman, L.D., Kane, D.L., Yoshikawa, K., Carr, A., Bolton, W.R., and Fraver, M., 2003, Hydrological variations among watersheds with varying degrees of permafrost: Proceedings of 8th International Conference on Permafrost, July 21–25, 2003, Zurich, Switzerland, p. 407– 411.

Hobbs, R.J., Higgs, E.S., and Hall, C.M., 2013, What do we know about, and what do we do about, novel ecosystems?, in Hobbs, R.J., Higgs, E.S., and Hall, C.M., eds., Novel ecosystems—Intervening in the new ecological world order: Oxford, United Kingdom, Wiley- Blackwell, p. 353–360.

Hobbs, R.J., Higgs, E.S., and Hall, C.M., eds., 2013, Novel ecosystems—Intervening in the new ecological world order: New York, Wiley, 368 p.

Hoberg, E.P., Galbreath, K.E., Cook, J.A., Kutz, S.J., and Polley, L., 2012, Northern host- parasite assemblages—History and biogeography on the borderlands of episodic climate and environmental transition: Advances in Parasitology, v. 79, p. 1–97.

Hoberg, E.P., Polley, L., Jenkins, E.J., Kutz, S.J., Veitch, A.M., and Elkin, B.T., 2008, Integrated approaches and empirical models for investigation of parasitic diseases in northern wildlife: Emerging Infectious Diseases, v. 14, p. 10–17.

Hocking, M.D., and Reynolds, J.D., 2011, Impacts of salmon on riparian plant diversity: Science, v. 331, p. 1,609–1,612.

Holland, S.S., 1976, Landforms of British Columbia—A physiographic outline: Victoria, Canada, British Columbia Ministry of Energy and Mines, Bulletin 48, accessed November 14, 2015, accessed November 10, 2017, at http://www.empr.gov.bc.ca/Mining/Geoscience/PublicationsCatalogue/BulletinInformation/BulletinsAfter1940/Documents/Bull48-txt.pdf.

Hollings T., Jones, M., Mooney, N., and McCallum, H., 2014, Trophic cascades following the disease-induced decline of an apex predator, the Tasmanian devil: Conservation Biology, v. 8, p. 63–75.

Hollingsworth, T.N., Walker, M.D., Chapin, F.S., III, and Parsons, A., 2006, Scale-dependent environmental controls over species composition in Alaskan black spruce communities: Canadian Journal of Forest Research, v. 36, p. 1,781–1,796.

Hopkins, D.M., 1982, Aspects of the paleogeography of Beringia during the Late Pleistocene, in Hopkins, D.M., Matthews, J.V., Jr, Schweger, C.E., and Young, S.B., eds., Paleoecology of Beringia: New York, Academic Press, p. 3–28.

Howitt, Richard, 2001, Rethinking resource management—Justice, sustainability, and Indigenous peoples: New York, Routledge, 446 p.

Hu, F.S., Brubaker, L.B., Gavin, D.G., Higuera, P.E., Lynch, J.A., Rupp, T.S., and Tinner, W., 2006, How climate and vegetation influence the fire regime of the Alaskan boreal biome—The Holocene perspective: Mitigation and Adaptation Strategies for Global Change, v. 11, no. 4, p. 829–846.

Hueffer, Karsten, O'Hara, T.M., and Follmann, E.H., 2011, Adaptation of mammalian host- pathogen interactions in a changing Arctic environment: Acta Veterinaria Scandinavica, v. 53, no. 17, 8 p.

Hueffer, Karsten, Parkinson, A.J., Gerlach, R., and Berner, J., 2013, Zoonotic infections in Alaska—Disease prevalence, potential impact of climate change and recommended actions for earlier disease detection, research, prevention and control: International Journal of Circumpolar Health, v. 72, no. 1, accessed November 10, 2017, at http://dx.doi.org/10.3402/ijch.v72i0.19562.

Hughes, J., and Macdonald, D.W., 2013, A review of the interactions between free-roaming dogs and wildlife: Biological Conservation, v. 157, p. 341–351.

Hulme, P.E. 2014. EDITORIAL—Bridging the knowing–doing gap—Know-who, know-what, know-why, know-how and know-when: Journal of Applied Ecology, v. 51, no. 5, p. 1,131–1,136.

Hunter, M.L., Jacobson, G.L., and Webb, T., 1988, Paleoecology and the coarse-filter approach to maintaining biological diversity: Conservation Biology, v. 2, no. 4, p. 375–385.

Huntington, H.P., 1992, Wildlife management and subsistence hunting in Alaska: Seattle, University of Washington Press, 177 p.

Huntington, H.P., Trainor, S.F., Natcher, D.C., Huntington, O.H., DeWilde, L., and Chapin, F.S., III, 2006, The significance of context in community-based research—Understanding discussions about wildfire in Huslia, Alaska: Ecology and Society, v. 11, no. 1, article 40, accessed November 10, 2017, at http://www.ecologyandsociety.org/vol11/iss1/art40/.

Huntley, B., 2005, North temperate responses, in Lovejoy, T.E., and Hanna, L., eds., Climate change and biodiversity: New Haven, Connecticut, Yale University Press, p. 109–124.

Indigenous and Northern Affairs Canada, 2010, First Nation and Inuit community well-being—Describing historical trends (1981–2006): Ottawa, Ontario, Canada, Indian and Northern Affairs Canada, Strategic Research and Analysis Directorate, 27 p., accessed March 28, 2015, at http://www.aadnc-aandc.gc.ca/eng/1100100016600/1100100016641.

Indigenous and Northern Affairs Canada, 2010, NWT contaminated sites, Sahtu settlement area: Indigenous and Northern Affairs Canada website, accessed December 17, 2015, at https://www.aadnc-aandc.gc.ca/eng/1100100026371/1100100026375#canol.

Indigenous Peoples Specialty Group, 2010, AAG Indigenous Peoples Specialty Group's declaration of key questions about research ethics with Indigenous communities: Indigenous Peoples Specialty Group of the Association of American Geographers document, 12 p., accessed April 2012 at http://www.indigenousgeography.net/IPSG/pdf/IPSGResearchEthicsFinal.pdf.

Industrial Forestry Service, Ltd., 2011, The effect of climate change on invasive species and their potential impacts in Whitehorse: Prepared for Yukon Invasive Species Council, Whitehorse, Yukon, Canada, by Industrial Forestry Service, Ltd., Prince George, British Columbia, Canada, 83 p., accessed September 12, 2014, at http://www.yukoninvasives.com/pdf_docs/ClimateChange_Report.pdf.

Jacobs, K., Garfin, G., and Lenart, M., 2005, More than just talk—Connecting science and decision-making: Environment, v. 47, no. 9, p. 6–21.

Jactel, H., Brockerhoff, E., and Duelli, P., 2005, A test of the biodiversity-stability theory— Meta-analysis of tree species diversity effects on insect pest infestations, and re-examination of responsible factors, in Scherer-Lorenzen, Michael, Korner, Christian, Schulze, Ernst-Detlef, eds., Forest diversity and function: New York, Springer, p. 235–262.

Jaenson, T., Jaenson, D., Eisen, L., Petersson, E., and Lindgren, E., 2012, Changes in the geographical distribution and abundance of the tick Ixodes ricinus during the past 30 years in Sweden: Parasites and Vectors, v. 5, no. 8, https://doi.org/10.1186/1756-3305-5-8.

Jafarov, E.E., Romanovsky, V.E., Genet, H., McGuire, A.D., and Marchenko, S.S., 2013, The effects of fire on the thermal stability of permafrost in lowland and upland black spruce forests of interior Alaska in a changing climate: Environmental Research Letters, v. 8, no. 3, doi:10.1088/1748-9326/8/3/035030.

Janowicz, J.R., 2004, Yukon overview—Watersheds and hydrologic regions, in Smith, C.A.S., Meikle, J.C., and Roots, C.F., eds., Ecoregions of the Yukon Territory—Biophysical properties of Yukon Landscapes: Agriculture and Agri-Food Canada, PARC Technical Bulletin No. 04-0115-18, 313 p.

Janowicz, J.R., 2008, Apparent recent trends in hydrologic response in permafrost regions of northwest Canada: Hydrology Research, v. 39, no. 4, p. 267–275.

Janowicz, J.R., 2010, Observed trends in the river ice regimes of northwest Canada: Hydrology Research, v. 41, no. 6, p. 462–470.

Janowicz, J.R., 2011, Streamflow responses and trends between permafrost and glacierized regimes in northwestern Canada: Oxford, United Kingdom, International Association of Hydrological Sciences Publication 346, p. 9–14.

Jaramillo-Correa, J.P., Beaulieu, J., and Bousquet, J., 2004, Variation in mitochondrial DNA reveals multiple glacial refugia in black spruce (Picea mariana), a transcontinental North American conifer: Molecular Ecology, v. 13, p. 2,735–2,747.

Jarema, S.I., Samson, J., McGill, B.J., and Humphries, M.M., 2009, Variation in abundance across a species' range predicts climate change responses in the range interior will exceed those at the edge—A case study with North American beaver: Global Change Biology, v. 15, no. 2, p. 508–522, doi:10.1111/j.1365-2486.2008.01732.x.

Jasanoff, S., 1995, Procedural choices in regulatory science: Technology in Society, v. 17, no. 3, p. 279–293.

Jenkins, E.J., Castrodale, L.J., de Rosemond, S.J., Dixon, B.R., Elmore, S.A., Gesy, K.M., Hoberg, E.P., Polley, L., Schurer, J.M., Simard, M., and Thompson, R.C., 2013, Tradition and transition—Parasitic zoonoses of people and animals in Alaska, northern Canada, and Greenland: Advanced Parasitology, v. 82, p. 33–204, doi:10.1016/B978-0-12-407706-5.00002-2.

Jenkins, E.J., Kutz, S., Veitch, A., Elkin, B., Chirino-Trejo, M., and Polley, L., 2000, Pneumonia as a cause of mortality in two Dall's sheep in the Mackenzie Mountains, Northwest Territories, Canada, in Carey, Jean, ed., Proceedings of the Twelfth Biennial Symposium Northern Wild Sheep and Goat Council, May 31–June 4, 2000, Whitehorse, Yukon, Canada, p. 40–53.

Jia, G.J., Epstein, H.E., and Walker, D.A., 2009, Vegetation greening in the Canadian Arctic related to decadal warming: Journal of Environmental Monitoring, v. 11, no. 12, p. 2,231– 2,238.

Johnson, C.J., 2011, Regulating and planning for cumulative effects—The Canadian experience, in Krausman, P.R., and Harris, L.K., eds., Cumulative effects in wildlife management—Impact mitigation: Boca Raton, Florida, CRC Press, p. 29–46.

Johnson, C.J., 2013, Identifying ecological thresholds for regulating human activity—Effective conservation or wishful thinking?: Biological Conservation, v. 168, p. 57–65.

Johnson, C.J., and St-Laurent, M-H., 2011, A unifying framework for understanding the impacts of human developments for wildlife, in Naugle, D.E., ed., Energy development and wildlife conservation in Western North America: Washington, D.C., Island Press, p. 27–54.

Johnson, C.J., Williamson-Ehlers, L., and Seip, D., 2015, Witnessing extinction—Cumulative impacts across landscapes and the future loss of an evolutionarily significant unit of woodland caribou in Canada: Biological Conservation, v. 186, p. 176–186.

Johnstone, J.F., and Chapin, F.S., III, 2003, Non-equilibrium succession dynamics indicate continued northern migration of lodgepole pine: Global Change Biology, v. 9, p. 1,401–1,409.

Johnstone, J.F., and Chapin, F.S., III, 2006b, Effects of soil burn severity on post-fire tree recruitment in boreal forest: Ecosystems, v. 9, no. 1, p. 14–31.

Johnstone, J.F., and Chapin, F.S., III, 2006a, Fire interval effects on successional trajectory in boreal forests of northwest Canada: Ecosystems, v. 9, no. 2, p. 268–277, doi:10.1007/s10021- 005–0061–2.

Johnstone, J.F., Chapin, F.S., III, Hollingsworth, T.N., Mack, M.C., Romanovsky, V.E., and Turetsky, M.R., 2010, Fire and resilience cycles in Alaska boreal forests—A conceptual synthesis: Canadian Journal of Forest Research, v. 40, p. 1,302–1,312, doi: 1310.1139/X1310-1061.

Johnstone, J. F., Hollingsworth, T.N., Chapin, F.S., III, and Mack, M.C., 2010, Changes in fire regime break the legacy lock on successional trajectories in Alaskan boreal forest: Global Change Biology, v. 16, p. 1,281–1,295.

Johnstone, J.F., and Kasischke, E.S., 2005, Stand-level effects of soil burn severity on postfire regeneration in a recently burned black spruce forest: Canadian Journal of Forest Research, v. 35, no. 9, p. 2,151–2,163, doi: 10.1139/x05-087.

Johnstone, J.F., McIntire, E.J.B., Pedersen, E.J., King, G., and Pisaric, M.J.F., 2010, A sensitive slope—Estimating landscape patterns of forest resilience in a changing climate: EcoSphere, v. 1, no. 6, p. 1–21, article 14, doi:10.1890/ES10-00102.1.

Johnstone, J.F., Rupp, T.S., Olsen, M., and Verbyla, D., 2011, Modeling impacts of fire severity on successional trajectories and future fire behaviour in Alaskan boreal forests: Landscape Ecology, v. 26, p. 487–500.

Joly, K., Duffy, P.A., and Rupp, T.S., 2012, Simulating the effects of climate change on fire regimes in Arctic biomes—Implications for caribou and moose habitat: Ecosphere, v. 3, no. 5, doi:10.1890/ES12-00012.1.

Jorgenson, M.T., Harden, Jennifer, Kanevskiy, Mikhail, O'Donnell, Jonathan, Wickland, Kim, and others, 2013, Reorganization of vegetation, hydrology and soil carbon after permafrost degradation across heterogeneous boreal landscapes: Environmental Research Letters, v. 8, no. 3, doi:10.1088/1748-9326/8/3/035017.

Jorgenson, M.T., and others, 2010, Resilience and vulnerability of permafrost to climate change: Canadian Journal of Forest Research, v. 40, p. 1,219–1,236.

Jos, P.H., and Watson, Annette, 2016, Privileging knowledge claims in collaborative regulatory management—An ethnography of marginalization: Administration and Society, https://doi.org/10.1177/0095399715623316.

Juday, G.P., and Alix, C., 2012, Consistent negative temperature sensitivity and positive influence of precipitation on growth of floodplain Picea glauca in Interior Alaska: Canadian Journal of Forest Research, v. 42, no. 3, p. 561–573.

Juday, G.P., Barber, V., and Duffy, P., 2005, Forests, land management and agriculture, in Artic climate impact assessment: Cambridge, United Kingdom, Cambridge University Press, p. 781–862.

Juday, G.P., Barber, V., Rupp, S., Zasada, J., and Wilmking, M.W., 2003, A 200-year perspective of climate variability and the response of white spruce in Interior Alaska, chap. 12 of Greenland, D., Goodin, D., and Smith, R., eds., Climate variability and ecosystem response at Long-Term Ecological Research (LTER) sites: New York, Oxford University Press, p. 226–250.

Kahan, D.M., and Braman, Donald, 2006, Cultural cognition and public policy: Yale Journal of Law and Public Policy, Faculty Scholarship Series, Paper 103, v. 24, p. 147–170, accessed November 10, 2017, at http://digitalcommons.law.yale.edu/fss_papers/103.

Kahan, D.M., Jenkins-Smith, H., and Braam, D., 2011, Cultural cognition of scientific consensus: Journal of Risk Research, v. 14, no. 2, p. 147–174.

Kane, D.L., and Stein, J., 1983, Field evidence of groundwater recharge in interior Alaska, in Proceedings of the Fourth International Conference on Permafrost: National Academy Press, p. 572–577.

Kane, D.L., and Yang, D., 2004, Overview of water balance determinations for high latitude watersheds, in Kane, D.L., and Yang, D., eds., Northern research basins water balance— Proceedings of a workshop held at Victoria, Canada, March 2004: International Association of Hydrological Sciences Series of Proceedings and Reports, no. 290, p. 1–12.

Kanevskiy, M., Shur, Y., Fortier, D., Jorgenson M.T., and Stephani, E., 2011, Cryostratigraphy of late Pleistocene syngenetic permafrost (yedoma) in northern Alaska, Itkillik River exposure: Quaternary Research, v. 75, p. 584–596.

Kashivakura, C.V., 2013, Detecting Dermacentor albipictus, the winter tick, at the northern extent of its distribution range—Hunter-based monitoring and serological assay development: Calgary, Alberta, Canada, University of Calgary, Faculty of Veterinary Medicine, M.S. thesis.

Kasischke, E.S., and Turetsky, M.R., 2006, Recent changes in the fire regime across the North American boreal region—Spatial and temporal patterns of burning across Canada and Alaska: Geophysical Research Letters, v. 33, no. 9, doi:10.1029/2006gl025677.

Kasischke, E.S., Verbyla, D.L., Rupp, T.S., McGuire, A.D., Murphy, K.A., Jandt, R., Barnes, J.L., Hoy, E.E., Duffy, P.A., Calef, M., and Turetsky, M.R., 2010, Alaska's changing fire regime—Implications for the vulnerability of its boreal forests: Canadian Journal of Forest Research, v. 40, no. 7, p. 1,313–1,324, accessed June 25, 2015, at http://www.nrcresearchpress.com/doi/abs/10.1139/X10-098.

Kawagley, A.O. 2006, A Yupiaq worldview—A pathway to ecology and spirit: Long Grove, Illinois, Waveland Press, 176 p.

Kawagley, A.O., and Barnhardt, R., 1998, Education Indigenous to place—Western science meets Native reality: Alaska Native Knowledge Network, 17 p, accessed November 10, 2017, at http://files.eric.ed.gov/fulltext/ED426823.pdf.

Kawaguchi, Y., and Nakano, S., 2001, Contribution of terrestrial invertebrates to the annual resource budget for salmonids in forest and grassland reaches of a headwater stream: Freshwater Biology, v. 46, p. 303–316.

Keane, R.E., Veblen, T., Ryan, K.C., Logan, J., Allen, C., and Hawkes, B., 2002, The cascading effects of fire exclusion in the Rocky Mountains in Baron, J.S., ed., Rocky Mountain futures— An ecological perspective: Washington, D.C., Island Press, p. 133–153.

Keeley, J.E., 2009, Fire intensity, fire severity and burn severity—A brief review and suggested usage: International Journal of Wildland Fire, v. 18, no. 1, p. 116–126.

Keenan, T.J., and Cwynar, L.C., 1992, Late quaternary history of black spruce and grasslands in southwest Yukon Territory: Canadian Journal of Botany, v. 70, p. 1,336–1,345.

Keevel, G., 2007, Mowing money on Yukon's highways: Yukon News, August 14, 2007, accessed September 12, 2014, at http://yukon-news.com/news/mowing-money-on-yukons-highways.

Kellomäki, S., 2016, Managing boreal forests in the context of climate change—Impacts, adaptation and climate change mitigation: Boca Raton, Florida, CRC Press, 365 p.

Kelly, E.N., Schindler, D.W., St. Louis, V.L., Donald, D.B., and Vladicka, K.E., 2006, Forest fire increases mercury accumulation by fishes via food web restructuring and increased mercury inputs: Proceedings of the National Academy of Sciences, v. 103, p. 19,380–19,385.

Kelly, R., Chipman, M.L., Higuera, P.E., Stefanova, I., Brubaker, L.B., and Hu, F.S., 2013, Recent burning of boreal forests exceeds fire regime limits of the past 10,000 years: Proceedings of the National Academy of Sciences, v. 110, p. 13,055–13,060, doi:10.1073/pnas.1305069110.

Kenai Cooperative Weed Management Area, 2014, Integrated pest management plan for eradicating elodea from the Kenai Peninsula—June 2014: Prepared by the Elodea Subcommittee of the Kenai Cooperative Weed Management Area, 47 p., accessed September 12, 2014, accessed November 10, 2017, at http://www.fws.gov/uploadedFiles/Integrated_Pest_Management_Plan_for_Eradicating_Elodea_ver4.pdf.

Kendall R.L., 1992, Public relations campaign strategies: planning for implementation. New York, NY: Harper Collins.

Kendall, Joan, 2001, Circles of disadvantage—Aboriginal poverty and underdevelopment in Canada: American Review of Canadian Studies, v. 31, no. 1–2, p. 43–59, http://dx.doi.org/10.1080/02722010109481581.

Kennedy, P.L., 2013, Perspective—Moving to the dark side, in Hobbs, R.J., Higgs, E.S., and Hall, C.M., eds., Novel ecosystems—Intervening in the new ecological world order: Oxford, United Kingdom, Wiley-Blackwell, p. 239–241

Kettle, N.P., and Dow, K., 2014, Cross-level differences and similarities in coastal climate change adaptation planning: Environmental Science and Policy, v. 44, p. 27–290, doi:10.1016/j.envsci.2014.08.013.

Kettle, N., and Trainor, S.F., 2015, The role of climate webinars in supporting extended boundary chain networks across Alaska: Climate Risk Management, v. 9, p. 6–19.

Klaminder, J., Yoo, K., Rydberg, J., and Giesler, R., 2008, An explorative study of mercury export from a thawing palsa mire: Journal of Geophysical Research, Biogeosciences, v. 113, iss. G4, doi:10.1029/2008JG000776.

Klein, E., Berg, E.E., and Dial, R., 2005, Wetland drying and succession across the Kenai Peninsula lowlands, south-central Alaska: Canadian Journal of Forest Research, v. 35, p. 1,931–1,941.

Klein, E.S., Berg, E.E., and Dial, R., 2011, Reply to comment by Gracz on "Wetland drying and succession across the Kenai Peninsula Lowlands, south-central Alaska": Canadian Journal of Forest Research, v. 41, no. 2, p. 429–433, https://doi.org/10.1139/X10-216.

Knapp, C.N., and Trainor, S.F., 2013, Adapting science to a warming world: Global Environmental Change, v. 23, no. 5, p. 1,296–1,306.

Knapp, C.N., and Trainor, S.F., 2015, Alaskan stakeholder-defined research needs in the context of climate change: Polar Geography, v. 38, no. 1, p. 42–69, http://dx.doi.org/10.1080/1088937X.2014.999844.

Knudtson, P., and Suzuki, D., 1992, Wisdom of the elders: Toronto, Stodart Publishing, Ltd., 274 p.

Kocher, S.D., Toman, E., Trainor, S.F., Wright, V., Briggs, J.S., Goebel, C.P., Montblanc, E.M., Oxarart, A., Pepin, D.L., Steelman, T.A., Thode, A., and Waldrop, T.A., 2012, How can we span the boundaries between wildland fire science and management in the United States?: Journal of Forestry, v. 110, no. 8, p. 421–428, doi: 10.5849/jof.11–085.

Kokelj, S.V., and Jorgenson, M.T., 2013, Advances in thermokarst research: Permafrost and Periglacial Processes, v. 24, p. 108–119.

Kokelj, S.V., and others, 2013, Thawing of massive ground ice in mega slumps drives increases in stream sediment and solute flux across a range of watershed scales: Journal of Geophysical Research—Earth Surface, v. 118, p. 681–692, doi:10.1002/jgrf20063.

Kokelj, S.V., Lantz, T.C., Tunnicliffe, J., Segal, R., and Laelle, D., 2017, Climate-driven thaw of permafrost preserved glacial landscapes, northwestern Canada: Geology, v. 45, p. 371–374.

Kovach, M.E., 2010, Indigenous methodologies—Characteristics, conversations, and contexts: Toronto, Ontario, Canada, University of Toronto Press, 216 p.

Krebs, C.J., Boutin, S., and Boonstra, R., 2001, Ecosystem dynamics of the boreal forest—The Kluane project: New York, Oxford University Press, 501 p.

Krupnik, I., and Jolly, D., eds., 2002, The Earth is faster now—Indigenous observations of Arctic environmental change: Fairbanks, Alaska, Arctic Research Consortium of the United States, 356 p.

Kruse, J.J., Smith, D.R., and Schiff, N.M., 2010, Monsoma pulveratum (Retzius) (Hymenoptera: Tenthredinidae: Allantinae), a palaearctic sawfly defoliator of alder in Alaska and new to the United States: Proceedings of the Entomological Society of Washington, v. 112, no. 2, p. 332– 335.

Kuhnlein, H.V., Receveur, O., Soueida, R., and Egeland, G.M., 2004, Arctic Indigenous peoples experience the nutrition transition with changing dietary patterns and obesity: Journal of Nutrition, v. 134, no. 6, p. 1,447–1,453.

Kurath, M. and Gisler, P., 2009, Informing, involving or engaging? Science communication, in the ages of atom-, bio- and nanotechnology: Public Understanding of Science, v. 18, no. 5, p. 559–573.

Kutz, S.J., Checkley, S., Verocai, G.G., Dumond, M., Hoberg, E.P., Peacock, R., Wu, J.P., Orsel, K., Seegers, K., Warren, A.L., and Abrams, A., 2013, Invasion, establishment, and range expansion of two parasitic nematodes in the Canadian Arctic: Global Change. Biology, v. 19, p. 3,254–3,262.

Kutz, S.J., Jenkins, E.J., Veitch, A.M., Ducrocq, J., Polley, L., Elkin, B., and Lair, S., 2009, The Arctic as a model for anticipating, preventing, and mitigating climate change impacts on host- parasite interactions: Veterinary Parasitology, v. 163, p. 217–228.

Laaksonen, S., Pusenius, J., Kumpula, J., Venäläinen, A., Kortet, R., Oksanen, A., and Hoberg, E., 2010, Climate change promotes the emergence of serious disease outbreaks of filarioid nematodes: EcoHealth, v. 7, p. 7–13.

Lambin, E.F., Turner, B.L., Geist, H.J., Agbola, S.B., Angelsen, A., Bruce, J.W., Coomes, O.T., Dirzo, R., Fischer, G., Folke, C., George, P.S., Homewood, K., Imbernon, J., Leemans, R., Li, X., Moran, E.F., Mortimore, M., Ramakrishnan, P.S., Richards, J.F., Skånes, H., Steffen, W., Stone, G.D., Svedin, U., Veldkamp, T.A., Vogel, C., and Xu, J., 2001, The causes of land-use and land-cover change—Moving beyond the myths: Global Environmental Change, v. 11, p. 261–269.

Lance, E.W., Howell, S.M., Lance, B.K., Howlin, S., Suring, L.H., and Goldstein, M.I., 2006, Spruce beetles and timber harvest in Alaska—Implications for northern red-backed voles: Forest Ecology Management, v. 222, nos. 1–3, p. 476–479.

Langdon, S.J., 2006, Tidal pulse fishing, in Menzies, C.R., and Butler, Caroline, eds., Traditional ecological knowledge and natural resource management: Lincoln, University of Nebraska Press, p. 21–46.

Larsen, C., 1997, Spatial and temporal variations in boreal forest fire frequency in northern Alberta: Journal of Biogeography, v. 24, no. 5, p. 663–673.

Larsen, P.H., and others, 2008, Estimating future costs for Alaska public infrastructure at risk from climate change: Global Environmental Change, v. 18, p. 442–457.

Latham, A.D.M., Latham, M.C., McCutchen, N.A., and Boutin, S. 2011. Invading white-tailed deer change wolf-caribou dynamics in northeastern Alberta: The Journal of Wildlife Management, v. 75, no. 1, p. 204–212.

Latour, B., 1987, Science in action—How to follow scientists and engineers through society: Cambridge, Massachusetts, Harvard University Press, 274 p.

Latour, B., 1993, We have never been modern: Cambridge, Massachusetts, Harvard University Press, 168 p.

Latour, B., and Woolgar, S., 1986, Laboratory life—The construction of scientific facts: Princeton, New Jersey, Princeton University Press, 296 p.

Law, K.S., and Stohl, A., 2007, Arctic air pollution—Origins and impacts: Science, v. 315, no. 5818, p. 1,537–1,540, accessed November 10, 2017, at http://www.jstor.org/stable/20035796.

Lawler, J.J., Ackerly, D., Albano, C., Anderson, M., Cross, M., Dobrowski, S., Gill, J., Heller, N., Pressey, R., Sanderson, E., and Weiss, S., 2015, The theory behind, and challenges of, conserving nature's stage in a time of rapid change: Conservation Biology, v. 29, no. 3, p. 618–629.

Lawler, J.L., Tear, T.H., Pyke, C., Shaw, M.R., Gonzalez, P., Kareiva, P., Hansen, L., Hannah, L., Klausmeyer, K., Aldous, A., Bienz, C., and Pearsall, S., 2010, Resource management in a changing and uncertain climate: Frontiers in Ecology and the Environment, v. 8, no. 1, p. 35– 43, doi:10.1890/070146.

Leckie, D., Paradine, D., Burt, W., Hardman, D., Eichel, F., Tinis, S., and Tammadge, D., 2006, NIR 2007 deforestation methods summary: Natural Resources Canada, Canadian Forest Service, Pacific Forestry Centre, Victoria, British Columbia, Report DRS-N-013, 12 p.

Lee, K.N., 1994, Compass and gyroscope—Integrating science and politics for the environment: Washington, D.C., Island Press, 255 p.

Leewis M.C., Reynolds, C.M., and Leigh, M.B., 2013, Long-term effects of nutrient addition and phytoremediation on diesel and crude oil contaminated soils in subarctic Alaska: Cold Regions Science and Technology, v. 96, p. 129–137.

Lemon, D.S., Warren, F.J., Lacroix, J., and Bush, E., eds., 2008, From impacts to adaptation— Canada in a changing climate 2008: Ottawa, Ontario, Government of Canada, p. 329–386, accessed June 25, 2015, at http://www.nrcan.gc.ca/environment/resources/publications/impacts-adaptation/reports/assessments/2008/10253.

Lemos, M.C., Kirchhoff, C.J., Kalafatis, S.E., Scavia, D., and Rood, R.B., 2014, Moving climate information off the shelf—Boundary chains and the role of RISAs as adaptive organizations: Weather, Climate, and Society, v. 6, no. 2, p. 273–285.

Lemos, M.C., Kirchhoff, C.J., and Ramprasad, V., 2012, Narrowing the climate information usability gap: Nature Climate Change, v. 2, no. 11, p. 789–794.

Leo, S.S.T., Samuel, W.M., Pybus, M.J., and Sperling, F.A.H., 2014, Origin of Demacentor albipictus (Acari:Ixodidae) on elk in the Yukon Canada. Journal of Wildlife Diseases, v. 50, no. 3, p. 544–551, https://doi.org/10.7589/2013-03-078.

Lewis, H.T., and Ferguson, T.A., 1988, Yards, corridors, and mosaics—How to burn a boreal forest: Human Ecology, v. 16, no. 1, p. 57–77.

Liebhold, A., and Kamata, N., 2000, Population dynamics of forest-defoliating insects: Population Ecology, v. 42, p. 205–278.

Lindenmayer, D.B., and Likens, G.E., 2010, The science and application of ecological monitoring: Biological Conservation, v. 143, no. 6, p. 1,317–1,328, doi:10.1016/j.biocon.2010.02.013.

Lindenmayer, D.B., Likens, G.E., Krebs, C.J., and Hobbs, R.J., 2010, Improved probability of detection of ecological "surprises": Proceedings of the National Academy of Sciences, v. 107, no. 51, p. 21,957–21,962, doi: 10.1073/pnas.1015696107.

Line, J., Brunner, G., Rosie, R., and Russell, K., 2008, Results of the 2007 invasive plants roadside inventory in Yukon: Whitehorse, Yukon Department of Environment, 38 p., accessed June 27, 2014, accessed November 10, 2017, at http://www.yukoninvasives.com/pdf_docs/InvasivePlantsRoadsideInventory_nov14_sm.pdf.

Littell, J.S., McKenzie, D., Kerns, B.K., Cushman, S., and Shaw, C.G., 2011, Managing uncertainty in climate-driven ecological models to inform adaptation to climate change: Ecosphere, v. 2, no. 9, article102, doi:10.1890/ES11-00114.1.

Lloyd, A.H., Edwards, M.E., Finney, B.P., Lynch, J.A., Barber, V.A., and Bigelow N.H., 2005, Development of the boreal forest, in Chapin, F.S., III, Oswood, M.W., Van Cleve, K., Viereck, L.A., and Verbyla, D.L., eds., Alaska's changing boreal forest: New York, Oxford University Press, p. 62–80.

Loarie, S.R., Duffy, P.B., Hamilton, H., Asner, G.P., Field, C.B., and Ackerly, D.D., 2009, The velocity of climate change: Nature, v. 462, p. 1,052–1,055.

Loo, T., 2007, Disturbing the peace—Environmental change and the scales of justice on a northern river: Environmental History, v. 12, no. 4, p. 895–919.

Loring, P.A., and Fazzino, D., 2014, From 'would' and 'will' to 'could' and 'can'—Climate change and environmental (in)justice in the North American Arctic: Anthropology News, v. 55, no. 4, doi:10.1111/j 1556–3502.2014.55402.x.pdf.

Loring, P.A., and Gerlach, S.C., 2009, Food, culture, and human health in Alaska—An integrative health approach to food security: Environmental Science and Policy, v. 12, no. 4, p. 466–478.

Loring, P.A., and Gerlach, S.C., 2010, Food security and conservation of Yukon River salmon— Are we asking too much of the Yukon River?: Sustainability, v. 2, no. 9, p. 2,965–2,98, doi:10.3390/su2092965.

Lottermoser, B., 2010, Mine wastes—Characterization, treatment and environmental impacts: New York, Springer, 304 p.

Louis, R.P., 2007, Can you hear us now?—Voices from the margin—Using Indigenous methodologies in geographic research: Geographical Research, v. 45, p. 130–139.

Luizza, M.W., Evangelista, P.H., Jarnevich, C.S., West, A., and Stewart, H., 2016, Integrating subsistence practice and species distribution modeling—Assessing invasive elodea's potential impact on Native Alaskan subsistence of Chinook salmon and whitefish: Environmental Management, doi:10.1007/s00267-016-0692-4.

Lynch, J.A., Clark, J.S., Bigelow, N.H., Edwards, M.E., and Finney, B.P., 2002, Geographic and temporal variations in fire history in boreal ecosystems of Alaska: Journal of Geophysical. Research, v. 107, no. D1, p. FFR 8-1-FFR 8-17, doi:10.1029/2001JD000332.

MacDonald, G.M., and Cwynar, L.C., 1986, A fossil pollen based reconstruction of the late Quaternary history of lodgepole pine (Pinus contorta ssp. latifolia) in the western interior of Canada: Canadian Journal of Forest Research, v. 15, p. 1,039–1,044.

Magness, D.R. and Morton, J., 2018, Using climate envelope models to identify potential ecological trajectories on the Kenai Peninsula, Alaska. PLoS ONE 13(12):e0208883.

Magness, D.R., Lovecraft, A.L., and Morton, J.M. 2012, Factors influencing individual management preferences for facilitating adaptation to climate change within the National Wildlife Refuge System: Wildlife Society Bulletin, v. 36, no. 3, p. 457–468.

Magness, D.R., Morton, J.M., Huettmann, F., Chapin, F.S., III, and McGuire, A.D., 2011, A climate-change adaptation framework to reduce continental-scale vulnerability across conservation reserves: Ecosphere, v. 2, no. 10, p. 1–23, article 212, doi:10.1890/ES11-00200.1.

Mahmoud, M., Liu, Y., Hartmann, H., Stewart, S., Wagener, T., Semmens, D., Stewart, R., Gupta, H., Dominguez, D., Dominguez, F., Hulse, D., Letcher, R., Rashleigh, B., Smith, C., Street, R., Ticehurst, J., Twery, M., van Delden, H., Waldick, R., White, D., and Winter, L., 2009, A formal framework for scenario development in support of environmental decision-making: Environmental Modeling Software, v. 24, no. 7, p. 798–808, doi:10.1016/j.envsoft.2008.11.010.

Maibach, E.W., 2017, Increasing public awareness and facilitating behavior change: Two guiding heuristics. In: Hannah, L and Lovejoy, T (eds), Climate Change and Biodiversity.

Maier, J.A.K., ver Hoef, J.M., McGuire, A.D., Bowyer, R.T., Saperstein, L. and Maier, H.A., 2005, Distribution and density of moose in relation to landscape characteristics—Effects of scale: Canadian Journal of Forest Research, v. 35, no. 9, p. 2,233–2,243.

Maletsky, L.D., Evans, W.P., Singletary, L. and Sicafuse, L.L., 2018, Joint Fire Science Program (JFSP) Fire Science Exchange Network: A national evaluation of initiative impacts: Journal of Forestry, v. 116, no. 4, p. 328–335.

Malmström, C.M., and Raffa, K.F., 2000, Biotic disturbance agents in the boreal forest— Considerations for vegetation change models: Global Change Biology, v. 6, no. S1, p. 35–48.

Manseau, M., Parlee, B., and Ayles, G.B., 2005, A place for traditional ecological knowledge in resource management in Berkes, F., Huebert, R., Fast, H., Manseau, M., and Diduck, A., eds., Breaking ice—Renewable resource and ocean management in the Canadian north: University of Calgary Press, Calgary, p. 141–164.

Margerum, R.D., 2011, Beyond consensus—Improving collaborative planning and management: Cambridge, Massachusetts, MIT Press, 416 p.

Markel, C., and Clark, D.A., 2012, Developing policy alternatives for the management of wood bison (Bison bison athabascae) in Kluane National Park and Reserve of Canada: The Northern Review, v. 36 (fall 2012), p. 53–76.

Marker, M., 2006, After the Makah whale hunt—Indigenous knowledge and limits to multicultural discourse: Urban Education, v. 41, no. 5, p. 482–505.

Markon, C.J., Trainor, S.F., and Chapin, F.S., III, eds., 2012, The United States National Climate Assessment—Alaska technical regional report: U.S. Geological Survey Circular 1379, 148 p.

Marles, R.J., Clavelle, C., Monteleone, L., Tays, N., and Burns, D., 2000, Aboriginal plant use in Canada's northwest boreal forest: Vancouver, Canada, University of British Columbia Press, 368 p.

Marsh, P., 1990, Snow hydrology, in Prowse, T.D., and Ommanney, C.S.L., eds., Northern hydrology—Canadian perspective: Environment Canada, National Hydrological Research Institute Science Report 1, p. 37–61.

Martin, J., Runge, M.C., Nichols, J.D., Lubow, B.C., and Kendall, W.L., 2009, Structured decision-making as a conceptual framework to identify thresholds for conservation and management: Ecological Applications, v. 19, no. 5, p. 1,079–1,090.

Mascaro, J., 2013, Perspective—From rivets to rivers, in Hobbs, R.J., Higgs, E.S., and Hall, C.M., eds., Novel ecosystems—Intervening in the new ecological world order: Oxford, United Kingdom, Wiley-Blackwell, p. 155–156.

Mascaro, J., Harris, J.A., Lach, L., Thompson, A., Perring, M.P., Richardson, D.M., and Ellis, E.C., 2013, Origins of the novel ecosystem concept, in Hobbs, R.J., Higgs, E.S., and Hall, C.M., eds., Novel ecosystems—Intervening in the new ecological world order: Oxford, United Kingdom, Wiley-Blackwell, p. 45–57.

Masterandrea, M.C., Field, C.B., Stocker, O.E., Ebi, K.L, Frame, D.J., and others, 2010, Guidance note for lead authors of the IPCC Fifth Assessment Report on the consistent treatment of uncertainties: Intergovernmental Panel on Climate Change Cross-Working Group meeting on Consistent Treatment of Uncertainties, Jasper Ridge, California, July 6–7, 2010, accessed November 10, 2017, at https://www.ipcc.ch/pdf/supporting-material/uncertainty-guidance-note.pdf.

Mattson, D., Karl, H., and Clark, S.G., 2012, Values in natural resource management and policy, in Karl, H., ed., Restoring lands—Coordinating science, policy, and action—Complexities of climate and governance: Dordrecht, Netherlands, Springer, p. 239–259, doi:10.1007/978-94-007-2549-2_12.

Mattson, D.J., and Clark, S.G., 2011, Human dignity in concept and practice: Policy Sciences, v. 44, no. 4, p. 303–319.

Mauger, S., Shaftel, R., Leppi, J.C., and Rinella, D.J., 2016, Summer temperature regimes in south-central Alaska streams—Watershed drivers of variation and potential implications for Pacific salmon: Canadian Journal of Fisheries and Aquatic Sciences, v. 74, no. 5, p. 702–715. https://doi.org/10.1139/cjfas-2016-0076.

McAfee, S.A., Guentchev, G., and Eischeid, J.K., 2013, Reconciling precipitation trends in Alaska—1, Station-based analyses: Journal of Geophysical Research—Atmospheres, v. 118, no. 14, p. 7,523–7,541.

McAfee S.A., Walsh, J., and Rupp, T.S., 2013, Statistically downscaled projections of snow/rain partitioning for Alaska: Hydrological Processes, v. 28, no. 12, p. 3,930–3,946.

McCulloch, L., and Woods, A., 2009, British Columbia's northern interior forests—Dothistroma stand establishment decision aid: BC Journal of Ecosystems and Management, v. 10, no. 1, p. 1–3.

McDowell Group, 2014, The economic impact of placer mining in Alaska: Prepared for Alaska Miners Association by McDowell Group, Inc., Juneau and Anchorage, Alaska, 22 p.

McGuire, A.D., Ruess, R.W., Lloyd, A., Yarie, J., Clein, J.S., and Juday, G.P., 2010, Vulnerability of white spruce tree growth in interior Alaska in response to climate variability—Dendrochronological, demographic, and experimental perspectives: Canadian Journal of Forest Research, v. 40, no. 7, p. 1,197–1,209.

McGuire, A.D., and others, 2009, Sensitivity of the carbon cycle in the Arctic to climate change: Ecological Monographs, v. 79, p. 523–555.

McIntosh, Roderick, Tainter, Joseph, and McIntosh, S.K., 2000, Climate, history, and human action, in McIntosh, Roderick, Tainter Joseph, and McIntosh, S.K., eds., The way the wind blows: New York, Columbia University Press, p. 1–42.

McNeeley, S., 2011, Examining barriers and opportunities for sustainable adaptation to climate change in interior Alaska: Climatic Change, v. 111, p. 835–857.

McNeeley, S.M., and Shulski, M.D., 2011, Anatomy of a closing window—Vulnerability to changing seasonality in interior Alaska: Global Environmental Change, v. 21, p. 464–473, doi:10.1016/j.gloenvcha.2011.02.003.

McNie, E.C., 2012, Delivering climate services—Organizational strategies and approaches for producing useful climate-science information: Weather, Climate, and Society, v. 5, no. 1, p. 14–26.

Meadow, A. M., D. B. Ferguson D.B., Guido, Z, Horangic, A., Owen, G., and Wall, T., 2015, Moving toward the deliberate coproduction of climate science knowledge: Weather, Climate, and Society, v. 7, no. 2, p. 179–191.

Menzel, A., and others, 2006, European phenological response to climate change matches the warming pattern: Global Change Biology, v. 12, no. 10, p. 1,969–1,976.

Millar, C.I., Stephenson, N.L., and Stephens, S.L., 2007, Climate change and forests of the future—Managing in the face of uncertainty: Ecological Applications, v. 17, no. 8 p. 2,145–2,151.

Millar, C.I., and Woolfenden, W., 1999, The role of climate change in interpreting historic variability: Ecological Applications, v. 9, p. 1,207–1,216.

Millennium Ecosystem Assessment, 2003, Ecosystems and human well-being: Washington, D.C., Island Press, 212 p.

Millennium Ecosystem Assessment, 2005, Ecosystems and human well-being—Synthesis: Washington D.C., Island Press, 160 p.

Miller, R.S., Farnsworth, M.L., and Malmberg, J.L., 2013, Diseases at the livestock–wildlife interface—Status, challenges, and opportunities in the United States: Preventive Veterinary Medicine, v. 110, no. 2, p. 119–132.

Minimata Convention on Mercury, 2015, Countries: Minimata Convention on Mercury website, accessed November 13, 2015, accessed November 10, 2017, at http://www.mercuryconvention.org/Countries/tabid/3428/Default.aspx.

Minister of Indian Affairs and Northern Development, 1998, Little Salmon/Carmacks First Nation final agreement: Ottawa, Canada, Minister of Indian Affairs and Northern Development, 325 p.

Miraglia, R., 1998, Traditional ecological knowledge handbook—A training manual and reference guide for designing, conducting and participating in research projects using traditional ecological knowledge: Anchorage, Alaska Department of Fish and Game, Division of Subsistence, 42 p.

Mitchell, D.C., 2003, Sold American—The Story of Alaska Natives and their land, 1867–1959: Fairbanks, University of Alaska Press, 564 p.

Moloney, K., and Unger, M., 2014, Transmedia storytelling in science communication—One subject, multiple media, unlimited stories, chap. 8 of Drake, J.L., and others, eds., New trends in earth-science outreach and engagement—Advances in natural and technological hazards research, Volume 38: Cham, Switzerland, Springer International Publishing, p. 109–120.

Moore, R.D., Fleming, S.W., Menounos, B., Wheate, R., Fountain, A., Stahl, K., Holm, K., and Jakob, M., 2009, Glacier change in western North America—Influences on hydrology, geomorphic hazards and water quality: Hydrological Processes, v. 23, no. 1, p. 42–61.

Mortimer-Sandilands, C., 2009, The cultural politics of ecological integrity—Nature and nation in Canada's National Parks, 1885–2000: International Journal of Canadian Studies/Revue Internationale d'Études Canadiennes, nos. 39–40, p. 161–189.

Morton, J.M., Berg, E., Newbould, D., MacLean, D., and O'Brien, L., 2006, Wilderness fire stewardship on the Kenai National Wildlife Refuge, Alaska: International Journal of Wilderness, v. 12, p. 14–17.

Morton, J.M., Magness, D.R., McCarty, M., Wigglesworth, D., Ruffner, R., Bernard, M., Walker, N., Fuller, H., Mauger, S., Bornemann, B., Fuller, L., and Smith. M., 2015, Kenai mountains to sea—A land conservation strategy to sustain our way of life on the Kenai Peninsula: Homer, Alaska, Kachemak Heritage Land Trust, 82 p.

Mote, P., Brekke, L., Duffy, P.B., and Maurer, E., 2011, Guidelines for constructing climate scenarios: Eos, Transactions American Geophysical Union, v. 92, no. 31, p. 257–258. doi:10.1029/2011EO310001.

Murphy, K., Huettmann, F., Fresco N., and Morton J., 2010, Connecting Alaska landscapes into the future—Results from an interagency climate modeling, land management and conservation project: U.S. Fish and Wildlife Service, Anchorage, Alaska, and University of Alaska, Fairbanks, 96 p.

Murton, J.B., 2013, Ground ice and cryostratigraphy, in Shroder, J.F., ed., Treatise on geomorphology: San Diego, Academic Press, 8, p. 173–201.

Muskwa-Kechika Advisory Board, 2016, Muskwa Kechika Management Area: Muskwa-Kechika Advisory Board Web site, accessed November 10, 2017, at http://www.muskwa-kechika.com/.

Nadasdy, P., 2003, Hunters and bureaucrats—Power, knowledge, and aboriginal-state relations in the southwest Yukon: Vancouver, University of British Columbia Press, 325 p.

Nader, Laura, ed., 1996, Naked science—Anthropological Inquiry into boundaries, power, and knowledge: New York, Routledge, 336 p.

Naiman, R.J., Bilby, R.E., Schindler, D.E., and Helfield, J.M., 2002, Pacific salmon, nutrients, and the dynamics of freshwater and riparian ecosystems: Ecosystems, v. 5, p. 399–417.

Naiman, R.J., Helfield, J.M., Bartz, K.K., Drake, D.C., and Honea, J.M., 2009, Pacific salmon, marine-derived nutrients, and the characteristics of aquatic and riparian ecosystems: American Fisheries Society Symposium, v. 69, p. 395–425.

Natcher, D.C., and Hickey, C.G., 2002, Putting the community back into community-based resource management—A criteria and indicators approach to sustainability: Human Organization, v. 61, no. 4, p. 350–363.

Natcher, D.C., Huntington, O., Huntington, H., Chapin, F.S., III, Trainor, S.F., and DeWilde, L.O., 2007, Notions of time and sentience—Methodological considerations for Arctic climate change research: Arctic Anthropology, v. 44, no. 2, p. 113–126, accessed on November 11, 2017, at http://www.jstor.org/stable/40316696.

Natcher, D.C., Davis, Susan, and Hickey, Clifford, 2005, Co-management—Managing relationships, not resources: Human Organization, v. 64, no. 3, p. 240–250.

National Academies of Sciences, Engineering, and Medicine, 2015, A review of the landscape conservation cooperatives: Washington, D.C., The National Academies Press, 150 p, accessed November 10, 2017, at http://www.nap.edu/catalog/21829/a-review-of-the-landscape-conservation-cooperatives.

National Academies of Sciences, Engineering, and Medicine, 2017, Communicating science effectively: a research agenda. Washington, DC: The National Academies Press. doi: 10.17226/23674.

National Park Service, 2013, Using scenarios to explore climate change—A handbook for practitioners: Fort Collins, Colorado, National Park Service, Natural Resource Stewardship and Science, Climate Change Response Program, 57 p.

National Park Service, 2015, What's so special about Wrangell-St. Elias: National Park Service website, accessed November 10, 2017, at https://www.nps.gov/wrst/whats-so-special.htm.

National Science Foundation, 2012, Principles for the conduct of research in the Arctic: National Science Foundation, Office of Polar Programs, accessed October 2012, at http://www.nsf.gov/od/opp/arctic/conduct.jsp.

Nelson, J.L., Zavaleta, E.S., and Chapin, F.S., III, 2008, Boreal fire effects on subsistence resources in Alaska and adjacent Canada: Ecosystems, v. 11, no. 1, p. 156–171.

Nelson, R., Howden, M., and Smith, M.S., 2008, Using adaptive governance to rethink the way science supports Australian drought policy: Environmental Science and Policy, v. 11, no. 7, p. 588–601.

Nichols, J.D., Koneff, M.D., Heglund, P.J., Knutson, M.G., Seamans, M.E., Lyons, J.E., Morton, J.M., Jones, M.T., Boomer, G.S., and Williams, B.K., 2011, Climate change, uncertainty, and natural resource management: The Journal of Wildlife Management, v. 75, p. 6–18.

Northern Climate Exchange, 2014, Compendium of Yukon climate change science 2003–2013: Whitehorse, Canada, Yukon College, Yukon Research Centre, 237 p., accessed November 11, 2017 at, https://www.yukoncollege.yk.ca/sites/default/files/inline-files/Compendium_updateMar2014_final_for_web.pdf.

Nossov, D.R., Hollingsworth, T.N., Ruess, R.W., and Kielland, K., 2011, Development of Alnus tenuifolia stands on an Alaskan floodplain—Patterns of recruitment, disease, and succession: Journal of Ecology, v. 99, p. 621–623, doi: 10.1111/j.1365–2745.2010.01792.x.

NWB LCC Strategic Plan, 2015, Northwest Boreal Landscape Conservation Cooperative Strategic Plan 2015–2025: Northwest Boreal Landscape Conservation Cooperative, 19 p, accessed April 10, 2018, at http://nwblcc.org/wp-content/uploads/2015/05/NWB-LCC-Strategic-Plan-V1.pdf.

O'Brien, K.L., and Leichenko, R.M., 2003, Winners and losers in the context of global change: Annals of the Association of American Geographers, v. 93, p. 89–103.

Ogden, N.H., Mechai, S., and Margos, G., 2013, Changing geographic ranges of ticks and tick- borne pathogens—Drivers, mechanisms, and consequences for pathogen diversity: Frontiers in Cellular and Infection Microbiology, v. 3, p. 1–11, doi:10.3389/fcimb.2013.00046.

Ontario Ministry of Natural Resources, 2012, Ontario invasive species strategic plan: Toronto, Ontario, Canada, Queen's Printer for Ontario, 58 p., accessed on November 11, 2017, at http://docs.ontario.ca/documents/2679-stdprod-097634.html.

Oswalt, C.M., Fei, S., Guo, Q., Iannone, B.V., III, Oswalt, S.N., Pijanowski, B.C., and Potter, K.M., 2015, A subcontinental view of forest plant invasions: NeoBiota, v. 24, p. 49–54, doi: 10.3897/neobiota.24.8378.

Pacific Climate Impacts Consortium, 2014, Climate summary for Skeena region: Victoria, British Columbia, Canada, Pacific Climate Impacts Consortium, University of Victoria, accessed June 25, 2015, at http://www.pacificclimate.org/sites/default/files/publications/Climate_Summary-Skeena.pdf.

Pachauri, R.K., and Reisinger, A., eds., 2007, Climate change 2007—Synthesis report— Contribution of working groups 1, II and III to the Fourth Assessment Report of the Intergovernmental Panel on Climate change: Geneva Switzerland, Intergovernmental Panel on Climate Change, 104 p., http://www.ipcc.ch/publications_and_data/publications_ipcc_fourth_assessment_report_synthesis_report.htm.

Pacyna, E.G., Pacyna, J.M., Steenhuisen, F., and Wilson, S., 2006, Global anthropogenic mercury emission inventory for 2000: Atmospheric Environment, v. 40, p. 4,048–4,063.

Padilla, E.S.R., 2010, Caribou leadership—A study of traditional knowledge, animal behavior, and policy: Fairbanks, University of Alaska, Ph.D. dissertation, 156 p.

Paine, R.T., 1995, A conversation on refining the concept of keystone species: Conservation Biology, v. 9, p. 962–964.

Parent, M.B., and Verbyla, D., 2010, The browning of Alaska's boreal forest: Remote Sensing, v. 2, p. 2,729–2,747.

Parker, B.H., 2003, A conceptual system modeling approach to ecological integrity planning—A case study of the Greater Kluane Region, Yukon: Parks Research Forum of Ontario 2003 Proceedings, p. 393–405.

Parlee, B., and Berkes, F., 2005, Health of the land, health of the people—A case study on Gwich'in berry harvesting in northern Canada: EcoHealth, v. 2, no. 2, p. 127–137.

Parlee, B.L., Geertsma, K., and Willier, A., 2012, Social-ecological thresholds in a changing boreal landscape—Insights from Cree knowledge of the Lesser Slave Lake Region of Alberta, Canada: Ecology and Society, v. 17, no. 2, article 20, accessed November 25, 2015, at http://www.ecologyandsociety.org/vol17/iss2/art20/.

Parlee, B., Goddard, E., Smith, M., and Lutsel K'e Dene First Nation, 2014, Tracking change—Traditional knowledge of wildlife health in northern Canada: Human Dimensions of Wildlife, v. 19, no. 1, p. 47–61.

Parlee, B., Manseau, M., and Lutsel K'e Dene First Nation, 2005, Understanding and communicating about ecological change—Denés⊠łiné indicators of ecosystem health, in Berkes, F., Huebert, R., Fast, H., Manseau, M., and Diduck, A., eds., Breaking ice—Integrated ocean management in the Canadian North: Calgary, Alberta, Canada, University of Calgary Press, p. 165–182.

Parmesan, C., and Yohe, G., 2003, A globally coherent fingerprint of climate change impacts across natural systems: Nature, v. 421, p. 37–42, doi:10.1038/nature01286.

Pastor, J., and Mladenoff, D.J., 1992, The southern boreal-northern hardwood forest border, in Shugart, H.H., Leemans, R., and Bonan, G.B., eds., A systems analysis of the global boreal forest: Cambridge, United Kingdom, Cambridge University Press, p. 216–240.

Pastor, J., Mladenoff, D., Haila, Y., Bryant, J., and Payette, S., 1996, Biodiversity and ecosystem processes in boreal regions, in Mooney, H.A., Cushman, J.H., Medina, E., Sala, O.E., and Schulze, E.D., eds., Functional roles of biodiversity—A global perspective: London, United Kingdom, John Wiley, p. 33–69.

Patra, R.W., Chapman, J.C., Lim, E.P., and Gehrke, P.C., 2007, The effects of three organic chemicals on the upper thermal tolerances of four freshwater fishes: Environmental Toxicology and Chemistry, v. 26, p. 1,454–1,459.

Pautasso, M., Holdenrieder, O., and Stenlid, J., 2005, Susceptibility to fungal pathogens of forests differing in tree diversity, in Forest diversity and function: Springer, p. 263–289.

Pearson, R.G., and Dawson, T.P., 2003, Predicting the impact of climate change on the distribution of species—Are bioclimate envelope models useful?: Global Ecology and Biogeography, v. 12, p. 361–371.

Peng, C., Ma, Z., Lei, X., Zhu, Q., Chen, H., Wang, W., Liu, S., Li, W., Fang X., and Zhou, X., 2011, A drought-induced pervasive increase in tree mortality across Canada's boreal forests: Nature Climate Change, v. 1, p. 467–471.

Peres, C.A., 2000, Effects of subsistence hunting on vertebrate community structure in Amazonian forests: Conservation Biology, v. 14, p. 240–253.

Peterson, G.D., Cumming, G.S., and Carpenter, S.R., 2003, Scenario planning—A tool for conservation in an uncertain world: Conservation Biology, v. 17, no. 2, p. 358–366.

Pewe, T.L., Hopkins, D.M., and Giddings, J.L., 1965, The Quaternary geology and archaeology of Alaska, in Wright, H.E.J., and Frey, D.G., eds., The Quaternary of the United States—A review volume for the VII Congress of the International Association for Quaternary Research: Princeton, New Jersey, Princeton University Press, p. 355–374.

Phillips, D., 1990, The climates of Canada: Canadian Government Publishing Centre, Ministry of Supply and Services, Ottawa, Ontario, Canada, 181 p., accessed November 11, 2017 at, http://publications.gc.ca/collections/collection_2014/ec/En56-1-1990-eng.pdf.

Pielke, R.A., 2007, The honest broker—Making sense of science in policy and politics: Cambridge, United Kingdom, Cambridge University Press, 188 p.

Podur, J., Martell, D.L., and Knight, K., 2002, Statistical quality control analysis of forest fire activity in Canada: Canadian Journal of Forest Research, v. 32, no. 2, p. 195–205.

Polley, L., and Thompson, R.C.A., 2009, Parasite zoonoses and climate change—Molecular tools for tracking shifting boundaries: Trends in Parasitology, v. 25, p. 285–291.

Post, E., and Forchhammer, M.C., 2008, Climate change reduces reproductive success of an Arctic herbivore through trophic mismatch: Philosophical Transactions of the Royal Society B, v. 363, no. 1501, p. 2,367–2,373.

Potter, C., 2014, Regional analysis of MODIS satellite greenness trends for ecosystems of interior Alaska: GIScience and Remote Sensing, v. 51, no. 4, p. 390–402, doi:10.1080/15481603.2014.933606.

Prowse, T.D., Furgal, C., Chouinard, R., Melling, H., Milburn, D., and Smith, S.L., 2009, Implications of climate change for economic development in northern Canada—Energy, resource, and transportation sectors: Ambio, v. 38, no. 5, p. 272–281.

Pyne, S.J., 1982, Fire in America—A cultural history of wildland and rural fire: Princeton, New Jersey, Princeton University Press, 654 p.

Quinn, T.P., Wetzel, L., Bishop, S., Overberg, K., and Rogers, D.E., 2001, Influence of breeding habitat on bear predation and age at maturity and sexual dimorphism of sockeye salmon populations: Canadian Journal of Zoology, v. 79, p. 1,782–1,793.

Rainville, R.A. and Gajewski, K., 2013, Holocene environmental history of the Aishihik Region, Yukon, Canada: Canadian Journal of Earth Sciences, v. 50, p. 397–405.

Randerson, J.T., Liu, H., Flanner, M.G., Chambers, S.D., Jin, Y., Hess, P.G., Pfister, G., Mack,M.C., Treseder, K.K., Welp, L.R., Chapin, F.S., Harden, J.W., Goulden, M.L., Lyons, E., Neff, J.C., Schuur,

E.A.G., and Zender, C.S., 2006, The impact of boreal forest fire on climate warming: Science, v. 314, no. 5802, p. 1,130–1,132, doi:10.1126/science.1132075.

Rasouli, K., Pomeroy, J.W., Janowicz, J.R., Carey, S.K., and Williams, T.J., 2014, Hydrological sensitivity of a northern mountain basin to climate change: Hydrological Process, v. 28, no. 14, p. 4,191–4,208.

Ray, L., 2011, Using Q-methodology to identify local perspectives on wildfires in two Koyukon Athabascan communities in rural Alaska: Sustainability—Science, Practice, & Policy, v .7, no. 2, p. 18–29.

Redford, K.H., 1992, The empty forest: BioScience, v. 42, no. 6, p. 412–422.

Reichard, S.H., and White, P., 2001, Horticulture as a pathway of invasive plant introductions in the United States: Bioscience, v. 51, no. 2, p. 103–113.

Rennert. K.J., Roe, G., and Putkonen, J., 2009, Soil thermal and ecological impacts of rain on snow events in the circumpolar Arctic: Journal of Climate, v. 22, no. 9, p. 2,302–2,315.

Resilience Alliance, 2010, Assessing resilience in social-ecological systems—Workbook for practitioners, Version 2.0: Resilience Alliance website, http://www.resalliance.org/3871.php.

Resource Development Council, 2010, Doyon to explore Yukon Flats for oil and gas: Resource Development Council website, accessed September 4, 2014, at http://www.akrdc.org/newsletters/2010/march/yukonflats.html.

Revich, B., Tokarevich, N., and Parkinson, A.J., 2012, Climate change and zoonotic infections in the Russian Arctic: International Journal of Circumpolar Health, 71, no. 1.

Rice, B., 1987, Changes in the Harding Icefield, Kenai Peninsula, Alaska: Fairbanks, University of Alaska, School of Agriculture and Land Resources Management, M.S. thesis, 116 p.

Richardson, D.M., and Gaertner, M., 2013, Plant invasions as builders and shapers of novel ecosystems, in Hobbs, R.J., Higgs, E.S., and Hall, C.M., eds., Novel ecosystems—Intervening in the new ecological world order: Oxford, United Kingdom, Wiley-Blackwell, p. 102–113.

Riordan, B., Verbyla, D., and McGuire, A.D., 2006, Shrinking ponds in subarctic Alaska based on 1950–2002 remotely sensed images: Journal of Geophysical Research, v. 111, no. G04002, doi:10.1029/2005JG000150.

Ritchie, J.C., and MacDonald, G.M., 1986, The patterns of post-glacial spread of white spruce: Journal of Biogeography, v. 13, p. 527–540.

Robinson, D., Hunsinger, E., Howell, D., and Sandberg, E., 2014, Alaska population projections 2012–2042: Juneau, State of Alaska, Department of Labor and Workforce Development, 108 p, accessed November 7, 2015, at http://laborstats.alaska.gov/pop/projected/pub/popproj.pdf.

Robson, D.B., Knight, J.D., Farrell, R.E., and Germida, J.J., 2003, Ability of cold-tolerant plants to grow in hydrocarbon-contaminated soil: International Journal of Phytoremediation, v. 5, p. 105–123.

Rohrs-Richey, J.K., 2010, Biotic pest damage of green alder (Alnus fruticosa)—Susceptibility to a stem disease (Valsa melanodiscus) and functional changes following insect herbivory: Fairbanks, University of Alaska, Ph.D. thesis.

Romanovsky, V. E., and others, 2013, Permafrost: National Oceanic and Atmospheric Administration website, http://www.arctic.noaa.gov/reportcard/permafrost.html.

Roon, D.A., 2011, Ecological effects of invasive European bird cherry (Prunus padus) on salmonid food webs in Anchorage, Alaska streams: Fairbanks, University of Alaska, M.S. thesis.

Roon, D.A., Wipfli, M.S., and Wurtz, T.L., 2014, Effects of invasive European bird cherry (Prunus padus) on leaf litter processing by aquatic invertebrate shredder communities in urban Alaskan streams: Hydrobiologia, v. 736, no. 1, p. 17–30, doi:10.1007/s10750-014-1881-x.

Roots, C., and Hart, C., 2004, Bedrock geology, in Smith, C.A.S., Meikle, J.C., and Roots, C.F., eds., Ecoregions of the Yukon Territory—Biophysical properties of Yukon landscapes: Summerland, Canada, Agriculture and Agri-Food Canada. PARC Technical Bulletin No. 04-01, p. 11–14.

Rosa, G.M., Anza, I., Moreira, P.L., Conde, J., Martins, F., Fisher, M.C., and Bosch, J., 2013, Evidence of chytrid-mediated population declines in common midwife toad in Serra da Estrela, Portugal: Animal Conservation, v. 16, p. 306–315.

Rose, M., and Hermanutz, L., 2004, Are boreal ecosystems susceptible to alien plant invasion?—Evidence from protected areas: Oecologia, v. 139, p. 467–477.

Ross, W.A., 1998, Cumulative effects assessment—Learning from Canadian case studies: Impact Assessment and Project Appraisal, v. 16, no. 4, p. 267–276.

Rowland, E.L., Davison, J.E., and Graumlich, L.J., 2011, Approaches to evaluating climate change impacts on species—A guide to initiating the adaptation planning process: Environmental Management, v. 47, no. 3, p. 322–337, doi:10.1007/s00267-010-9608-x.

Rowland, E.R., Cross, M.S., and Hartmann H., 2014, Considering multiple futures—Scenario planning to address uncertainty in natural resource conservation: U.S. Fish and Wildlife Service, 162 p.

Rozell, N., 2009, Invasion of the redlegged frogs: University of Alaska, Fairbanks, website, accessed March 29, 2016, at http://www.uaf.edu/files/news/a_news/20090306140320.html.

Ruess, R.W., McFarland, J.M., Trummer, L.M., and Rohrs-Richey, J.K., 2009, Disease-mediated declines in N-fixation inputs by Alnus tenuifolia to early-successional floodplains in interior and south-central Alaska: Ecosystems, v. 12, no. 3, p. 489–502.

Ruggerone, G.T., Hanson, R., and Rogers, D.E., 2000, Selective predation by brown bears (Ursus arctos) foraging on spawning sockeye salmon (Oncorhynchus nerka): Canadian Journal of Zoology, v. 78, p. 974–981.

Ruhl, J.B., 2008, Climate change and the Endangered Species Act—Building bridges to the no- analog future: Boston University Law Review, v. 88, p. 1–62.

Rupp, T.S., Chapin, F.S., III, and Starfield, A.M., 2001, Modeling the influence of topographic barriers on treeline advance at the forest-tundra ecotone in northwestern Alaska: Climate Change, v. 48, p. 399–416.

Rupp, T.S., Olson, M., Adams, L.G., Dale, B.W., Joly, K., Henkelman, J., Collins, W.B., and Starfield, A.M., 2006, Simulating the influences of various fire regimes on caribou winter habitat: Ecological Applications, v. 16, no. 5, p. 1,730–1,743.

Rupp, T.S., and Springsteen, A., 2009, Projected climate change scenarios for the Bureau of Land Management Eastern Interior Management Area, Alaska, 2001–2099: Bureau of Land Management, 10 p., accessed November 14, 2015, at https://eplanning.blm.gov/epl-front- office/projects/lup/1100/43915/47266/Eastern_Interior_Management_Area_%28EIMA%29_Climate_Change_508.pdf.

Rutherford, M.B., Gibeau, M.L., Clark, S.G., and Chamberlain, E.C., 2009, Interdisciplinary problem solving workshops for grizzly bear conservation in Banff National Park, Canada: Policy Sciences, v. 42, no. 2, p. 163–187.

Sabatier, P.A., and Jenkins-Smith, H.C., 1999, The advocacy coalition framework—An assessment, in Sabatier, P.A., ed., Theories of the policy process: Boulder, Colorado, Westview Press, p. 117–168.

Salafsky, N., Salzer, D., Stattersfield, A.J., Hilton-Taylor, C., Neugarten, R., Butchart, S.H.M., Collen, B., Cox, N., Master, L.L., O'Connor, S., and Wilkie, D., 2008, A standard lexicon for biodiversity conservation—Unified classifications of threats and actions: Conservation Biology, v. 22, p. 897–911, doi: 10.1111/j.1523-1739.2008.00937.x.

Salmón, E., 2000, Kincentric ecology—Indigenous perceptions of the human-nature relationship: Ecological Applications, v. 10, no. 5, p. 1,327–1,332.

Salmon, R.A., Priestley, R.K., and Goven, J., 2015, The reflexive scientist: an approach to transforming public engagement: Journal of Environmental Studies and Sciences, v. 7, no. 1, p. 1–16.

Saltmarsh, D.M., Bowser, M.L., Morton, J.M., Lang, S., Shain, D., and Dial, R., 2016, Distribution and abundance of exotic earthworms within a boreal forest system in southcentral Alaska: NeoBiota, v. 28, p. 67–86, doi: 10.3897/neobiota.28.5503.

Sanderson, L.A., McLaughlin, J.A., and Antunes, P.M., 2012, The last great forest—A review of the status of invasive species in the North American boreal forest: Forestry, v. 85, p. 329–340.

Sarewitz, D., 2004, How science makes environmental controversies worse: Environmental Science and Policy, v. 7, p. 385–403.

Scenarios Network for Alaska and Arctic Planning, 2014, Historical monthly temperature and precipitation 2km CRU TS 3.1—Projected monthly temperature and precipitation 2km CMIP5/AR5—Projected derived temperature products 2km CMIP 5/AR5: University of Alaska, Fairbanks SNAP data portal, accessed May 2014, at http://www.snap.uaf.edu/data.php.

Scenarios Network for Alaska and Arctic Planning and Ecological Wildlife Habitat Data Analysis for the Land and Seascape, 2012, Predicting future potential climate-biomes for the Yukon, Northwest Territories, and Alaska: Fairbanks, University of Alaska, Scenarios Network for Arctic Planning (SNAP), and Ecological Wildlife Habitat Data Analysis for the Land and Seascape, 105 p., http://www.snap.uaf.edu/attachments/Cliomes-FINAL.pdf.

Schaefer, J., 2003, Long-term range recession and the persistence of caribou in the taiga: Conservation Biology, v. 17, no. 5, p. 1,435–1,439.

Scheffer, M., Carpenter, S.R., Foley, J.A., Folke, C., and Walke, B., 2001, Catastrophic shifts in ecosystems: Nature, v. 413, p. 591–596.

Scheuerell, M.D., Moore, J.W., Schindler, D.E., and Harvey, C.J., 2007, Varying effects of anadromous sockeye salmon on the trophic ecology of two species of resident salmonids in southwest Alaska: Freshwater Biology, v. 52, p. 1,944–1,956.

Schiedek, D., Sundelin, B., Readman, J.W., and Macdonald, R.W., 2007, Interactions between climate change and contaminants: Marine Pollution Bulletin, v. 54, p. 1,845–1,856.

Schirato, T., and Yell, S., 1997, Communication and Cultural Literacy: An Introduction, Sydney: Allen & Unwin.

Schmiegelow, F., 2007, On benchmarking natural systems: Canadian Silviculture, v. 5, no. 1, p. 3–7.

Schmiegelow, F.K.A., Cumming, S.G., Lisgo, K.A., Leroux, S.J., and Krawchuk, M.A., 2014, Catalyzing large landscape conservation in Canada's boreal systems—The BEACONs project experience, in Levitt, J.N., ed., Conservation catalysts: Cambridge, Massachusetts, Harvard University Press, Lincoln Institute of Land Policy, p. 97–122.

Schneider, R.R., and Bayne, E.M., 2015, Reserve design under climate change—From land facets back to ecosystem representation: PLoS ONE, v. 10, no. 5, p. e0126918, doi:10.1371/journal.pone.0126918.

Schneider, R.R., Stelfox, J.B., Boutin, S., and Wasel, S., 2003, Managing the cumulative impacts of land uses in the Western Canadian sedimentary basin—A modelling approach: Ecology and Society, v. 7, no. 1, article 8, accessed November 25, 2015, at http://www.ecologyandsociety.org/vol7/iss1/art8/.

Schweger, C.E., Froese, D., White, J.M., and Westgate, J.A., 2011, Pre-glacial and interglacial pollen records over the last 3 Ma from northwest Canada—Why do Holocene forests differ from those of previous interglaciations?: Quaternary Science Review, v. 30, p. 2,122–2,133.

Schwörer, T., Federer, R.N., and Ferren, H.J., II, 2014, Invasive species management programs in Alaska—A survey of statewide expenditures, 2007–11: Arctic, v. 67, no. 1, p. 20–27.

Seastedt, T.R., Hobbs, R.J., and Suding, K.N., 2008, Management of novel ecosystems—Are novel approaches required?: Frontiers in Ecology and Environment, v. 6, no. 10, p. 547–553, doi:10.1890/070046.

Shaffer, M.L., 2014, Policy challenges for wildlife management in a changing climate, in Sample, V.A., and Bixler, R.P., eds., Forest conservation and management in the Anthropocene—Conference proceedings: U.S. Forest Service, Rocky Mountain Research Station, Fort Collins, Colorado, RMRS-P-71, p. 427–441.

Sherriff, R.L., Berg, E.E., and Miller, A.E., 2011, Climate variability and spruce beetle (Dendroctonus rufipennis) outbreaks in south-central and southwest Alaska: Ecology, v. 92, no. 7, p. 1,459–1,470.

Sherry, E., and Myers, H., 2002, Traditional environmental knowledge in practice: Society and Natural Resources, v. 15, no. 4, p. 345–358.

Shiklomanov, N.I., Streletskiy, D.A., and Nelson, F.E., 2012, Northern Hemisphere component of the global Circumpolar Active Layer Monitoring (CALM) program, in Proceedings of the 10th International Conference on Permafrost, June 25–28, 2012, Salekhard, Russia: The Northern Publisher, v. 1, p. 377–382.

Shulski, M., and Wendler, G., 2007, The climate of Alaska: Fairbanks, University of Alaska Press, 216 p.

Shur, Y., and Zheskhova T., 2003, Cryogenic structure of a glacio-lacustrine deposit, in Proceedings of the Eighth International Conference on Permafrost: Lisse, The Netherlands, A.A. Balkema Publishers, p. 1,051–1,056 p.

Sillmann, J., Kharin, V., Zwiers, F., Zhang, X., and Bronaugh, D., 2013, Climate extremes indices in the CMIP5 multimodel ensemble: Part 2—Future climate projections: Journal of Geophysical Research, Atmospheres, v. 118, p. 2,473–2,493.

Simon, N.P.P., Schwab, F.E., Baggs, E.M., and Cowan, G., 1998, Distribution of small mammals among successional and mature forest types in western Labrador: Canadian Field-Naturalist, v. 112, no. 3, p. 441–445.

Singer, A.C., Thompson, I.P., and Bailey, M.J., 2004, The tritrophic trinity—A source of pollutant-degrading enzymes and its implications for phytoremediation: Current Opinion in Microbiology, v. 7, no. 239–244.

Singletary, L., Evans, B., Sicafuse, L., and Maletksy, L., 2015, Evaluation resource guide for Joint Fire Science Program Consortia: University of Nevada, Cooperative Extension and Joint Fire Science Program, 57 p., accessed December 2, 2017, at https://www.firescience.gov/documents/JFSP_Evaluation_Resource_Guide_2015.pdf.

Slater, H., Gouin, T., and Leigh, M.B., 2011, Assessing the potential for rhizoremediation of PCB contaminated soils in northern regions using native tree species: Chemosphere, v. 84, p. 199–206.

Slocombe, D.S., Hartley, L., and Noonan, M., 2016, Environmental assessment and land claims, devolution and co-management—Evolving opportunities and challenges in Yukon, in Hanna, K.S., ed., Environmental impact assessment—Practice and participation (3rd ed.): Toronto, Ontario, Canada, Oxford University Press. p. 220–240.

Smith, B., 2004, Applying the knowledge, experience, and values of Yukon Indian people, Inuvialuit, and others in conservation decisions—Summaries of 55 Yukon projects, 1985– 2003: Whitehorse, Yukon Environment, Report MR-04-01, 206 p.

Smith, S.L., and others, 2010, Thermal state of permafrost on North America—A contribution to the International Polar Year: Permafrost and Periglacial Processes, v. 21, p. 117–135.

Smith, S.L., Throop, J., and Lewkowicz, A.G., 2012, Recent changes in climate and permafrost temperatures at forested and polar desert sites in northern Canada: Canadian Journal of Earth Sciences, v. 49, p. 914–924.

Smith, S.L., Wolfe, S.A., Riseborough, D.W., and Nixon, F.M., 2009, Active-layer characteristics and summer climatic indices, Mackenzie Valley, Northwest Territories, Canada: Permafrost and Periglacial Processes, v. 20, p. 201–220.

Snaddon, J., Petrokofsky, G., Jepson, P., Willis, K.J., 2013, Biodiversity technologies—Tools as change agents: Biology Letters, v. 9, 3 p., http://rsbl.royalsocietypublishing.org/content/roybiolett/9/1/20121029.full.pdf.

Snover, A.K., Mantua, N.J., Littell, J.S., Alexander, M.A., Mcclure, M.W., and Nye, J., 2013, Choosing and using climate-change scenarios for ecological-impact assessments and conservation decisions: Conservation Biology, v. 27, no. 6, p. 1,147–1,157, doi:10.1111/cobi.12163.

Soja, A.J., Tchebakova, N.M., French, N.H.F., Flannigan, M.D., Shugart, H.H., Stocks, B.J., Sukhinin, A.I., Parfenova, E.I., Chapin, F.S., III, and Stackhouse, P.W., Jr., 2007, Climate- induced boreal forest change—Predictions versus current observations: Global and Planetary Change, v. 56, no. 3, p. 274–296.

Southcentral Alaska Northern Pike Control Committee, 2014, Management plan for invasive northern pike in Alaska: Southcentral Alaska Northern Pike Control Committee, 58 p, accessed September 12, 2014, at http://www.adfg.alaska.gov/static/species/nonnative/invasive/pike/pdfs/invasive_pike_manage ment_plan.pdf.

Spellman, B.T., and Wurtz, T.L., 2011, Invasive sweetclover (Melilotus albus) impacts seedling recruitment along floodplains in Alaska: Biological Invasions, v. 13, p. 1,779–1,790.

Spellman, K.V., Mulder, C.P.H., and Carlson, M.L., 2013, Effects of white sweetclover invasion on the pollination and berry production of Vaccinium vitis-idaea in Alaska [poster]: Proceedings of the 9th Ecological Society of America Annual Meeting, Minneapolis, Minnesota, August 4–9, 2013, accessed on November 11, 2017, at https://www.researchgate.net/publication/267289553_Effects_of_white_sweetclover_invasion_on_the_pollination_and_berry_production_of_Vaccinium_sp_in_Alaska.

Spellman, K.V., Mulder, C.P.H., and Hollingsworth, T.N., 2014, Susceptibility of burned black spruce (Picea mariana) forests to non-native plant invasions in interior Alaska: Biological Invasions, v. 16, no. 9, p. 1,879–1,895, doi:10.1007/s10530-013-0633-6.

Spence, M.D., 2000, Dispossessing the wilderness—Indian removal and the making of the national parks: New York, Oxford University Press, 200 p.

Squires, A.J., Westbrook, C.J., and Dubé, M.G., 2010, An approach for assessing cumulative effects in a model river, the Athabasca River Basin: Integrated Environmental Assessment and Management, v. 6, no. 1, p. 119–134.

Standish, R.J., Hobbs, R.J., Mayfield, M.M., Bestelmeyer, B.T., Suding, K.N., Battaglia, L.L., Eviner, V., Hawkes, C.V., Temperton, V.M., Cramer, V.A., Harris, J.A., Funk, J.L., and Thomas, P.A., 2014, Resilience in ecology—Abstraction, distraction or where the action is?: Biological Conservation, v. 177, p. 43–51.

Stafford, J., Wendler, G., and Curtis, J., 2000, Temperature and precipitation of Alaska—50 year trend analysis: Theoretical and Applied Climatology, v. 67, p. 33–44.

Stanley, R.G., Ahlbrandt, T.S., Charpentier, R.R., Cook, T.A., Crews, J.M., Klett, T.R., Lillis, P.G., Morin, R.L., Phillips, J.D., Pollastro, R.M., Rowan, E.L., Saltus, R.W., Schenk, C.J., Simpson, M.K., Till, A.B., and Troutman, S.M., 2004, Oil and gas assessment of Yukon Flats, east-central Alaska, 2004, U.S. Geological Survey Fact Sheet 2004–3121, 2 p.

Staples, Kiri, Chavez-Ortiz, Manuel, Barrett, M.J., and Clark, D.A., 2013, Fixing land-use planning in the Yukon before it really breaks—A case study of the Peel Watershed: The Northern Review, v. 37, no. 3, p. 143–165, http://journals.sfu.ca/nr/index.php/nr/article/view/278.

Starzomski, B.M., 2013, Novel ecosystems and climate change, in Hobbs, R.J., Higgs, E.S., and Hall, C.M., eds., Novel ecosystems—Intervening in the new ecological world order: Oxford, United Kingdom, Wiley-Blackwell, p. 88–101.

Steffen, W., Crutzen, P.J., and McNeil, J.R., 2007, The Anthropocene—Are humans now overwhelming the great forces of nature?: Ambio, v. 36, no. 8, p. 614–621.

Stein, B.A., Glick, P., Edelson, N., and Staudt, A., eds., 2014, Climate-smart conservation— Putting adaptation principles into practice: Washington, D.C., National Wildlife Federation, 262 p.

Stein, B.A., Scott, C., and Benton, N., 2008, Federal lands and endangered species—The role of military and other Federal lands in sustaining biodiversity: BioScience, v. 58, no. 4, p. 339–347.

Steneck, R.S., Hughes, T.P., Cinner, J.E., Adger, W.N., Arnold, S.N., Berkes, F., Boudreau, S.A., Brown, K., Folke, C., Gunderson, L., Olsson, P., Scheffer, M., Stephenson, E., Walker, B., Wilson, J., and Worm, B., 2011, Creation of a gilded trap by the high economic value of the Maine lobster fishery: Conservation Biology, v. 25, no. 5, p. 904–912, doi:10.1111/j.1523-1739.2011.01717.x.

Stephenson, P.H., 1995, A persistent spirit—Towards understanding aboriginal health in British Columbia: Victoria, Department of Geography, University of Victoria, 390 p.

Stephenson, B., and Rogers, R.R., 2007, Wood bison restoration in Alaska—A review of environmental and regulatory issues and proposed decisions for project implementation: Alaska Department of Fish and Game, 91 p., http://www.wc.adfg.state.ak.us/management/game/wood_bison/er_no_appendices.pdf.

Stevenson, M.G., and Webb, J., 2003, Just another stakeholder?—First Nations and sustainable forest management in Canada, in Burton, P.J., Messier, C., Smith, D.W., and Adamowicz, W.L., eds., Towards sustainable management of the boreal forest: Ottawa, Ontario, Canada, NRC Research Press, p. 65–112.

St. Martin, K., 2001, Making space for community resource management in fisheries: Annals of the Association of American Geographers, v. 91, no. 1, p. 122–142.

St. Martin, K., 2006, The impact of "community" on fisheries management in the US Northeast: Geoforum, v. 372, p. 169–184.

Stocker, T.F., Qin, D., Plattner, G.-K., Alexander, L.V., Allen, S.K., Bindoff, N.L., Bréon, F.-M., Church, J.A., Cubasch, U., Emori, S., Forster, P., Friedlingstein, P., Gillett, N., Gregory, J.M., Hartmann, D.L., Jansen, E., Kirtman, B., Knutti, R., Krishna Kumar, K., Lemke, P., Marotzke, J., Masson-Delmotte, V., Meehl, G.A., Mokhov, I.I., Piao, S., Ramaswamy, V., Randall, D., Rhein, M., Rojas, M., Sabine, C., Shindell, D., Talley, L.D., Vaughan, D.G., and Xie, S.-P., 2013, Technical summary, in Stocker, T.F., Qin, D., Plattner, G.-K., Tignor, M., Allen, S.K., Boschung, J., Nauels, A., Xia, Y., Bex, V., and Midgley, P.M., eds., Climate change 2013— The physical science basis—Contribution of working group I to the Fifth Assessment Report of the Intergovernmental Panel on Climate Change: Cambridge, United Kingdom, Cambridge University Press.

Stockholm Convention, 2015, Status of ratifications: Chatelaine, Switzerland, Secretariat of the Stockholm Convention Clearing House website, accessed November 13, 2015, at http://chm.pops.int/Countries/StatusofRatifications/Overview/tabid/3484/Default.aspx.

Stokstad, E., 2008, Canada's experimental lakes: Science, v. 322, no. 5906, p. 1,316–1,319. doi:10.1126/science.322.5906.1316.

Strauss, S.E., Tetroe, J., and Graham, I., 2009, Defining knowledge translation: Canadian Medical Association Journal, v. 181, nos. 3–4, p. 165–168, doi:10.1503/cmaj.081229.

Streletskiy, D.A., Shiklomanov, N.I., Little, J.D., Nelson, F.E., Brown, J., Nyland, K.E., and Klene, A.E., 2017, Thaw subsidence in undisturbed tundra landscapes, Barrow, Alaska, 1962– 2015: Permafrost and Periglacial Processes, v. 28, p. 566–572.

Streicker, J., 2016, Yukon climate change indicators and key findings 2015: Northern Climate ExChange, Yukon Research Centre, Yukon College, 84 p., accessed November 13, 2017, at https://www.yukoncollege.yk.ca/sites/default/files/inline-files/Indicator_Report_Final_web.pdf.

Swanston, C.J., and Janowiak, M., eds., 2012, Forest adaptation resources—Climate change tools and approaches for land managers: U.S. Forest Service General Technical Report NRS- 87, 121 p., accessed November 14, 2015, at http://www.nrs.fs.fed.us/pubs/40543.

Telford, A., Cavers, S., Ennos, R.A., and Cottrell, J.E., 2014, Can we protect forests by harnessing variation in resistance to pests and pathogens?: Forestry, v. 88, no. 1, p. 3–12, doi:10.1093/forestry/cpu012.

Thompson, J.L., Fallon-Lambert, K., Foster, D., Blumstein, M., Broadbent, E., and. Zambrano, A.A., 2014, Changes to the land—Four scenarios for the future of the Massachusetts landscape: Cambridge, Massachusetts, Harvard University, Harvard Forest, 32 p., accessed November 14, 2015, at http://harvardforest.fas.harvard.edu/changes-to-the-land.

Thompson, J.R., Wiek, A., Swanson, F.J., Carpenter, S.R., Fresco, N., Hollingsworth, T., Spies, T.A., and Foster, D.R., 2012, Scenario studies as a synthetic and integrative research activity for long-term ecological research: BioScience, v. 62, no. 4, p. 367–376, doi: 10.1525/bio.2012.62.4.8.

Thormann, M.N., Bayley, S.E., and Szumigalski, A.R., 1998, Effects of hydrologic changes on aboveground production and surface water chemistry in two boreal peatlands in Alberta— Implications for global warming: Hydrobiologia, v. 362, no. 1, p. 171–183.

Timoney, K., and Lee, P., 2001, Environmental management in resource-rich Alberta, Canada— First world jurisdiction, third world analogue: Journal of Environmental Management, v. 63, no. 4, p. 387–405.

Timoney, K., Peterson, G., Fargey, P., Peterson, M., McCanny, S., and Wein, R., 1997, Spring ice-jam flooding of the Peace-Athabasca Delta—Evidence of a climatic oscillation: Climatic Change, v. 35, no. 4, p. 463–483.

Trainor, S.F., and Leigh, M.B., 2013, In a time of change—The art of fire: Alaska Fire Science Consortium, Bonanza Creek Long Term Ecological Research Program, Final Report—Joint Fire Science Program Project ID: 11-S-2–2, 10 p., accessed October 29, 2017, at https://www.firescience.gov/projects/11-S-2–2/project/11-S-2-2_final_report.pdf.

Trainor, S.F., Kettle, N., Gamble, J.B., and Taylor, K., 2016, Not another webinar!—Analysis of regional webinars as an effective knowledge to action network for climate science application Alaska, in Parris, A., and Garfin, G., eds., Climate in context—Science and society partnering for adaptation: Hoboken, New Jersey, Wiley, p. 117–142.

Tribbia, J., and Moser, S.C., 2008, More than information—What coastal managers need to plan for climate change: Environmental Science and Policy, v. 11, no. 4, p. 315–328.

Tremont, J.D., 1987, Surface-transportation networks of the Alaskan North Slope: U.S. Department of the Interior, Minerals Management Service, Anchorage, Alaska, Alaska Outer Continental Shelf Report No. OCS/MMS-87/0010, 96 p, accessed December 14, 2014, at http://www.arlis.org/docs/vol1/18833832.pdf.

Tuhiwai Smith, Linda, 2012, Decolonizing methodologies—Research and Indigenous peoples: Zed Books, Ltd., 240 p.

Turetsky, M.R., Harden, J.W., Friedli, H.R., Flannigan, Mike, Payne, Nicholas, Crock, James, and Radke, Lawrence, 2006, Wildfires threaten mercury stocks in northern soils: Geophysical Research Letters, v. 33, no. 16, L16043.1, 6 p., doi:10.1029/2005GL025595.

Turner, N.J., and Davis, A., 1993, "When everything was scarce"—The role of plants as famine foods in northwestern North America: Journal of Ethnobiology, v. 13, no. 2, p. 171–201.

Turner, N.J., Ignace, M.B., and Ignace, R., 2000, Traditional ecological knowledge and wisdom of aboriginal peoples in British Columbia: Ecological Applications, v. 10, no. 5, p. 1,275–1,287.

Tyrrell, M., and Clark, D., 2014, What happened to climate change?—CITES and the reconfiguration of polar bear conservation discourse: Global Environmental Change, v. 24, p. 363–372.

Ullsten, O., Speth, J.G., and Chapin, F.S., III, 2004, Options for enhancing the resilience of northern countries to rapid social and environmental change—A message to policy makers: AMBIO—A Journal of the Human Environment, v. 33, no. 6, p. 343–343.

Urquhart, D., 2012, The null hypothesis—Co-management doesn't work: Rangifer, v. 20, p. 103–112.

U.S. Department of Transportation, 2001, Alaska transportation profile: U.S. Department of Transportation, Bureau of Transportation Statistics, 122 p., accessed November 7, 2015, at http://www.rita.dot.gov/bts/sites/rita.dot.gov.bts/files/publications/state_transportation_statistics/alaska/pdf/entire.pdf.

U.S. Forest Service, 2011, Forest health conditions in Alaska—2010: Anchorage, Alaska, U.S. Forest Service Alaska Region, FHP Protection Report R10-PR-23, 65 p.

U.S. Forest Service, 2012, Forest health conditions in Alaska—2011: Anchorage, Alaska, U.S. Forest Service Alaska Region, FHP Protection Report R10-PR-25, 68 p.

U.S. Forest Service, 2014, Forest health conditions in Alaska—2013: Anchorage, Alaska, U.S. Forest Service Alaska Region, FHP Protection Report R10-PR-035, 89 p. accessed October, 26, 2017, at https://www.fs.usda.gov/Internet/FSE_DOCUMENTS/stelprd3797075.pdf.

U.S. Forest Service, 2015, Forest health conditions in Alaska—2014: Anchorage, Alaska, U.S. Forest Service Alaska Region, FHP Protection Report R10-PR-036, 88 p.

Van Hemert, C., Pearce, J.M., and Handel, C.M., 2014, Wildlife health in a rapidly changing North—Focus on avian disease: Frontiers in Ecology and the Environment, v. 12, p. 548–556.

van Oort, H., McLellan, B.N., and Serrouya, R., 2010, Fragmentation, dispersal and metapopulation function in remnant populations of endangered mountain caribou: Animal Conservation, v. 14, no. 3, p. 215–224.

Veldkamp, A., and Lambin, E.F., 2001, Predicting land-use change: Agriculture, Ecosystems and Environment, v. 85, nos. 1–3, p. 1–6.

Viereck, L.A., 1982, Effects of fire and firelines on active layer thickness and soil temperatures in interior Alaska: Proceedings of the Fourth Canadian Permafrost Conference, p. 123–135.

Visser, M.E., 2008, Keeping up with a warming world; assessing the rate of adaptation to climate change: Proceedings of the Royal Society B, v. 275, no. 1635, p. 649–659, doi:10.1098/rspb.2007.0997.

Vitousek, P.M., Mooney, H.A., Lubchenco, J., and Melillo, J.M., 1997, Human domination of Earth's ecosystems: Science, v. 277, p. 494–499.

Walker, B., Holling, C.S., Carpenter, S.R., and Kinzig, K., 2004, Resilience, adaptability, and transformability in social-ecological systems: Ecology and Society, v. 9, no. 2, article 5.

Walker, B., and Salt, D., 2012, Resilience practice—Building capacity to absorb disturbance and maintain function: Washington, D.C., Island Press, 248 p.

Wall, T.U, McNie, E., and Garfin, G.M., 2017, Use-inspired science: Making science usable by and useful to decision makers: Frontiers in Ecology and the Environment, v. 15, no. 10, p. 551–559.

Wallenius, T.H., 1999, Yield variations of some common wild berries in Finland in 1956–1996: Annales Botanici Fennici, v. 36, no. 4, p. 299–314.

Walsh, J.E., Chapman, W.L., Romanovsky, V., Christensen, J.H., and Stendel, M., 2008, Global climate model performance over Alaska and Greenland: Journal of Climate, v. 21, no. 23, p. 6,156–6,174.

Walters, C.J., and Holling, C.S., 1990: Large-scale management experiments and learning by doing: Ecology, v. 71, no. 6, p. 2,060–2,068.

Walvoord, M.A., and Striegl, R.G., 2007, Increased groundwater to stream discharge from permafrost thawing in the Yukon River Basin—Potential impacts on lateral export of carbon and nitrogen: Geophysical Research Letters, v. 34, L12402, doi:10.1029/2007GL030216.

Wang, T., Hamann, A., Spittlehouse, D.L., and Murdock, T.Q., 2012, ClimateWNA—High- resolution spatial climate data for Western North America: Journal of Applied Meteorology and Climatology, v. 51, no. 1, p. 16–29.

Ware, C., Bergstrom, D.M., Muller, E., and Alsos, I.G., 2012, Humans introduce viable seeds to the Arctic on footwear: Biological Invasions, v. 12, p. 567–577.

Warren, Becky, and Maynard, Jill, 2009, Interior regional energy plan—Phase I: Fairbanks, University of Alaska, Alaska Center for Energy and Power, 58 p., accessed November 11, 2017, at http://www.chena.org/wp-content/uploads/climate/interior-ak/InteriorRegionalEnergyPlan-Final_Draft-1209.pdf.

Watson, Annette, 2013, Misunderstanding the "Nature" of co-management—A geography of regulatory science and Indigenous knowledges (IK): Environmental Management, v. 52, no. 5, p. 1,085–1,102.

Watson, Annette, and Huntington, O.H., 2008, They're here—I can feel them—The epistemic spaces of Indigenous and Western knowledges: Social and Cultural Geography, v. 9, no., 3, p. 257–281.

Watson, Annette, and Huntington, Orville, 2014, Transgressions of the man on the moon—Climate change, Indigenous expertise, and the posthumanist ethics of place and space: GeoJournal, v. 79, no. 6, p. 721–736, accessed November 10, 2017, at http://link.springer.com/article/10.1007/s10708–014-9547-9.

Watson, Annette, Huntington, O.H., Machuca, N., Hoke, J., and Ned, S., 2014, Traditional ecological knowledge of moose, related wildlife species, and climate change in Allakaket/Alatna, Alaska: Report to the National Park Service, contract #P12PS22483.

Weber, M.G., and Flannigan, M.D., 1997, Canadian boreal forest ecosystem structure and function in a changing climate—Impact on fire regimes: Environmental Reviews, v. 5, nos. 3– 4, p. 145–166.

Weeks, D., Malone, P. and Welling, L., 2011, Climate change scenario planning—A tool for managing parks into uncertain futures: Park Science, v. 28, no. 1 (spring), accessed November 14, 2015, accessed November 10, 2017, at https://www.nature.nps.gov/ParkScience/index.cfm?ArticleID=475.

Weiher, E., and Keddy, P., eds., 1999, Ecological assembly rules—Perspectives, advances, retreats: Cambridge, United Kingdom, Cambridge University Press, 418 p.

Wendler, G., and Shulski, M., 2009, A century of climate change for Fairbanks, Alaska: Arctic, v. 62, no. 3, p. 295–300.

Wenzel, G., 2004, From TEK to IQ—Inuit Qajimajatuqangit and Inuit cultural ecology: Arctic Anthropology, v. 41, no. 2, p. 238–250.

Werner, R.A., Holsten, E.H., Matsuoka, S.M., and Burnside, R.E., 2006, Spruce beetles and forest ecosystems in south-central Alaska—A review of 30 years of research: Forest Ecology and Management, v. 227, no. 3, p. 195–206.

Werner, R.A., Raffa, K.F., and Illman, B.L., 2006, Dynamics of phytophagous insects and their pathogens in Alaskan boreal forests, in Alaska's changing boreal forests: New York, Oxford University Press, p. 133–146.

Wessels, K.J., Freitag, S., and van Jaarseld, A.S., 1999, The use of land facets as biodiversity surrogates during reserve selection at a local scale: Biological Conservation, v. 89, no. 1, p. 21–28.

Western Region Climate Center, 2018a, NCDC 1981–2010 Normals for Fairbanks WSO Airport, Alaska: Western Region Climate Center, accessed April 10, 2018 at https://wrcc.dri.edu/cgi-bin/cliMAIN.pl?akfair.

Western Region Climate Center, 2018b, Climate of Alaska: Western Regional Climate Center, accessed April 10, 2018, at https://wrcc.dri.edu/narratives/ALASKA.htm.

Wheeler, N.C., and Critchfield, W.B., 1985, The distribution and botanical characteristics of lodgepole pine—Biogeographical and management implications, in Baumgartner, D.M., Krebill, R.G., Arnott, J.T., and Weetman, G.F., eds., Lodgepole pine—The species and its management: Symposium proceedings, May 8–10, 1984, Spokane, Washington, and repeated May 14–16, 1984, Vancouver, British Columbia, Canada, p. 1–13.

Wholey, J.S., Hatry, H.P., and Newcomer, K.E., eds., 2004, Analyzing evaluation data, part three of Handbook of practical program evaluation (2d ed.): San Francisco, Jossey-Bass, p. 413–542.

Wilkinson, C.E., Hocking, M.D., Reimchen, T.E., 2005, Uptake of salmon-derived nitrogen by mosses and liverworts in coastal British Columbia: Oikos, v. 108, p. 85–98.

Wilkinson, K.M., Clark, S.G. and Burch, W.R., 2007, Other voices, other ways, better practices—Bridging local and professional environmental knowledge: New Haven, Connecticut, Yale University, School of Forestry & Environmental Studies, Report No. 14, 69 p, accessed November 10, 2017, at https://environment.yale.edu/publication-series/documents/downloads/a-g/FES-Report-14.pdf.

Williams, J.W., and Jackson, S.T., 2009, Novel climates, no-analog communities, and ecological surprises: Frontiers in Ecology and the Environment, v. 5, p. 475–482.

Wilmking, M., and Myers-Smith, I., 2008, Changing climate sensitivity of black spruce (Picea mariana Mill.) in a peatland-forest landscape in Interior Alaska: Dendrochronologia, v. 25, p. 167–175.

Wilson, R.R., Liebezeit, J.R., and Loya, W.M., 2013, Accounting for uncertainty in oil and gas development impacts to wildlife in Alaska: Conservation Letters, v. 6, no. 5, p. 350–358.

Winterstein, Mark, Hollingsworth, T.N., and Parker, Carolyn, 2016, A range extension of Carex Sartwellii in interior Alaska: Canadian Field Naturalist, v. 130, no. 3, DOI: http://dx.doi.org/10.22621/cfn.v130i3.1878.

Wipfli, M.S., 1997, Terrestrial invertebrates as salmonid prey and nitrogen sources in streams: Contrasting old-growth and young-growth riparian forests in southeastern Alaska, USA: Canadian Journal of Fisheries and Aquatic Sciences, v. 54, p. 1,259–1,269.

Wipfli, M.S., and Baxter, C.V., 2010, Linking ecosystems, food webs, and fish production—Subsidies in salmonid watersheds: Fisheries, v. 35, p. 373–387.

Wipfli, M.S., and Gregovich, D.P., 2002, Export of invertebrates and detritus from fishless headwater streams in southeastern Alaska—Implications for downstream salmonid production: Freshwater Biology, v. 47, p. 957–969.

Wipfli, M.S., Hudson, J., and Caouette, J., 1998, Influence of salmon carcasses on stream productivity—Response of biofilm and benthic macroinvertebrates in southeastern Alaska, USA: Canadian Journal of Fisheries and Aquatic Science, v. 55, p. 1,503–1,511.

Wipfli, M.S., Hudson, J.P., Caouette, J.P., and Chaloner, D.T., 2003, Marine subsidies in freshwater ecosystems—Salmon carcasses increase the growth rates of stream-resident salmonids: Transactions of the American Fisheries Society, v. 132, p. 371–381.

Wipfli, M.S., Hudson, J.P., Chaloner, D.T., and Caouette, J.P., 1999, Influence of salmon spawner densities on stream productivity in southeast Alaska: Canadian Journal of Fisheries and Aquatic Sciences, v. 56, p. 1,600–1,611.

Wiseman, J., Biggs, C., Rickards, L., and Edwards, T., 2011, Scenarios for climate adaptation— Guidebook for practitioners: Carlton, Australia, University of Melbourne, Victoria Centre for Climate Adaptation Research (VICCAR), 76 p., accessed November 14, 2015, at http://www.vcccar.org.au/publication/final-report/scenarios-for-climate-adaptation-final-report.

Wolfe, B.B., Humphries, M.M., Pisaric, M.F.J., Balasubramaniam, A.M., Burn, C.R., Chan, Laurie, Cooley, Dorothy, Froese, D.G., Graupe, Shel, Hall, R.I., Lantz, Trevor, Porter, T.J., Roy-Leveillee, Pascale, Turner, K.W., Wesche, S.D., and Williams, Megan, 2011, Environmental change and traditional use of the Old Crow Flats in Northern Canada—An IPY opportunity to meet the challenges of the new northern research paradigm: Arctic, v. 64, no. 1, p. 128–135.

Wolfe, R.J., 2006, Playing with fish and other lessons from the North: Tucson, University of Arizona Press, 152 p.

Wolken, J.M., Hollingsworth, T.N., Rupp, T.S, Chapin, F.S., III, and others, 2011, Evidence and implications of recent and projected climate change in Alaska's forest ecosystems: Ecosphere, v. 2, no. 11, p. 1–35, article 124, doi:10.1890/ES11-00288.1.

Woo, M.K., 1986, Permafrost hydrology in North America: Atmosphere-Ocean, v. 24, no. 3, p. 201–234.

Woods, A., Coates, K.D., and Hamann, A., 2005, Is an unprecedented Dothistroma needle blight epidemic related to climate change?: Bioscience, v. 55, no. 9, p. 761–769.

Woodward, A., and Beever, E.A., 2011, Conceptual ecological models to support detection of ecological change on Alaska National Wildlife Refuges: U.S. Geological Survey Open-File Report 2011–1085, 136 p., accessed November 14, 2015, at http://pubs.usgs.gov/of/2011/1085/.

Wotton, B., Nock, C., and Flannigan, M., 2010, Forest fire occurrence and climate change in Canada: International Journal of Wildland Fire, v. 19, no. 3, p. 253–271.

Wray, K., and Parlee, B., 2013, Ways we respect caribou—Teetł'it Gwich'in rules: Arctic, v. 66, no. 1, p. 68–78, doi:10.14430/arctic4267.

Wyborn, C., 2015, Co-productive governance: a relational framework for adaptive governance: Global Environmental Change, v. 30, p. 56–57.

Xiao, J., and Zhuang, Q., 2007, Drought effects on large fire activity in Canadian and Alaskan forests: Environmental Research Letters, v. 2, no. 4, p. 044003.

Xu, L., Myneni, R.B., Chapin, F.S., III, Callaghan, T.V., Pinzon, J.E., Tucker, C.J., Zhu, Z., Bi, J., Ciais, P., Tømmervik, H., Euskirchen, E.S., Forbes, B.C., Piao, S.L., Anderson, B.T., Ganguly, S., Nemani, R.R., Goetz, S.J., Beck, P.S.A., Bunn, A.G., Cao, C., and Stroeve, J.C., 2013, Temperature and vegetation seasonality diminishment over northern lands: Nature Climate Change, v. 3., p. 581–586.

Yannic, G., Pellissier, L., Ortego, J., and others, 2014, Genetic diversity in caribou linked to past and future climate change: Nature Climate Change, v. 4, p. 132–137.

Yellowstone to Yukon Conservation Initiative, 2012, Muskwa-Kechika Management Area biodiversity conservation and climate change assessment—Summary report: Canmore, Alberta, Canada, Yellowstone to Yukon Conservation Initiative, 112 p., accessed November 16, 2015, accessed November 10, 2017, at http://y2y.net/publications/copy_of_technical-reports.

Yue, S., Pilon, P., and Phinney, B., 2003, Canadian streamflow trend detection: impacts of serial and cross-correlation: Hydrological Sciences, v. 48, no. 1, p. 51–63.

Yukon Bureau of Statistics, 2017, Populations and dwellings, census 2016: Government of Yukon, Whitehorse, 4 p., accessed October 24, 2017, at http://www.eco.gov.yk.ca/stats/pdf/2016PopulationDwellings.pdf.

Yukon Geological Survey, 2012, Yukon exploration and geology overview 2011, in MacFarlane, K.E., ed., NTS mapsheet(s)—Yukon wide: Yukon Geological Survey, 98 p, accessed November 10, 2017, at http://data.geology.gov.yk.ca/Reference/50373.

Yukon Government, 2009, Energy strategy for Yukon: Whitehorse, Yukon Government, Energy, Mines and Resources, 28 p., accessed November 10, 2017, at http://www.energy.gov.yk.ca/energy_strategy.html.

Yukon Government, 2017, Yukon oil and gas annual report 2016: Whitehorse, Yukon Government, Energy, Mines and Resources, 11 p., accessed November 13, 2017, at http://www.emr.gov.yk.ca/oilandgas/pdf/yukon-oil-and-gas-annual-report-2016.pdf.

Zemp, M., Hoezle, M., and Haeberli, W., 2007, Six decades of glacier mass-balance observations—A review of the worldwide monitoring network: Annals of Glaciology, v. 50, no. 50, p. 101–111.

Zhang, X., Harvey, K.D., Hogg, W.D., and Yuzyk, T.R., 2001, Trends in Canadian streamflow: Water Resources Research, v. 37, no. 4, p. 987–998.

INDEX

Note: page numbers followed by b, f, or t refer to boxes, figures, or tables respectively. Those followed by n refer to notes, with note number.